Mary Ann McDowell • Sima Rafati
Editors

Neglected Tropical Diseases - Middle East and North Africa

Springer

Editors
Mary Ann McDowell
Department of Biological Sciences
Eck Institute for Global Health
University of Notre Dame
Notre Dame
Indiana
USA

Sima Rafati
Molecular Immunology and Vaccine Research
Pasteur Institute of Iran
Tehran
Iran

ISSN 2194-8275 ISSN 2194-8283 (electronic)
ISBN 978-3-7091-1983-9 ISBN 978-3-7091-1613-5 (eBook)
DOI 10.1007/978-3-7091-1613-5
Springer Wien Heidelberg New York Dordrecht London

Softcover reprint of the hardcover 1st edition 2014

Printed on acid-free paper

Springer is part of Springer Science+Business Media (www.springer.com)

Neglected Tropical Diseases

Series Editors

Dr. Peter Hotez
Department of Pediatrics
Baylor College of Medicine
Texas Children's Hospital
Texas, USA

For further volumes:
http://www.springer.com/series/11165

Human beings are members of a whole,
In creation of one essence and soul.
If one member is afflicted with pain,
Other members uneasy will remain.
If you've no sympathy for human pain,
The name of human you cannot retain.
Saadi Shirazi 1210–1291

بنی آدم اعضای یکدیگرند

که در آفرینش ز یک گوهرند

چو عضوی به درد آورد روزگار

دگر عضوها را نماند قرار

تو کز محنت دیگران بی غمی

نشاید که نامت نهند ادمی

Preface

The Neglected Tropical Diseases (NTDs) are a group of 17 viral, bacterial, protozoan, and helminth infections that disproportionately affect the most vulnerable populations. The concept of NTDs begun to take shape following the 2000 Millennium Development Goals put forth by the United Nations, specifically Goal # 6 to combat HIV/AIDs, malaria, and other diseases. Although the "other diseases" category spurred substantial debate, it is now accepted that the 17 diseases classified as NTDs by the World Health Organization (WHO) represent some of this "other" category. Compared to HIV/AIDs, malaria, and tuberculosis at 42.1 %, NTDs have generally been ignored (i.e., neglected), receiving only 0.6 % of official assistance for health (Liese and Schubert 2009). These diseases do not cause substantial global mortality; however, morbidity can rival HIV/AIDs and malaria (Murray et al. 2012; Vos et al. 2012). Importantly, NTDs are some of the most common diseases on the globe, thrive in impoverished regions, and perpetuate the cycle of poverty, causing mental impairment in children and hindering socioeconomic development.

The Middle East and North Africa (MENA) is highly endemic for several NTDs. This region is economically diverse, encompassing both oil-rich and resource-poor nations. 340 million people live in the region, of which 12 % (~50 million) live on less than $2 per day (World Bank 2010). While global efforts to eliminate some NTDs have been successful in MENA countries, many have not received such attention. Soil-transmitted helminth infections are the most prevalent NTDs in the MENA; however, modifications in human behavior, recent environmental changes, and political turmoil have increased the risk for many others. The present volume, *Neglected Tropical Diseases in the Middle East and North Africa*, covers the most prevalent NTDs in the MENA region, including chapters on dengue virus, rabies, brucellosis, leprosy, trachoma, toxoplasmosis, cutaneous and visceral leishmaniasis, fascioliasis, schistosomiasis, and soil-transmitted helminth infections. The authors of individual chapters are experts in their respective fields, either MENA-endemic scientists or non-endemic researchers with an intimate knowledge of these diseases in the MENA region. This book emphasizes disease burden, clinical manifestation, and current control approaches and outlines the major obstacles for reducing the burden of NTDs in the MENA. In most cases, social determinants, including human migration, political instability, urbanization, and agricultural practices, are all drivers in preventing control of these devastating diseases. In particular, the recent political landscape in the region has had devastating impacts on public health management leading to a breakdown in control efforts and an increase in outbreaks. Increased surveillance efforts, including advanced training, improved diagnosis methods, and mandatory reporting, are needed for most of the NTDs in the region. Successful strategies to combat the burden of NTDs will undoubtedly require strong political commitment and intimate international collaboration involving research, policy, and veterinary and human health implementation sectors.

The hope is that this volume will stimulate increased awareness and commitment from research institutions, funding agencies, and governments to eliminating the devastation caused by NTDs in the MENA region. Ultimately, we desire strengthened cooperative efforts of all the MENA nations for controlling the burden of NTDs and international commitment to stabilize the political situation in the region.

We express our deep appreciation to the editorial staff of SpringerVerlag, in particular Claudia Panuschka and Ursula Gramm, for their organization and editorial expertise. We also are thankful to Dr. Peter Hotez for the opportunity to contribute to such an important project. Finally, we are extremely grateful to all the contributing authors for their valuable contributions, cooperation, and patience to this project. We value their time and insight.

It is with our deepest regret that one author, Professor Rashida Barakat, passed from this world before this project came to fruition. Her contribution to this volume and her tireless efforts towards schistosomiasis control in Egypt will be forever remembered. It is to this distinguished scholar and mentor that we dedicate this book.

Notre Dame, IN, USA Mary Ann McDowell
Tehran, Iran Sima Rafati

References

Liese BH, Schubert L (2009) Official development assistance for health-how neglected are neglected tropical diseases? An analysis of health financing. Int health 1:141–147

Murray CJ, Vos T, Lozano R, Naghavi M, Flaxman AD, et al. (2012) Disability-adjusted life years (DALYs) for 291 diseases and injuries in 21 regions, 1990-2010: a systematic analysis for the Global Burden of Disease Study 2010. Lancet 380:2197–2223

Vos T, Flaxman AD, Naghavi M, Lozano R, Michaud C, et al. (2012) Years lived with disability (YLDs) for 1160 sequelae of 289 diseases and injuries 1990-2010: a systematic analysis for the Global Burden of Disease Study 2010. Lancet 380:2163–2196

World Bank (2010) Data. http://data.worldbank.org/region/MNA

Contents

Soil-Transmitted Helminth (STH) Infections in the MENA Region

Mohammad Bagher Rokni, Wael M. Lotfy, Peter J. Hotez, and Nilanthi R. de Silva

Contents

M.B. Rokni (✉)
Department of Medical Parasitology and Mycology, School of Public Health, Tehran University of Medical Sciences, P.O. Box 14155-6446, Tehran, Iran
e-mail: roknimoh@tums.ac.ir

W.M. Lotfy
Department of Parasitology, Medical Research Institute, Alexandria University, Alexandria, Egypt

P.J. Hotez
National School of Tropical Medicine, Baylor College of Medicine, Houston, TX 77030, USA
e-mail: hotez@bcm.edu

N.R. de Silva
Department of Parasitology, University of Kelaniya, PO Box 6, Ragama, Sri Lanka
e-mail: nrdesilva@gmail.com

M.A. McDowell and S. Rafati (eds.), *Neglected Tropical Diseases - Middle East and North Africa*, Neglected Tropical Diseases, DOI 10.1007/978-3-7091-1613-5_1,

Abstract Soil-transmitted helminths (STHs), or geohelminths, including Ascaris lumbricoides, Trichuris trichiura, Ancylostoma duodenale, Necator americanus, and Strongyloides stercoralis are a group of intestinal nematode parasites transmitted through contaminated soil, vegetables, and water. STH infections are leading neglected tropical diseases that affect the Middle East and North Africa (MENA) region. To conduct this review, publication databases related to STHs in the MENA countries were surveyed. Search engines utilized were PUBMED, SCIRUS, GOOGLE SCHOLAR, IMEMR, The Global Atlas of Helminth Infections, and local sources. Overall the MENA region accounts for less than 10 % of the global number of cases of STH infection. Results show that ascariasis has the lowest rate of infection as 0.02 % in Oman and 56 % as the highest rate in the Palestinian Territories. More than 20 million infections with ascariasis are in the MENA region. As for trichuriasis, a range of 0 % infection in Lebanon to 45 % infection in Somalia was found; overall about seven infections are in the MENA region. For hookworm infection, five to ten million infections are in the MENA region with infection rates ranging between 0 % in Algeria and 50 % in Kuwait (50 %). Minimal data for strongyloidiasis were available, but the estimates range from 0.0 % in Algeria to 15.5 % in the Palestinian Territories. Overall, anthelminthic drug coverage through periodic deworming is extremely low in the MENA region. Here we describe various risk factors in transmission of STHs in the region and report the prevalence of contamination of vegetables with the parasites' eggs. Moreover, different aspects of clinical manifestations, control, prevention, and treatment will be discussed.

Keywords Soil-transmitted helminths • MENA region • Helminths • Prevalence

Background

Soil-transmitted helminths (STHs), also called geohelminths, are a group of intestinal nematode parasites transmitted primarily through contaminated soil, vegetable, and water. They exhibit direct life cycles that require no intermediate hosts or vectors. The soil provides conditions under which development of unembryonated eggs to the infective stage can take place, with human infection occurring through ingestion of eggs (*Ascaris lumbricoides*, *Trichuris trichiura*) or larvae (*Ancylostoma duodenale*) or through direct larval penetration (*Necator americanus*, *Ancylostoma duodenale*, *Strongyloides stercoralis*. The STHs together represent

Table 1 Global disease burden of soil-transmitted helminths modified from Murray et al. (2013), Lozano et al. (2013), and Fürst et al. (2012)

Diseases	Population at risk (millions)	People infected (millions)	People with morbidity (millions)	Deaths (thousands)	DALYs (thousands)
Hookworm infection	3195	576–740	150	0 –65	3,231
Ascariasis	4211	807–1,221	350	3–60	1,315
Trichuriasis	3212	604–795	220	0–10	638
Strongyloidiasis	Not determined	30–100	Not determined	Not determined	Not determined

the most common parasitic infections of humans worldwide. They have particular public health relevance because of significant child morbidities (WHO 2002, 2012). The extraordinary numbers of STH infections, which approach two billions, are a reflection of a remarkably successful adaptation to survival of the eggs or larval stages in the environment and parasitism in humans lasting years (de Silva et al. 2003). They present an enormous global disease burden, resulting in more than five million disability adjusted life years lost annually according to a new Global Burden of Disease 2010 Study (Murray et al. 2013), with ascariasis also causing a significant number of deaths in young children (Lozano et al. 2013). Moreover, STH infections are regarded as one of the world's leading causes of intellectual and physical growth deficits and disabilities (Bethony et al. 2006).

In the MENA region the STHs disproportionately affect the estimated 65 million people living on less than $2 per day (Hotez et al. 2012). The basis by which poverty promotes endemicity of STH infections has not been well studied in the MENA region, although presumably this situation reflects high rates of inadequate sanitation and access to clean water especially in impoverished rural and some urban areas. In 2003, de Silva et al. estimated that the MENA area harbor about 23 million cases of ascariasis which constitute about 2 % of the global disease burden, 7 million cases of trichuriasis which equals about 1 % of the global burden, and 10 million cases of hookworm infections which equals about 1 % of the global burden (de Silva et al. 2003), numbers that were modified in Hotez et al. (Hotez et al. 2012). The World Health Organization (WHO) estimates that approximately 9 % of children at risk for STH infections live in its Eastern Mediterranean Region (WHO 2013). The most updated data in Table 1 depict various features of STHs.

The MENA Region

According to World Bank, the MENA region includes countries of the Middle East and North Africa: Middle East: Afghanistan, Bahrain, Iran, Iraq, Israel, Jordan, Kuwait, Lebanon, Oman, Pakistan, Qatar, Saudi Arabia, Syria, United Arab

Emirates, West Bank and Gaza; North Africa: Algeria, Djibouti, Egypt, Libya, Malta, Morocco, Somalia, Sudan, Tunisia, and Yemen. Almost 400 million people, approximately 5 % of the world's population, live in the MENA region, led by Egypt (80 million), Iran (75 million), Algeria (36 million), and Morocco and Iraq (31–32 million each) as the most populated countries (Hotez et al. 2012).

Methodology

For this review, publication databases related to STHS in the MENA countries were surveyed. Search engines utilized were PUBMED, SCIRUS, GOOGLE SCHOLAR, IMEMR, Global Atlas of Helminth Infections (GAHI), and local sources. Database searches were not restricted by date. In addition to full text papers, abstracts were also reviewed. The panel of key words utilized was "*Ascaris lumbricoides*," "*Trichuris trichiura*," "*Strongyloides stercoralis*," "*Necator americanus*," "*Ancylostoma duodenale*," "Hook worms," and related terminology of diseases, e.g., ascariasis. In addition to papers in English, papers published in other languages such as French, Persian, and Arabic were considered as well.

Ascaris lumbricoides Linnaeus, 1758

Etiology, Life Cycle, and Major Clinical Manifestations

Ascaris lumbricoides, is one of the leading infectious agents of humans affecting up to one billion or more (Holland 2009). The Egyptian medical papyri documented human infection of *As. lumbricoides* since the time of pharaohs (Grove 1990; Hoeppli 1959). This worm is possibly the earliest recorded human helminths; it is referred to in texts from Mesopotamia, Greece, Rome, and China.

Humans acquire *As. lumbricoides* via ingestion of embryonated eggs in both rural and some impoverished urban environments. The eggs are almost ubiquitous in the soil of poor environments and they tenaciously adhere to inert substances. After ingestion, the liberated larvae migrate from the intestine through the liver, heart, and lungs where they cause wheezing and Loeffler's pneumonitis, which clinically resembles asthma. Indeed in Saudi Arabia seasonal asthma has been linked to endemic *As. lumbricoides* infection (Gelpi and Mustafa 1967). After *As. lumbricoides* larvae return to the gut they develop into adult worms, typically in the jejunum and ileum where they can grow to 15–40 cm in length. The adult worms result in vitamin A malabsorption and other malnutrition syndromes that result in growth stunting and cognitive deficits. Adult worms in the ileum of small children can cause acute intestinal obstruction, which can result in an estimated 2,700 deaths annually (Lozano et al. 2013).

Prevalence of Ascariasis in the MENA Region

Table 2 illustrates the prevalences of STHs including ascariasis in different countries of the MENA region. Accordingly, the lowest rate of infection was shown in Oman as 0.02 % and the highest rate in Palestine with 56 % infection.

Studies conducted by (de Silva et al. 2003) estimated that the MENA region had about 23 million infections with ascariasis, accounting for approximately 2–3 % of the global estimates of disease burden. Ascariasis may represent the most common neglected tropical disease in the MENA region. Among the MENA countries, Egypt leads in the number of cases of ascariasis with about 8.3 million infections. Yemen is the country with the second highest number of infections (5.8 million infections), followed by Iran with 5.1 million infections, and Morocco with 1.3 million infections (Hotez et al. 2012). In Egypt, the prevalence of ascariasis in Lower Egypt was higher than in Upper Egypt (El-Gammal et al. 1995). Recent studies in either Upper or Lower Egypt showed discrepancies in prevalence rates in different samples and localities (Bakr et al. 2009; El-Masry et al. 2007; El-Nofely and Shaalan 1999; El-Sahn et al. 1997; Hamed et al. 2013; Ibrahium 2011; Mahfouz et al. 1997; Mohammad et al. 2012).

In Yemen *As. lumbricoides* has been reported as 0.42–15.9 % (Baswaid 2008; Nasher 1988). In Oman during 1994–1998 ascariasis has been reported from 0.02 to 1.1 % (Ibrahium 2011). A study in Lebanon during two periods of 1997–1998 and 2007–2008 shows that although the rate of infection has decreased, the risk of infection still persists (Table 2) (Araj et al. 2011). In Algeria, from 1,042 individuals examined, 0.4 % were infected with ascariasis (Benouis et al. 2013).

In Libya, in primary schoolchildren among 1,039 stool specimens, *As. lumbricoides* was detected in 0.1 % of cases (Sadaga and Kassem 2007). From 126 samples of fresh vegetables, 85 (86 %) cases were contaminated with *Ascaris* eggs (Abougrain et al. 2010).

Trichuris trichiura (Linnaeus, 1771) Stiles, 1901

Etiology, Life Cycle, and Major Clinical Manifestations

Tr. trichiura or **Trichocephalus** *trichiura* is commonly known as whipworm because of its characteristic morphology. The ecological and environmental requirements of *Tr. trichiura* eggs are similar to those of *As. lumbricoides*. After egg ingestion the *Tr. trichiura* larvae migrate to the cecum and ascending colon where they develop to adult worms approximately 3–5 cm in length. In cases of moderate and heavy infections, the host's inflammatory response to adult whipworms results in *Trichuris* colitis, which resembles inflammatory bowel disease. In very heavy infections (more than 500 worms), worms spread throughout the colon to the rectum, where they cause hemorrhages, mucopurulent stools, symptoms of dysentery (*Trichuris* dysentery syndrome) with rectal prolapse.

Table 2 Summary of published reports on prevalence of STHs in countries of the MENA region

Country	AL (%)	SS (%)	TT (%)	HW (%)	Kind of study	Subjects (No.)	Year of study	Reference
Afghanistan	40.9	–	9.9	0.7	CBS	Schoolchildren (1,001)	2003	Gabrielli et al. (2005)
Algeria	0.4	–	–	–	HBS	All (1,042)	1010–1011	Benouis et al. (2013)
Algeria	1.4	0.0	–	0.0	CBS	All (11,601)	1984–1988	Bachta et al. (1990)
Iran	1.5	–	0.1	<0.1	CBS	All (53,995)	1999–2000	Sayyari et al. (2005)
Iran	0.57	0.5	–	–	CBS	Food handlers (62,007)	2000–2009	Saki et al. (2012)
Israel	20.3	4.5	19.2	54.2	CBS	Ethiopian immigrants (5,412)	1990	Nahmias et al. (1991)
Jordan	4.9	–	1.1	2.5	CBS	Food handlers (283)	1990	al-Lahham et al. (1990)
Kuwait	5.0	2.0	28	50	HBS	Housemaids (100)	–	Grover et al. (2008)
Lebanon	2.0	0.1	0.2	0.05	HBS	(14771)	1997–1998	Araj et al. (2011)
Lebanon	1.0	0.5	0.0	0.1	HBS	(7,477)	2007–2008	Araj et al. (2011)
Morocco	13.3	–	13.3	–	CBS	Children (610)	–	Amahmid and Bouhoum (2005)
Oman	0.02–1.1	–	0.2	–	CBS	School children (2,213)	1994–1998	Idris et al. (2001)
Pakistan	22.8	–	(2.5)	–	HBS	All (237)	2008	Ahmed et al. (2012)
Pakistan	1.9	–	0.6	4.6	CBS	Farmers & Textile laborers (1,704)	2002–2003	Ensink et al. (2005)
Palestine	56	15.5	2.7	–	CBS	All (1,000)	2011–2002	Al-Zain and Al-Hindi (2005)
Palestine	0.0–1.3	0.02–0.24	–	0.01–0.12	HBS	All (123,290)	2000–2009	Bdir and Adwan (2010)
Qatar	0.3	–	0.5	2.1	HBS	Immigrants (9,208)	2005–2008	Abu-Madi et al. (2010)
Qatar	2.1	–	10.8	7.9	HBS	Immigrants (1,737)	2005–2006	Abu-Madi et al. (2008)
Qatar	2.5	0.5	3.5	8.3	HBS	Immigrants (1,538)	2008	Abu-Madi et al. (2011)
Saudi Arabia	25.9	0.9	26.7	27.4	HBS	Expatriate workers (1,019)	1994	Abahussain and Abahussain (2005)

Saudi Arabia	0.66	0.07	0.36	0.036	HBS	All (10,427)	2005–2007	Eligail et al. (2010)
Saudi Arabia	15.9	–	12.5	15.4	HBS (All)	(23,278)	2006–2008	Imam et al. (2012)
Saudi Arabia	12.6	3.0	13.1	13.2	HBS (All)	(63,892)	1996–2003	Alkhalife (2006)
Saudi Arabia	0.05	–	0.03	0.02	HBS (All)	(12,054)	2004–2009	Zaglool et al. (2011)
Somalia	17	–	45	–	CBS (Children and mothers)	(517)	–	Peltola et al. (1988) from 517 cases
United Arab Emirate	6.6	–	6.2	2.4	HBS (Immigrants)	(60,268)	–	Ibrahim et al. (1993)
Yemen	1.7	–	–	1.7	HBS (Restaurant workers)	(500)	2007	Baswaid (2008)

AL, *Ascaris lumbricoides;* SS, *Strongyloides stercoralis*; TT, *Trichuris trichiura;* HW, Hook worms; CBS, Community-based study; HBS, Hospital-based study

Allergic manifestations such as urticaria, rhinitis, and eosinophilia are frequently seen (Chandrasekhara et al. 2007). A significant number of patients, especially children with longstanding massive infections, have dysenteric syndrome presenting with chronic mucous diarrhea, rectal prolapse, anemia from chronic blood loss and iron deficiency, clubbing of fingers, protein-energy malnutrition, and growth retardation. Deficits in cognitive function and stunting have been observed in infected children, hindering educational achievement and psychomotor development. In rare cases the parasite may cause intussusception that mimics acute appendicitis (Alkhulaiwi et al. 1996; To et al. 2006).

Prevalence of Trichuriasis in the MENA Region (Table 2)

According to Table 2, a range of 0.0–0.2 % infection in Lebanon to 45 % infection in Somalia with this parasite has been demonstrated. *Trichuris trichiura* is the second most common roundworm of humans. In 2003, it was estimated that 800 million people were infected worldwide (de Silva et al. 2003). Approximately 7–9 million infections are in the MENA region which equals about 1 % of the global estimates (de Silva et al. 2003; Hotez et al. 2012). In this region, trichuriasis infections are highest in Morocco with about 3.2 million infections, followed by Egypt with about 1.7 million infections, then Iran with 1.6 million infections, and Yemen with 1.5 million infections (Hotez et al. 2012).

It was reported that in a slum area of Alexandria (Egypt), dwellers experienced an increase in trichuriasis (Curtale et al. 1998). This species was reported among the parasites mechanically transmitted by cockroaches and flies in Egypt (El-Sherbini and El-Sherbini 2011; El-Sherbini and Gneidy 2012).

Hookworms

Different species of the family Ancylostomatidae are commonly known as hookworms. They colonize the intestinal tract of man and other mammal hosts. Among the different known hookworms, only two species, *Ancylostoma duodenale* and *Necator americanus*, are known to have a major public health importance (de Silva et al. 2003).

Hookworms are the third most common STH of humans. In 2003, de Silva et al. estimated that 740 million people were infected worldwide, and the greatest prevalence estimates of infection were in sub-Saharan Africa (29 %); East Asia and the Pacific Islands (26 %); China (16 %); and South Asia (16 %). Worldwide, *Ne. americanus* is the predominant hookworm species, while *An. duodenale* is more geographically restricted. In Egypt pure *An. duodenale* infections have been noted (el Shazly et al. 1998). More typically, mixed infections with both species may be common in the Middle East, although very few epidemiological studies have attempted to differentiate hookworm species (Brooker et al. 2004; Eid et al. 2008). The eggs of the two species cannot be distinguished morphologically. To identify these hookworm species the larvae must be examined. Larvae cannot be found in stool specimens unless the specimen is left at ambient temperature for a day or more (de Silva et al. 2003).

Ancylostoma duodenale (Dubini, 1843) Creplin, 1845

Necator americanus *(Stiles 1902) Stiles, 1906*

Etiology, Life Cycle, and Major Clinical Manifestations

Fully developed adult *Ne. americanus* worms are smaller than *An. duodenale* worms (females are up to 11 and 13 mm long, respectively; males are slightly smaller), and consequently cause more blood loss from their hosts, 0.2 vs. 0.05 ml/worm/day (Loukas and Prociv 2001). However, *Ne. americanus* is more widespread worldwide and, therefore, more significant as a cause of disease burden (Hotez et al. 2005). *Ne americanus* tends to be much more tropical in distribution (Schad 1991), whereas *An. duodenale* is often better adapted to more northerly latitudes in the subtropics, including areas of the Middle East.

Humans acquire hookworm infection through larval skin penetration, although *An. duodenale* is also infective via the oral route (Bethony et al. 2006). Upon skin penetration the infective larvae migrate through the lungs and are coughed and swallowed. Acute upper gastrointestinal discomfort occurs as the larvae enter the gastrointestinal tract and resume their development to become adult worms. This period often coincides with the onset of eosinophilia. Adult hookworms in the small intestine produce blood loss leading to iron and protein losses. Iron deficiency anemia results when host iron reserves are depleted. In recent systematic reviews, moderate and heavy hookworm infections were linked to anemia in children (Smith and Brooker 2010), whereas even light infections could cause anemia in some populations of both pregnant and nonpregnant adults (Smith and Brooker 2010; Brooker et al. 2008). In heavy infections protein malnutrition can also occur.

Prevalence of Hookworms in the MENA Region (Table 2)

About five to ten million infections are in the MENA region which accounts for approximately 1 % of the global disease burden (de Silva et al. 2003; Hotez et al. 2012). In the MENA region, hookworm infections are highest in Egypt with about 3.6 million infections, followed by Iran with about 0.4 million infections, then Saudi Arabia with 0.4 million infections, and Oman with 0.2 million infections (Hotez et al. 2012).

According to Table 2, the lowest and the highest rate of infection with hookworms belong to Algeria (0.0 %) and Kuwait (50 %). However, among immigrant populations, a prevalence of 54.2 % was reported in Ethiopian immigrants to Israel (Nahmias et al. 1991).

An. duodenale infection is reported from Morocco (Jiménez-Albarrán and Odda 1994), Algeria (Pampiglione and Hadjerès 1965), Tunisia (Al-Binali et al. 2006), Egypt (Kuntz et al. 1956; Mohamed et al. 1985; Mohamed et al. 1988; Wells and Blagg 1956), Palestine (Scott et al. 1934), Israel (Avins et al. 1971; Brauman et al. 1982), Jordan (Altaif 2011), Lebanon (Yenikomshian and Berberian 1932), Syria (Yenikomshian and Berberian 1932), Saudi Arabia (Zaglool et al. 2011; Al-Binali et al. 2006; Abdel-Hafez et al. 1986), Yemen (Farag 1985), Oman (Patel and Khandekar 2006), and Iraq (Farhan 2012; Kadir and Salman 1999; Niazi et al. 1975).

In Egypt, laborers in agriculture are significantly at risk of *An. duodenale* infection (Bakr et al. 2009). In contrast to ascariasis, ancylostomiasis is more prevalent in Upper Egypt than Lower Egypt (Augustine et al. 1930; Miller et al. 1980; Scott 1939). *An. duodenale* was reported among the parasites mechanically transmitted by cockroaches and flies in Egypt (El-Sherbini and El-Sherbini 2011; El-Sherbini and Gneidy 2012).

Strongyloides stercolaris (Bavay, 1876) Stiles et Hassall, 1902

Because of the difficulties in diagnosing *St. stercoralis* infections, we know the least about strongyloidiasis in terms of its prevalence and global disease burden. The GBD 2010 Study, for instance, did not provide an estimate in DALYs of human strongyloidiasis. Some estimates indicate, however, that this STH affects between 10 and 40 % of the population worldwide (Schär et al. 2013). Because of its ability to produce autoinfection and multiply in the body, the infection can persist for decades, even more than 75 years in some cases (Schär et al. 2013). Under selected conditions patients who receive corticosteroids (e.g., during solid organ transplantation or for treatment of autoimmune disease or malignancy) or in HTLV-1 coinfections, *St. stercoralis* can cause a hyperinfection syndrome associated with severe morbidity and mortality.

Etiology, Life Cycle, and Major Clinical Manifestations

The life cycle of *St. stercoralis* embraces two free-living and parasitic cycles.

Under the normal environmental conditions, the first-stage rhabditiform larvae (L1) after passing in the stool will transform to adult free-living worms and start a free-living cycle. Under the unfavorable conditions, the rhabditiform larvae become third-stage filariform larvae (L3) and switch to parasitic cycle to infect humans through penetrating the skin and after an internal migration reside in small

Table 3 Country-wide prevalence rates for *Strongyloides stercoralis* for countries of MENA region (Schär et al. 2013)

Country	Community-based surveys	Hospital-based surveys	Refugees and immigrants
Egypt	2.5	11.1	–
Iran	0.3	0.6	–
Iraq	–	24.2	–
Israel	94.9	–	–
Jordan	0.03	–	–
Kuwait	–	16.3	–
Oman	3	–	–
Palestine	–	4.2	–
Saudi Arabia	–	12.5	7.1
Sudan	3.7	–	–
Tunisia	–	0.5	–

intestine (Schär et al. 2013). The adult worms in the small intestine produce eggs which hatch and release L1 larvae in gut. Under conditions still not well defined the L1 can continue their developmental program while still in the intestine to become L3 leading to autoinfection. When autoinfection is dysregulated, both hyperinfection and disseminated infection can result.

Strongyloidiasis is a wide spectrum disease in the context of clinical manifestations including enteritis resulting from the adult worms in the small intestine to life-threatening hyperinfection syndrome and disseminated disease, which can occur in patients receiving exogenous corticosteroids or in patients with HTLV-1 coinfection. The basis by which steroids or HTLV-1 can trigger *St stercoralis*-induced hyperinfection is not known.

The most important feature is its role as a fatal disease in immunocompromised individuals and patients receiving corticosteroid therapy (Schär et al. 2013).

Prevalence of Strongyloides stercoralis in the MENA Region

In general, information on infection rates and prevalence of the parasite is scarce in the region. Table 3 provided by Schar et al. depicts the most updated and comprehensive information on the prevalence of the disease in some countries of the MENA region (Schär et al. 2013). Table 2 provided by the present authors shows the lowest and the highest rate of infection as 0.0 % in Algeria and 15.5 % in Palestinian.

The reality is that in comparison to other STHs, the diagnosis of this disease is more difficult. Many surveys conducted in the region were based on only one stool sample examination which is of low sensitivity. Normally more samples should be examined over consecutive days from one subject to reach a decisive conclusion (Marti and Koella 1993).

Environmental and Human Factors that Promote STH Infections in the MENA Region

Some of the major influences of STH infections in the MENA region include the following elements.

Animals

The major STH infections of humans are not generally thought as zoonotic diseases. Thus, *As. lumbricoides* is mainly a human (anthroponotic) parasite. However, the results of a study carried out in Egypt suggested that dogs could act as reservoir hosts of *As. lumbricoides* and environmental contaminators that increase risk of infection in humans (Shalaby et al. 2010). In Saudi Arabia, the Arabian sacred baboon, *Papio hamadryas arabicus*, may play a role in transmission of *Ascaris* spp. to neighboring human communities (Nasher 1988).

Foreign Workers

Several countries in the MENA region employ foreign workers as household servants and in the construction sector. Many from Asian and African countries are believed to be infected with different parasitic diseases because they are coming from countries of low socioeconomic levels and inadequate medical care and may serve as disease reservoirs (Table 2). Although in many nations in the MENA region the immigrant workers are required to pass selected clinical examinations it is not uncommon for them to become reinfected upon return to their native homeland (Abu-Madi et al. 2010).

Geophagia

Geophagia (Soil-eating habit), which is a kind of pica, has been reported from many countries of the region. Soil is an important source of infection which contains many parasites' eggs especially STHS (Geissler et al. 1998). In Pakistan, geophagia was reported as 24.8 % in participant children (Mehraj et al. 2008).

Wastewater

The role of wastewater in spreading STH infections has been confirmed in two surveys in Morocco. For example, the prevalence of ascariasis in the schoolchildren from an area contaminated with wastewater versus a control group was 32.8 vs. 1.45 %, respectively (Bouhoum and Amahmid 2000). Similarly, Habbari et al. in another part of the country reported the prevalence as 20.5 vs. versus 3.8 % (Habbari et al. 2000).

In Tunisia, the concentration of *Ascaris* sp. was 455 eggs per liter and 46 eggs per liter in raw wastewater and treated wastewater, respectively (Ayed et al. 2009), which indicates that even treated wastewater might still serve as a source of infection for ascariasis.

The role of sewage is especially critical in Afghanistan, where tens of millions of Ascaris eggs were recovered from sources of water that were simultaneously contaminated with sewage and also used for irrigation and wells holding drinking water (Safi and Buerkert 2012).

Vegetables

An important vehicle of ascariasis transmission is through the ingestion of eggs adhering to vegetables. Different studies in countries of MENA region show that eggs of STHs are present in vegetables. It is important to note that consuming raw vegetables is a traditional habit in several different nations of the region. Another issue is that vegetables are washed in water, which after treatment are still contaminated with parasites' eggs (Bolbol 1992; Amin 1988).

In Saudi Arabia, the presence of *As. lumbricoides* eggs in treated municipal wastewater of the Riyadh metropolitan area was reported (Bolbol 1992). In two different studies in Egypt and Saudi Arabia, it was found that *As. lumbricoides* eggs were common in leafy vegetables and the use of tap water does little to remove them (Al-Binali et al. 2006; Fawzi et al. 2004). Study of vegetables in Saudi Arabia showed the contamination with *Ascaris* eggs as 26.3 % and *Ancylostoma* 11.8 %. Altogether, 16.2 % of all samples were found contaminated with different parasites' eggs (Al-Megrm 2010).

In Bahrain, sludge produced in the central sewage treatment plant was found contaminated with *As. lumbricoides* eggs. It is used for agricultural purposes and poses a threat to public health (Amin 1988).

In Algeria, of 36 tomato, 36 cucumber, 27 lettuce, and 27 cress samples examined, eggs of *Ascaris* spp. were detected in 19 %, 75 %, 96 %, and 96 %, respectively (Abougrain et al. 2010).

In Iran, a prevalence of 2 % and 1 % infection with *As. lumbricoides* eggs in vegetables has been reported in two studies (Daryani et al. 2008; Garedaghi et al. 2011). From 44 farms and 20 markets, 40 farms and all 20 markets had parasitic contamination in Tehran (Gharavi et al. 2002). In addition, 5.4 % and 2.17 % of examined samples of vegetables had *Ascaris* and hookworm eggs, respectively (Nazemi et al. 2012).

Fertilizer

Primitive agricultural practices using human feces as fertilizer are responsible for the high prevalence of ascariasis in certain regions of the world. Raw wastewater reuse can lead to a high risk of ascariasis (El Kettani and Azzouzi 2006). In a field study in Marrakech, Morocco, where raw sewage is used to fertilize crop fields, *As. lumbricoides* eggs were detected at the rate of 0.18 eggs/kg in potatoes, 0.27 eggs/kg in turnip, 4.63 eggs/kg in mint, 0.7 eggs/kg in carrots, and 1.64 eggs/kg in radish (Habbari et al. 1999).

In addition to aforementioned cases, more risk factors of critical significance in the region are as follows: (1) Lower socioeconomic status; (2) Lack of access to clean water; (3) Poor hygienic environment; (4) Poor education; (5) Overcrowded conditions; (6) Poverty; (7) Not washing hands after defecation; (8) Immigrants

Polyparasitism

Although understudied, STH coinfection with other intestinal parasites is believed to be common in the MENA region. In Pakistan, in 13.9 % of patients are coinfected with ascariasis and *Giardia* (Mehraj et al. 2008). In addition, 1.68 % of mixed infestations of *As. lumbricoides* and *Tr. trichiura* as well as 0.84 % coinfection with *As. lumbricoides* and *Giardia lamblia* were reported in Pakistan (Ahmed et al. 2012). In Iran polyparasitism was reported in 2 % of schoolchildren from Tehran, 8.2 % of food sellers in Kashan, 5.8 % of students from Hormozgan, and 0.8 % of those attending day-care centers in Tehran (Rokni 2008). Another study reports 5.53 % and 0.6 % of coinfection for two and three parasites, respectively (Mowlavi et al. 2008).

Treatment, Control, and Preventive Measures

In Israel, the cure rate for necatoriasis by treatment with 400 mg of albendazole was 84.4 %; besides, albendazole, 400 mg for 3 days, cured 92 % of the cases with *St. stercoralis* infection (Nahmias et al. 1991).

In 2001 the World Health Assembly adopted a resolution for frequent and periodic deworming through mass drug administration (also known as "preventive chemotherapy") as a means to control STH infections. While the initial recommendation was for school-aged children, the WHO has since recommended that coverage should be extended to preschool-aged children at risk for acquiring infection. Typically this is conducted by mass drug administration with single dose albendazole (400 mg) or mebendazole (500 mg). The WHO's major region that covers the MENA area is known as the Eastern Mediterranean Region. Unfortunately coverage of both preschool-aged and school-aged children in this area has been low. For 2011, WHO estimated that only 16–17 % of the more than 25 million preschool-aged children at risk for STH infection receive deworming treatments, while less than 1 % of the more than 55 million school-aged children have received deworming (WHO 2013). Overall of the roughly 80 million children in the region, only 5–6 % received either albendazole or mebendazole treatments in 2011 (WHO 2013).

Despite this overall poor coverage for STH infections there are some success stories. In Oman, a set of single annual dose of albendazole 400 mg, health education and promotion of environmental health, succeeded to reduce the prevalence of *Ne. americanus* in two parts of the country from 40 to 1.3 % and from 6 to 0 %, respectively, among rural and urban school children (Idris et al. 2001). In the Sultanate of Oman the climate is in favor of hookworm transmission and in some parts of the country infection with hook worms varies from 13 to 60 % (Idris et al. 1995). Fallah et al. reported that parasites' eggs were detected in 32.6 % of unwashed, 1.3 % of traditionally washed, and not in any standard washed samples of vegetables ($P < 0.001$) (Fallah et al. 2012). Therefore it clearly shows the importance of washing vegetables with standard methods which is easily conveyable to the people of the region.

Iran stands out among the countries which have experienced a remarkable decrease on the prevalence and incidence of helminthic diseases. Recent estimates show the prevalences of ascariasis, strongyloidiasis, hookworm as 0.1 %, 0.3 %, and <1 % of the population (Rokni 2008). Ascariasis has decreased significantly from 86.3 % in 1961 to merely 0.3 % in 1995 (Rokni 2008). In another study in Iran, among food handlers, the infection with ascariasis decreased from 1.55 % in 2000 to 0.48 % in 2009 (Saki et al. 2012). Moreover, out of 1494 examined subjects from nomad people in southwest of Iran, ascariasis and strongyloidiasis was detected in 0.13 % and 0.6 % of cases (Mowlavi et al. 2008). It is another reason that the rate of infection with STHs is decreased remarkably in Iran, because lots of risk factors are there among nomads in terms of infection with STHs.

In Libya, a significant difference was noted between parasitic infections and parent's education ($P = 0.000$), socioeconomic status of the family ($P = 0.000$), family size and number of rooms in houses ($P = 0.000$), and source of water for human consumption ($P = 0.05$) (Sadaga and Kassem 2007).

Concluding Remarks and Future Priorities

Despite the facts that most of the MENA region is comprised of middle income rather than low-income countries, and most parts are environmentally unfavorable to transmission, STH infections are still widespread. Ascariasis in particular is pervasive in much of the region. Undoubtedly, a major reason for rampant STH infections is the near absence of deworming in the schools and for school-aged children and limited targeting of preschool aged children.

The basis for low anthelminthic drug coverage needs to be better understood. The MENA region is notable for its political turmoil in many of its countries. Unfortunately this phenomenon has imposed a vast negative impact on different aspects of the human affairs including public health and parasitic diseases. STHs, as shown in this article, have critical role on increasing the sorrows of the people there. International cooperation among the ministers of health to promote universal deworming of school-aged children is a high priority and there is urgency to establish a climate of political will in this area.

In addition, it is likely that the actual amount of infection in the region is underestimated. One reason is the number of stool examinations that are routinely conducted. Many researchers prefer to conduct single stool examination, which results to underestimate the true prevalence. One or two samples are often not enough to yield a trustful estimation (Marti and Koella 1993).

As for strongyloidiasis, it is necessary to recommend researchers to utilize higher sensitivity diagnostic methods, such as Koga Agar plate culture, the Baermann, ELISA, and advise them to examine at least three samples of stool to be sure of the output. Unfortunately for this reason in the MENA region, most studies conducted on STHs lack of any data on *St. stercoralis*.

Ultimately, the MENA region still needs time to overcome some serious political hardships and the looming and actual prospect of conflict. As political agitations increasingly become more common there is a risk that public health might become compromised, as we have recently seen in Syria with resultant emergence of cutaneous leishmaniasis (Aleppo ulcer). Increasingly it might be necessary to rely on international organizations such as the WHO and their Eastern Mediterranean Regional Office (EMRO) in order to intervene and assist in surveillance and mass drug administration.

Acknowledgements The authors declare that there is no conflict of interest.

References

Abahussain NA, Abahussain N (2005) Prevalence of intestinal parasites among expatriate workers in Al-Khobar, Saudi Arabia. Middle East J Fam Med 3:17–21

Abdel-Hafez MM, El-Kady N, Bolbol AS, Baknina MH (1986) Prevalence of intestinal parasitic infections in Riyadh district, Saudi Arabia. Ann Trop Med Parasitol 80:631–634

Abougrain AK, Nahaisi MH, Madi NS, Saied MM, Ghenghesh KS (2010) Parasitological contamination in salad vegetables in Tripoli-Libya. Food Control 21:760–762
Abu-Madi MA, Behnke JM, Ismail A (2008) Patterns of infection with intestinal parasites in Qatar among food handlers and housemaids from different geographical regions of origin. Acta Trop 106:213–220
Abu-Madi MA, Behnke JM, Doiphode SH (2010) Changing trends in intestinal parasitic infections among long-term-residents and settled immigrants in Qatar. Parasit Vectors 3:1–13
Abu-Madi MA, Behnke JM, Ismail A, Al-Olaqi N, Al-Zaher K, El-Ibrahim R (2011) Comparison of intestinal parasitic infection in newly arrived and resident workers in Qatar. Parasit Vectors 4:1–10
Ahmed K, Jan M, Imran R, Shuja N, Shah G (2012) Prevalence of intestinal parasitic pathogens among gastroenteritis patients in district Gilgit, Gilgit-Baltistan, Pakistan. Pak J Zool 44:1059–1063
Al-Binali AM, Bello CS, El-Shewy K, Abdulla SE (2006) The prevalence of parasites in commonly used leafy vegetables in South Western, Saudi Arabia. Saudi Med J 27:613–616
Alkhalife IS (2006) Retrospective analysis of intestinal parasitic infections diagnosed at a University Hospital in Central, Saudi Arabia. Saudi Med J 27:1714
Alkhulaiwi AA, Afzal M, Rasul K, Mohammad AA, Baez-Giangreco A (1996) Ileocolic intussusception with *Trichuris trichiura* infestation: A case report. Ann Saudi Med 16:570–572
al-Lahham AB, Abu-Saud M, Shehabi AA (1990) Prevalence of Salmonella, Shigella and intestinal parasites in food handlers in Irbid, Jordan. J Diarrhoeal Dis Res 8:160–162
Al-Megrm WI (2010) Prevalence of intestinal parasites in leafy vegetables in Riyadh, Saudi Arabia. Int J Zool Res 6:137–142
Altaif KI (2011) Prevalence of intestinal parasitic infestation in Ma'an governorate, Jordan. Asian Pac J Trop Dis 1:110–112. doi:10.1016/S2222-1808(11)60047-9
Al-Zain B, Al-Hindi A (2005) Distribution of *Strongyloides stercoralis* and other intestinal parasites in household in Beit-lahia city, Gaza Strip, Palestine. Ann Alquds Med 1:48–52
Amahmid O, Bouhoum K (2005) Assessment of the health hazards associated with wastewater reuse: transmission of geohelminthic infections (Marrakech, Morocco). Int J Environ Health Res 15:127–133
Amin OM (1988) Pathogenic micro-organisms and helminths in sewage products, Arabian Gulf, country of Bahrain. Am J Public Health 78:314–315
Araj G, Musharrafieh U, Haydar A, Ghawi A, Itani R, Saliba R (2011) Trends and prevalence of intestinal parasites at a tertiary care center in Lebanon over a decade. Le Journal médical libanais. Leban Med J 59:143
Augustine DL, Helmy M, Nazmi M (1930) Ancylostomiasis and ascariasis in Egypt. Am J Hyg 11:136–148
Avins L, Sadowsky C, Gunders AE (1971) Hookworm infection in Yavne. Harefuah 81:374–376
Ayed LB, Schijven J, Alouini Z, Jemli M, Sabbahi S (2009) Presence of parasitic protozoa and helminth in sewage and efficiency of sewage treatment in Tunisia. Parasitol Res 105:393–406
Bachta E, Zenaidi N, Belkaid M, Tabet-Derraz O (1990) Intestinal parasitic infections in and around Algiers 1984-88. Bulletin de la Société de Pathologie Exotique 83:510–516
Bakr IM, Arafa NA, Ahmed MA, Mostafa MH, Mohamed MK (2009) Prevalence of intestinal parasitosis in a rural population in Egypt, and its relation to socio-demographic characteristics. J Egypt Soc Parasitol 39:371–381
Baswaid S (2008) Parasitic infections among restaurant workers in Mukalla (Hadhramout/Yemen). Iranian J Parasitol 3:37–41
Bdir S, Adwan G (2010) Prevalence of intestinal parasitic infections in Jenin Governorate, Palestine: a 10–year retrospective study. Asian Pac J Trop Med 3:745–747
Benouis A, Bekkouche Z, Benmansour Z (2013) Epidemiological study of human intestinal parasitosis in the hospital of Oran (Algeria). Int J Innovat Appl Stud 2:613–620

Bethony J, Brooker S, Albonico M, Geiger SM, Loukas A, Diemert D, Hotez PJ (2006) Soil-transmitted helminth infections: ascariasis, trichuriasis, and hookworm. Lancet 367:1521–1532

Bolbol AS (1992) Risk of contamination of human and agricultural environment with parasites through reuse of treated municipal wastewater in Riyadh, Saudi Arabia. J Hyg Epidemiol Microbiol Immunol 36:330–337

Bouhoum K, Amahmid O (2000) Health effect of wastewater reuse in agriculture. Schriftenreihe des Vereins für Wasser-, Boden-und Lufthygiene 105: 241

Brauman A, Gilboa Y, Pick A, Gil I, Gluskin I, Karpuch H (1982) Anemia due to hookworm infestation: an epidemic in a closed institution. Harefuah 103:346–347

Brooker S, Bethony J, Hotez PJ (2004) Human hookworm infection in the 21st century. Adv Parasitol 58:197–288

Brooker S, Hotez PJ, Bundy DA (2008) Hookworm-related anaemia among pregnant women: a systematic review. PLoS Negl Trop Dis 2:e291

Chandrasekhara V, Arslanlar S, Sreenarasimhaiah J (2007) Whipworm infection resulting in eosinophilic colitis with occult intestinal bleeding. Gastrointest Endosc 65:709–710

Curtale F, Shamy MY, Zaki A, Abdel-Fattah M, Rocchi G (1998) Different patterns of intestinal helminth infection among young workers in urban and rural areas of Alexandria Governorate, Egypt. Parassitologia 40:251–254

Daryani A, Ettehad G, Sharif M, Ghorbani L, Ziaei H (2008) Prevalence of intestinal parasites in vegetables consumed in Ardabil, Iran. Food Contr 19:790–794

de Silva NR, Brooker S, Hotez PJ, Montresor A, Engels D, Savioli L (2003) Soil-transmitted helminth infections: updating the global picture. Trends Parasitol 19:547–551

Eid HO, Hefny AF, Joshi S, Abu-Zidan FM (2008) Non-traumatic perforation of the small bowel. Afr Health Sci 8:36–39

El Kettani S, Azzouzi E (2006) Prevalence of helminths in a rural population using wastewater for agricultural purposes at Settat (Morocco). Sante 16:245–251

el Shazly A, Handousa A, Ibrahem M (1998) Histochemical and pathological studies on biopsied materials from patients with Ancylostoma duodenale infection. J Egypt Soc Parasitol 28:665–672

El-Gammal N, Sayed El-Ahl S, Osman FH, Salem HS (1995) Comparative study of parasitic infections among school children in two rural areas in upper Egypt (Demo village) and lower Egypt (Malames village). Egypt J Community Med 3:25–30

Eligail A, Masawi A, Al-Jaser N, Abdelrahman K, Shah A (2010) Audit of stool analysis results to ensure the prevalence of common types of intestinal parasites in Riyadh region, Saudi Arabia. Saudi J Biol Sci 17:1–4

El-Masry HM, Ahmed YA, Hassan AA, Zaky S, Abd-Allah ES, El-Moselhy EA, Baraka YA, Abdel-Rahem MA (2007) Prevalence, risk factors and impacts of Schistosomal and intestinal parasitic infections among rural school children in Sohag Governorate. Egypt J Hosp Med 29:616–630

El-Nofely A, Shaalan A (1999) Effect of Ascaris infection on the nutritional status and I.Q. of children. Int J Anthropol 14:55–59

El-Sahn FF, Deghedi BM, Mahdy NH, El Sahn A (1997) The impact of intestinal parasitic infections on the nutritional status of primary school children in Alexandria, Egypt. J Egypt Public Health Assoc 72:113–151

El-Sherbini GT, El-Sherbini ET (2011) The role of cockroaches and flies in mechanical transmission of medical important parasites. J Entomol Nematol 3:98–104

El-Sherbini GT, Gneidy MR (2012) Cockroaches and flies in mechanical transmission of medical important parasites in Khaldyia village, El-Fayoum, Governorate, Egypt. J Egypt Soc Parasitol 42:165–174

Ensink JH, van der Hoek W, Mukhtar M, Tahir Z, Amerasinghe FP (2005) High risk of hookworm infection among wastewater farmers in Pakistan. Trans R Soc Trop Med Hyg 99:809–818

Fallah AA, Pirali-Kheirabadi K, Shirvani F, Saei-Dehkordi SS (2012) Prevalence of parasitic contamination in vegetables used for raw consumption in Shahrekord, Iran: Influence of season and washing procedure. Food Contr 25:617–620
Farag HF (1985) Intestinal parasitosis in the population of the Yemen Arab Republic. Trop Geogr Med 37:29–31
Farhan AO (2012) Prevalence of Intestinal Parasitic Infestations in Al-Anbar Province, West of Iraq. J Univ Anbar Pure Sci 6:12–15
Fawzi M, El-Sahn AA, Ibrahim HF, Shehata AI (2004) Vegetable-transmitted parasites among inhabitants of El-Prince, Alexandria and its relation to housewives' knowledge and practices. J Egypt Pub Health Assoc 79:13–29
Fürst T, Keiser J, Utzinger J (2012) Global burden of human food-borne trematodiasis: a systematic review and meta-analysis. Lancet Infect Dis 12:210–221
Gabrielli AF, Ramsan M, Naumann C, Tsogzolmaa D, Bojang B, Khoshal MH, Connolly M, Stothard JR, Montresor A, Savioli L (2005) Soil-transmitted helminths and haemoglobin status among Afghan children in World Food Programme assisted schools. J Helminthol 79:381–384. doi:10.1079/JOH2005316
Garedaghi Y, Farhang HH, Pooryagoobi S (2011) Parasitic contamination of fresh vegetables consumed in Tabriz, Iran. Res J Biol Sci 6:518–522
Geissler PW, Mwaniki D, Thiong'o F, Friis H (1998) Geophagy as a risk factor for geohelminth infections: a longitudinal study of Kenyan primary schoolchildren. Trans R Soc Trop Med Hyg 92:7–11
Gelpi AP, Mustafa A (1967) Seasonal pneumonitis with eosinophilia. A study of larval ascariasis in Saudi Arabs. Am J Trop Med Hyg 16:646–657
Gharavi M, Jahani M, Rokni M (2002) Parasitic contamination of vegetables from farms and markets in Tehran. Iranian J Pub Health 31:83–86
Grove DI (1990) A history of human helminthology. CAB International, Wallingford
Grover S, Alanezi A, Radovanovic Z (2008) Frequency of intestinal nematode infections among housemaids in Kuwait as a hint for their mass treatment. Med Princ Pract 4:99–103
Habbari K, Tifnouti A, Bitton G, Mandil A (1999) Helminthic infections associated with the use of raw wastewater for agricultural purposes in Beni Mellal, Morocco. East Mediterr Health J 5:912–921
Habbari K, Tifnouti A, Bitton G, Mandil A (2000) Geohelminthic infections associated with raw wastewater reuse for agricultural purposes in Beni-Mellal, Morocco. Parasitol Int 48:249–254
Hamed AF, Yousef FMA, Omran EK, Moustafa A (2013) Common parasitic infestation among rural population in Sohag Governorate, Egypt. J Am Sci 9:596–601
Hoeppli R (1959) Parasites and parasitic infections in early science and medicine. University of Malaya Press, Singapore
Holland C (2009) Predisposition to ascariasis: patterns, mechanisms and implications. Parasitology 136:1537–1547
Hotez PJ, Bethony J, Bottazzi ME, Brooker S, Buss P (2005) Hookworm: "the great infection of mankind". PLoS Med 2:e67
Hotez PJ, Savioli L, Fenwick A (2012) Neglected tropical diseases of the Middle East and North Africa: review of their prevalence, distribution, and opportunities for control. PLoS Negl Trop Dis 6:e1475. doi:10.1371/journal.pntd.0001475
Ibrahim O, Bener A, Shalabi A (1993) Prevalence of intestinal parasites among expatriated workers in Al-Ain, United Arab Emirates. Ann Saudi Med 13:126
Ibrahium AA (2011) Prevalence and predisposing factors regarding intestinal parasitic infections among rural primary school pupils at Minia Governorate, Egypt. J Public Health Africa 2:e29
Idris M, Ruppel A, Shaban M, Mehta F, Al-Aufy S, Badesab B, Mohamed A (1995) Study on intestinal parasitic infections in Dhofar region. Community Health and Disease Surveillance Newsletter 6: 1–4

Idris M, Shaban M, Fatahallah M (2001) Effective control of hookworm infection in school children from Dhofar, Sultanate of Oman: a four-year experience with albendazole mass chemotherapy. Acta Trop 80:139–143

Imam AM, Altayyar A, Eltayeb E, Almushawa Y (2012) Frequency and seasonality of intestinal parasitism in Qassim region, Saudi Arabia. Pak J Med Sci 28

Jiménez-Albarrán M, Odda R (1994) A coprological study of intestinal infections in Northern Morocco (provinces of Tangier, Tetuán and Larache). Rev Sanid Hig Publica 68:405–418

Kadir MA, Salman YG (1999) Prevalence of intestinal parasites among primary school children in Al-Taameem province, Iraq. Annals Coll Med Mosul 25: 94–8

Kuntz RE, Lawless DK, Strome CP (1956) Intestinal parasites in an Egyptian village of the Nile Valley with emphasis on the protozoa. Am J Trop Med Hyg 5:1010–1014

Loukas A, Prociv P (2001) Immune responses in hookworm infections. Clin Microbiol Rev 14:689–703

Lozano R, Naghavi M, Foreman K, Lim S, Shibuya K, Aboyans V, Abraham J, Adair T, Aggarwal R, Ahn SY (2013) Global and regional mortality from 235 causes of death for 20 age groups in 1990 and 2010: a systematic analysis for the Global Burden of Disease Study 2010. Lancet 380:2095–2128

Mahfouz AA, El-Morshedy H, Farghaly A, Khalil A (1997) Ecological determinants of intestinal parasitic infections among pre-school children in an urban squatter settlement of Egypt. J Trop Pediatr 43:341–344

Marti H, Koella JC (1993) Multiple stool examinations for ova and parasites and rate of false-negative results. J Clin Microbiol 31:3044–3045

Mehraj V, Hatcher J, Akhtar S, Rafique G, Beg MA (2008) Prevalence and factors associated with intestinal parasitic infection among children in an urban slum of Karachi. PLoS One 3:e3680

Miller FD, Hussein M, Mancy KH, Hilbert MS (1980) Human intestinal parasitic infections and environmental health factors in rural Egyptian communities. A Report of the U.S.-Egyptian River Nile and Lake Nasser Research Project. Environmental Research Laboratory, Office of Research and Development, U.S. Environmental Protection Agency, Athens, GA

Mohamed NH, Bishara SA, Soffar SA, Fawzy AF (1985) Prevalence of parasitic diseases in Tahrir Province and Shubra El Kheima. J Egypt Soc Parasitol 15:535–542

Mohamed NH, Fawzy AF, Sarwat MA, Farrag AM (1988) Parasitic diseases in El Korean, Sharkyia Governorate with evaluation of some methods used in stool examination. J Egypt Soc Parasitol 18:305–311

Mohammad KA, Mohammad AA, Abu El-Nour MF, Saad MY, Timsah AG (2012) The prevalence and associated risk factors of intestinal parasitic infections among school children living in rural and urban communities in Damietta Governorate, Egypt. Academia Arena 4:90–97

Mowlavi GR, Mir Ahmadi H, Rezaeian M, Beigom kia E, Ebrahimi Daryani N, Rokni MB, Golestan B, Shafiei R, Fereshtehnejad SM, Keramati MR (2008) Prevalence of Intestinal parasites in tribal parts of Khuzestan province during 2005-07. Govaresh 12: 219–228

Murray CJ, Vos T, Lozano R, Naghavi M, Flaxman AD, Michaud C, Ezzati M, Shibuya K, Salomon JA, Abdalla S (2013) Disability-adjusted life years (DALYs) for 291 diseases and injuries in 21 regions, 1990–2010: a systematic analysis for the Global Burden of Disease Study 2010. Lancet 380:2197–2223

Nahmias J, Greenberg Z, Djerrasi L, Giladi L (1991) Mass treatment of intestinal parasites among Ethiopian immigrants. Isr J Med Sci 27:278

Nasher AK (1988) Zoonotic parasite infections of the Arabian sacred baboon *Papio hamadryas arabicus Thomas* in Asir Province, Saudi Arabia. Ann Parasitol Hum Comp 63:448–454

Nazemi S, Raei M, Amiri M, Chaman R (2012) Parasitic contamination of raw vegetables in Shahroud, Semnan. Zahedan J Res Med Sci 14:84–86

Niazi AD, Al-Issa TB, Al-Dorki KA, Al-Hussaini M, Al-Khalissi A, Khamis F (1975) Pilot study on prevalence of ankylostomiasis in Iraq. Bull Endem Dis 16:105–142

Pampiglione S, Hadjerès S (1965) On a new focus of anchylostomiasis discovered in eastern Algeria. Riv Parassitol 26:241–250

Patel PK, Khandekar R (2006) Intestinal parasitic infections among school children of the Dhahira Region of Oman. Saudi Med J 27:627–632

Peltola H, Kataja M, Mohamed ON, Kyrönsepp H (1988) Intestinal parasitism of children and mothers in rural Somalia. Pediatr Infect Dis J 7:488–491

Rokni M (2008) The present status of human helminthic diseases in Iran. Ann Trop Med Parasitol 102:283–295

Sadaga G, Kassem H (2007) Prevalence of intestinal parasites among primary schoolchildren in Derna District, Libya. J Egypt Soc Parasitol 37:205

Safi Z, Buerkert A (2012) Heavy metal and microbial loads in sewage irrigated vegetables of Kabul, Afghanistan. J Agri Rural Dev Tropics Subtropics 112:29–36

Saki J, Khademvatan S, Masoumi K, Chafghani M (2012) Prevalence of intestinal parasitic infections among food handlers in Khuzestan, Southwest of Iran: a 10-year retrospective study. Afr J Microbiol Res 6:2475–2480

Sayyari A, Imanzadeh F, Bagheri Yazdi S, Karami H, Yaghoobi M (2005) Prevalence of intestinal parasitic infections in the Islamic Republic of Iran. East Mediterr Health J 11:377–383

Schad GA (1991) The parasite. In: Gilles HM, Ball PAJ (eds) Hookworm infections: human parasitic diseases, vol 4, Elsevier. Amsterdam, The Netherlands, pp 121–154

Schär F, Trostdorf U, Giardina F, Khieu V, Muth S, Marti H, Vounatsou P, Odermatt P (2013) Strongyloides stercoralis: global distribution and risk factors. PLoS Negl Trop Dis 7:e2288

Scott JA (1939) Observations on infection with the common round worm, *Ascaris lumbricoides*, in Egypt. Am J Epidemiol 30-SectionD: 83–116

Scott JA, Ayoub GE, Reitler R (1934) Hookworm disease in the coastal plain of Palestine. Am J Epidemiol 19:601–628

Shalaby HA, Abdel-Shafy S, Derbala AA (2010) The role of dogs in transmission of *Ascaris lumbricoides* for humans. Parasitol Res 106:1021–1026. doi:10.1007/s00436-010-1755-8

Smith JL, Brooker S (2010) Impact of hookworm infection and deworming on anaemia in non-pregnant populations: a systematic review. Trop Med Int Health 15:776–795

To KK, Cheng VC, Hung IF, Tang BS, Wong SS, Yuen KY (2006) Caecal-caecal intussusception caused by *Trichuris trichiura* in a young healthy adult. Scand J Infect Dis 38:813–815

Wells WH, Blagg W (1956) A survey of human intestinal parasites in a fishing village of Northern Egypt. Am J Trop Med Hyg 5:266–268

WHO (2002) Prevention and control of schistosomiasis and soil-transmitted helminthiasis. Report of a WHO Expert Committee. WHO Technical Report Series No. 912 (i-vi). World Health Organization, Geneva, Switzerland

WHO (2012) Eliminating soil-transmitted helminthiasis as a public health problem in children: progress report 2001-2010 and strategic plan 2011-2020. World Health Organization, Geneva

WHO (2013) Soil-transmitted helminthiases: number of children treated in 2011. Weekly Epidemiol Record 88: 145–52

Yenikomshian HA, Berberian DA (1932) A preliminary report on the incidence of hookworm disease in Syria and the Lebanon. Trans R Soc Trop Med Hyg 25:399–406

Zaglool DA, Khodari YA, Gazzaz ZJ, Dhafar KO, Shaker HA, Farooq MU (2011) Prevalence of intestinal parasites among patients of Al-Noor Specialist Hospital, Makkah, Saudi Arabia. Oman Med J 26:182

Human Schistosomiasis in the Middle East and North Africa Region

Rashida Barakat, Hala El Morshedy, and Azza Farghaly

Contents

Abstract Human schistosomiasis is one of the most common Neglected Tropical Diseases (NTDs); it is an intravascular parasite caused by the trematode blood fluke (*Schistosoma*). Most human infections are caused by *S. haematobium*, *S. mansoni*, and *S. japonicum*. An estimated total of 237 people are infected worldwide, and 732 million people are at risk of being infected. In the Middle East and North Africa (MENA) Region alone, 12.7 million individuals are infected. The link between poverty and high prevalence is evident, where approximately ten million of infected individuals are clustered in Egypt and Yemen. However, during the past 20 years significant changes had occurred in the region. Schistosomiasis was eliminated from Islamic Republic of Iran, Oman, Lebanon, and Tunisia. Transmission has been greatly reduced in Egypt, Morocco, Saudi Arabia, Iraq, Jordan, and Syria. Evidence from the Egyptian experience indicated that a nonintegrated intervention strategy,

Author was deceased at the time of publication.

R. Barakat • A. Farghaly
High Institute of Public Health, Alexandria University, Alexandria, Egypt
e-mail: barakat@dataxprs.com.eg; azza.farghaly@yahoo.com

H. El Morshedy (✉)
Faculty of Medicine, Princess Nora Bint Abdul Rahman University, Riyadh, KSA
e-mail: elmorshedyh@hotmail.com

M.A. McDowell and S. Rafati (eds.), *Neglected Tropical Diseases - Middle East and North Africa*, Neglected Tropical Diseases, DOI 10.1007/978-3-7091-1613-5_2,

such as annual drug delivery for morbidity control, did not succeed to alter transmission.

Large-scale mass chemotherapy is the first step to reducing the burden of *Schistosoma*-related disease; yet, such programs may not significantly alter parasite transmission in high-risk areas. Snail control, integrated with drug treatment proved to be most efficacious in preventing and controlling schistosomiasis in Saudi Arabia, Morocco, and Egypt. Intersectoral collaboration between health, agriculture, and education is an extremely important part of advocacy in any program and is an essential part of any successful program in countries which achieved elimination or near elimination progress. This necessitates political commitments for decades. Despite the notable success in schistosomiasis control in the region, achievements are jeopardized by the current political instability, therefore resurgence of high prevalence, high intensity of infection, and severe morbidity might ensue.

Yemen is the only country in the Middle East not to have eliminated the disease. The significant World Bank funds being allocated to the 6-year control program in Yemen will continue provided the political situation there returns to a level that allows drug distribution.

Keywords Human schistosomiasis • The MENA Region • Epidemiology • Control

Human Schistosomiasis

Human schistosomiasis is one of the most common NTDs; it is an intravascular parasite caused by the trematode blood fluke (*Schistosoma*). Most human infections are caused by *S. haematobium*, *S. mansoni*, and *S. japonicum*; less prevalent species include *S. mekeongi* and *S. intercalatum*. Schistosomiasis is endemic in 77 countries in tropical and subtropical regions; estimates of infected individuals worldwide are 237 million; another 600–779 are at risk of being infected (Chitsulo et al. 2000; Steinmann et al. 2006; WHO Weekly Epidemiological Report 2012). The life cycle of the parasite is characterized by alteration of generation; asexual reproduction occurs in the snail intermediate host and sexual reproduction occurs in humans (Fig. 1). The pathology of schistosomiasis is due to egg-mediated immune response in the form of granuloma formation followed by fibrosis which results in obstructive manifestations in the gastrointestinal tract (GIT) in case of intestinal schistosomiasis and in the urinary tract in the case of *S. haematobium* (Nash et al. 1982; Wynn et al. 2004; Wilson et al. 2007). However, eggs can be disseminated to other organs, e.g., the brain, the spinal cord, genital organs, and the lungs leading to severe morbidity (Gryseels et al. 2006). Squamous cell carcinoma is one of the serious complications of urinary schistosomiasis in Egypt and North Africa (Fedewa et al. 2009). In infected children, studies of physical and intellectual functions indicate significant reductions in physical fitness and spontaneous activity

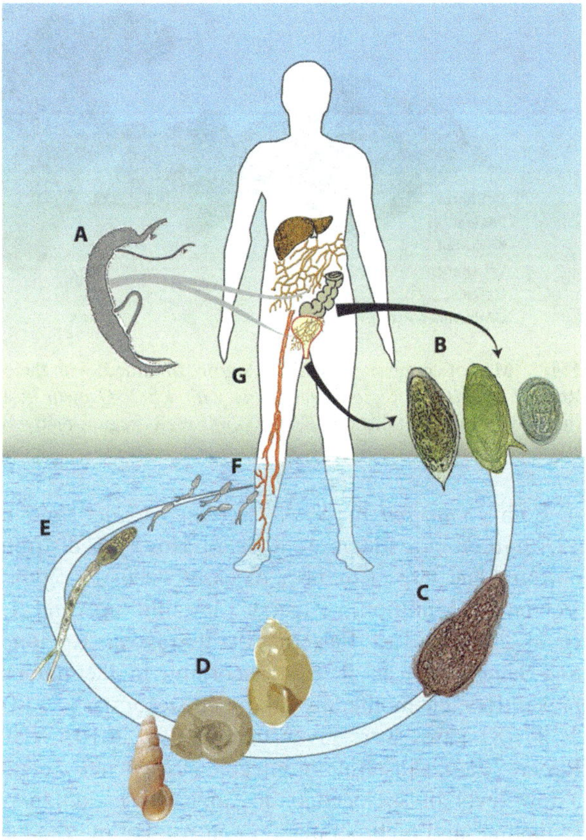

Fig. 1 Life cycle of human schistosomiasis (A) adult worms, (B) eggs (left to right, *S. haematobium*, *S. mansoni*, *S. japonicum*), (C) miracidium, (D) intermediate snail host (left to right, *Oncomelania*, *Biomphalaria*, *Bulinus*), (E) cercaria. Used with permission from the Institute of Tropical Medicine Antwerp

among children (Latham et al. 1990). Linear growth and nutrition are impaired, resulting in stunting and underweight status among infected children (Assis et al. 1998; Coutinho et al. 2006). Poor performance in standardized intelligence and achievement tests has also been associated with schistosomiasis (Nazel et al. 1999; Jukes et al. 2002; Ezeamama et al. 2005, 2012).

Overview of the Middle East and North Africa Region

The MENA Region includes 21 countries inhabited by 336.5 million people (Fig. 2). Population density in MENA accounts for 5 % of the world's total population. The highest density is found in Egypt (82 million), followed by Iran (75 million), Algeria (36 million), Morocco and Iraq (31–32 million each), Kingdom of Saudi Arabia (KSA) (26 million), and Yemen (25 million) (Population Reference Bureau 2013).

Fig. 2 Map of countries in MENA Region. Modified, with the permission of the publisher, from Roy "*IFM: MENA Region Recovering with 4.5 % Growth in 2010,*" Offshore Capitalist, 2010 (http://offshore-capitalist.com/2010/04/imf-mena-region-recovering-with-4-5-growth-in-2010/)

Both *S. mansoni* and *S. haematobium* are endemic in the Region; approximately 12.7 million individuals are infected. However, distribution of infected cases is not uniform; the largest number of cases occur in Egypt (7.2 million), followed by Yemen (2.9 million), Algeria (2.3 million), and Libya (0.3 million) (Hotez et al. 2012). The Eastern Mediterranean Region ranked second after the Sub-Sahara African Region according to the number of individuals requiring preventive chemotherapy for schistosomiasis (14,493,641); however, only 2,137,787 cases were given treatment in 2010 (WHO Weekly Epidemiological Report 2012). Clustering of infected cases in a few countries of the Region is due to a low level of socioeconomic standards including poverty, bad environmental sanitation, and high population density. Estimates since 2011 indicate that 2.4 % of the population lives below the World Bank poverty figure of US$1.25 per day and 12 % lives below US$2 per day. It is noteworthy that most of the countries in the Region are classified by the World Bank as low-middle income countries (World Bank (n.d.) Data:Middle East and North Africa).

However, during the past 20 years significant changes have occurred in the Region. Schistosomiasis was eliminated from the Islamic Republic of Iran, Oman, Lebanon, and Tunisia. Transmission has been greatly reduced in Egypt, Iraq, Jordan, Morocco, Saudi Arabia, and Syria, while in Yemen schistosomiasis is considered a major health problem (Table 1 and Fig. 3) (Fenwick et al. 2006; Rollinson et al. 2012; International Association for Medical Assistance to Travelers 2012). In this chapter the focus will be on countries with a large population size which achieved notable progress in the control of schistosomiasis, e.g., Egypt, Morocco, and Saudi Arabia, and countries where schistosomiasis has remained as a major health problem such as the Yemen.

Table 1 Condition of schistosomiasis in MENA Region according to country

Country	Condition	Snails
Algeria	*S. haematobium* is absent from most of the country, risk of infection is localized in the province of Boumèrdes	*B. truncatus*
Egypt	Both species are endemic; *S. haematobium* is endemic throughout southern Egypt including Fayoum. *S. mansoni* is endemic in the Nile Delta and Suez Canal Region. There are limited foci in southern Egypt. Control activities have reduced the infection; estimated prevalence in 2003 is <3 % in most villages and 0.3 % in 2012	*B. truncatus* *B. alexandrina*
Iran	*S. haematobium* is eliminated	*B. truncatus*
Iraq	*S. haematobium* is present along the Euphrates and Tigris river system, estimated prevalence in 2003 and 2010 is 0.1 %	*B. truncatus*
Jordan	Eliminated, non-endemic	–
Lebanon	*S. haematobium* elimination was declared	*B. truncatus*
Libya	Both *S. haematobium* and *S. mansoni* are endemic. Areas of risk are limited and specified in: 1. Near the Mediterranean coast in Darnah and in oasis south of Misratah. 2. In the central part of Fezzan. 3. Near the southwestern border of Algeria. Estimate of prevalence in 2003 and 2010 is 5 %	*B. truncatus* *B. globsus* *B. alexandrina*
Morocco	Transmission of *S. haematobium* is interrupted based on serological study of the remaining low risk foci	*B. truncatus*
Oman	*S. mansoni* is absent from most of the country, estimated prevalence in 2003 and 2010 is 0.1 % and 0.01 % respectively	*B. alexandrina*
Saudi Arabia	Both S. *haematobium* and *S. mansoni* are endemic. Control has reduced the infection to 0.1 % in 2003 and 0.02 % in 2010. Areas of risk are restricted to Asir in the southwestern region	*B. truncatus* *B. buccarii* *B. wright* *B. phfeifferi*
Syria	*S. haematobium* low risk areas are located along the river system in the northern parts of the country. Estimated prevalence in 2003 and 2010 is <0.1 % and <10 %, respectively	*B. truncatus*
Tunisia	Elimination of *S. haematobium* was declared	*B. truncatus*
Turkey	Eliminated, non-endemic	–
Yemen	Both *S. haematobium* and *S. mansoni* are endemic in the whole country including urban communities, estimated prevalence in 2003 and 2010 is 14.6 % and 14.3 %, respectively	*B. truncatus* *B. phfeifferi* *B. alexandrina*

Human Schistosomiasis in Egypt

History

The history of urinary schistosomiasis in Egypt is long-standing, i.e., since the time of ancient Egypt. The first description of the disease was found in Kahun papyrus (1900 B.C.) referring to hematuria as a manifestation of a disease known as a-a-a disease (Shokeir and Hussein 1999) (Fig. 4). The parasite etiology of the disease is proposed in the Ebers papyrus in (1550 B.C.), while the eggs of *S. haematobium* have been

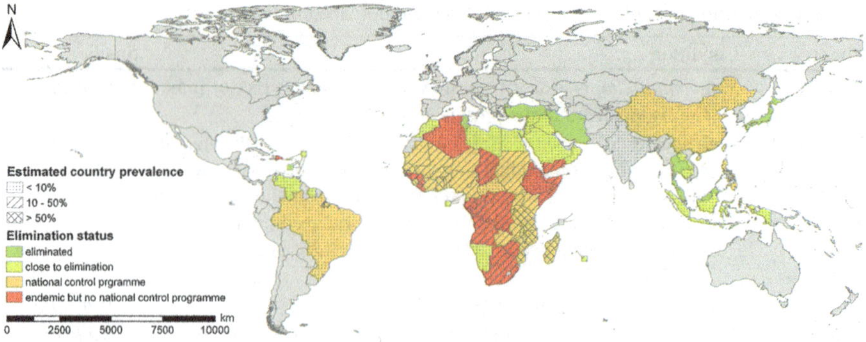

Fig. 3 World map, highlighting countries where schistosomiasis has been eliminated (*green color*), is close to elimination (*yellow color*), ongoing control program (*orange color*). *Red color*, where schistosomiasis is endemic, but national control programs have yet to be implemented. *Hachure* indicate prevalence in countries, *pointed hachure* (low prevalence <10 %), *dashed hachure* (moderate prevalence 10–50 %), and *crossed hachure* high prevalence >50 %). Reprinted from (Rollinson et al. 2012) Copyright (2012), with permission from Elsevier

Fig. 4 Hieroglyphic script of haematuria referring to schistosomiasis (a-a-a disease) as it appears in the Kahun papyrus. From http://www.scribd.com/doc/16105584/Urology-doc-Circumcision

found in the tissues taken from mummies of the twentieth dynasty, 1250–1000 B.C. (Ruffer 1910; Ebbel 1937). The morbidity of urinary schistosomiasis was mentioned in the Chester Beatty papyrus describing bladder cancer as a disease complication (Badr 1981). The causative organism was termed *Bilharzia* after Theodor Bilharz, who described the parasite from human autopsy in Kasr El Ainy hospital in Cairo in 1851 (Bilharz 1853). The discovery of the life cycle and presence of both *S. haematobium* and *S. mansoni* in Egypt was first reported by Leiper (1915). Recently, in the early 1990s, Schistosome antigens were detected in tissues from mummies using the ELISA technique (Deelder et al. 1990; Miller et al. 1992).

Distribution of Schistosome Species in Egypt

The comprehensive epidemiology of schistosomiasis was first described by Scott (1937). He mapped the distribution of both *S. haematobium* and *S. mansoni* in the Nile Delta representing Northern Egypt and the Nile Valley representing Southern Egypt. The Nile Valley comprises Middle Egypt which extends from Giza to

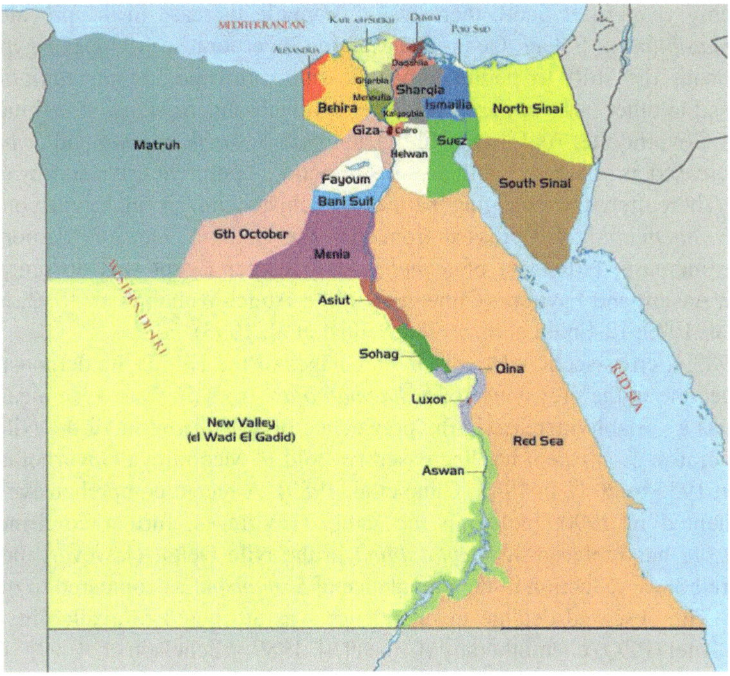

Fig. 5 Map of Egyptian Governorates. Food and Agriculture Organization of the United Nations, 2011, Mohamed A. El-Nahrawy, Country pasture/Forage Resource Profile. http://www.fao.org/ag/AGP/AGPC/doc/Counprof/PDF%20files/Egypt.pdf. Reproduced with permission

Menya Governorate and Upper Egypt extending from Assiut to Aswan. *S. haematobium* was endemic throughout Egypt; prevalence in the Nile Delta and in the parts of the Nile Valley with perennial irrigation was high (60 %), while low prevalence (6 %) was reported in regions of basin irrigation in the Nile Valley. Distribution of *S. mansoni* was limited to the Delta region; prevalence was high (60 %) in the northern and eastern parts of the Delta, whereas low prevalence (6 %) occurred in southern parts of the Delta (Fig. 5).

The snail intermediate host of *S. mansoni* and *Biomphalaria alexandrina* was abundant in the Nile Delta only, whereas *Bulinus truncatus*, the snail intermediate host of *S. haematobium*, was found in the Nile Delta and the Nile Valley; its density was higher in areas of the Valley experiencing perennial irrigation than in areas with basin irrigation. The breeding places of snails were swept by the annual flood of the River Nile and desiccated during the hot summer in areas of basin irrigation resulting in interruption of disease transmission. Accordingly, Scott concluded that a shift to perennial irrigation in parts of the valley resulted in high prevalence of *S. haematobium*. Similarly, several studies confirmed that a shift to perennial irrigation was associated with high prevalence of *S. haematobium* in the Nile Valley as far as Nouba, the farthest south in Upper Egypt (Azim 1935; Khalil and Azim 1938; El-Zawahry 1964).

Twenty years after Scott, there was a dramatic increase in the prevalence of *S. haematobium* in Sohag, Qena, and Aswan Governorates in Upper Egypt coinciding with the shift to perennial irrigation; meanwhile, *S. haematobium* had decreased in other parts of the valley. On the other hand, *S. mansoni* had increased in Giza Governorate. As for the Nile Delta, both *S. haematobium* and *S. mansoni* had decreased as compared to the results of Scott's comprehensive survey (Wright 1973). Thereafter, the changing pattern of schistosomiasis in Egypt continued, while *S. haematobium* decreased from the Nile Delta; *S. mansoni* demonstrated relative increase. Extension of *S. mansoni* to Upper Egypt was documented in Menya, Assiut, and Fayoum (Cline et al. 1989; Abdel-Wahab et al. 1993; Michelson et al. 1993; El-Enien et al. 1993; Medhat et al. 1993).

In1983, a cross-sectional study of 71 villages of the Nile Delta demonstrated a sharp decline in the prevalence of *S. haematobium* from 56 % to 5 %; meanwhile, there was a variable increase in the prevalence of *S. mansoni* in all the Nile Delta Governorates. *S. mansoni* had increased twofold in Menoufeya Governorate from 10 % in 1935 to 20 % in 1983 (Cline et al. 1989). A recent comprehensive survey implemented in 1990, including the same 71 villages, further confirmed that *S. mansoni* has replaced *S. haematobium* in the Nile Delta. However, there was an overall 38 % reduction in the prevalence of *S. mansoni* as compared to the 1983 survey. The observed decline in prevalence was attributed to availability of the praziquantel (PZQ) chemotherapy (Cline et al. 1989; Michelson et al. 1993). In the Nile Valley governorates, *S. haematobium* decreased except in Sohag, Qena, and Aswan (Miller et al. 1981; El-Khoby et al. 2000a).

In 1990, a comprehensive house-to-house survey covered 251 villages from 9 governorates representing the Nile Delta, and the Nile Valley was implemented by the epidemiology teams of the Schistosomiasis Research Project (SRP) (El-Khoby et al. 2000b). The results demonstrated an average *S. mansoni* prevalence of 36.45 %, while *S. haematobium* vanished from most of the governorates. The highest prevalence was reported from Ismailia, Suez Canal zone (1.8 %), and the lowest was in Qalubia (0.08 %) (Barakat et al. 1995; Habib et al. 2000; El-Hawey et al. 2000; Abdel-Wahab et al. 2000; Nooman et al. 2000). In Upper Egypt, the average prevalence of *S. haematobium* was 7.8 %, while the highest prevalence was 13.7 % and the lowest prevalence was 4.8 %. On the other hand, *S. mansoni* was rare, except in Fayoum where the recorded prevalence was 4.3 % (Abdel-Wahab et al. 2000; El-Khoby et al. 2000a; Hammam et al. 2000a, b).

An Update of the Epidemiology Profile of Human Schistosomiasis in Egypt

The SRP encompassed 89,180 individuals from 251 villages representing rural communities in Egypt. The project provided estimates of prevalence and intensity of infection of both Schistosome species, evaluated risk factors, investigated morbidity for the first time using portable ultrasonography and documented changing patterns of schistosomiasis in Egypt. *S. haematobium* infection was clustered in

Upper Egypt, prevalence of infection ranged from 4.8 % in Qena to 13.7 % in Fayoum. Overall, infection peaked at 15.7 % among the age group 10–14 years; males were 1.9 times more infected than females. The average intensity of infection was below 10 eggs per 10 ml of urine and ranged from 1.53 eggs to 9.95 eggs per 10 ml of urine; the intensity of infection was higher among males in all age groups. The force of infection was higher in ezbas (hamlets) than in large villages, due to poverty, lack of potable water supply, bad environmental sanitation and inaccessibility to health services which are more common in ezbas than in large villages. Microscopic hematuria increased the risk of infection (OR = 12.4). Among infected cases, bladder lesions were detected in 2 % only, whereas first grade periportal pipe stem fibrosis (PPF) and splenomegaly were found in 14.7 % and 13.4 % respectively (Abdel-Wahab et al. 2000; El-Khoby et al. 2000a; Hammam et al. 2000a, b).

The overall prevalence of *S. mansoni* in the Nile Delta and Ismailia (Suez Canal zone) was 36.4 %. The highest prevalence was in Ismailia Governorate (42.9 %), followed by Kafr El-Sheikh (39.2 %), Menoufeya (28.5 %), and the least in Qalubia (17.5 %). Infection peaked at 48.3 %; males were 1.6 times as likely to be infected as females. The overall intensity of infection was below 100 eggs per gram of stool (epg) and ranged from 62.6 epg in Qalubia Governorate to 93.3 epg in Ismailia Governorate. Similar to *S. haematobium* individuals living in ezbas were at higher risk of becoming infected (OR = 1.9) as compared to those living in large villages. Examination with portable ultrasonography demonstrated splenomegaly in 20.8 % while grade II and grade III PPF was found in 50.3 % (Barakat et al. 1995; El-Khoby et al. 2000a; El-Hawey et al. 2000; Habib et al. 2000; Abdel-Wahab et al. 2000; Nooman et al. 2000).

The results of the SRP comprehensive survey representing schistosomiasis endemic communities in Egypt provided accurate data about the epidemiology profile and changing pattern of schistosomiasis, while *S. haematobium* continues to decline sharply in the Nile Delta, *S. mansoni* transmission is continuing at an appreciable level, in addition to evidence of extension of *S. mansoni* to Upper Egypt (Talaat et al 1999; El-Khoby et al. 2000a; Lotfy 2009; Barakat 2012). Moreover, the pattern of morbidity sequelae of *S. haematobium* has changed over the past 26 years. Transitional cell carcinoma of the urinary bladder has replaced the squamous cell carcinoma previously associated with a history of urinary schistosomiasis (Felix et al. 2008; Fedewa et al. 2009; Salem et al. 2011). Age and sex distribution of infection of this study is close to data of individual governorates in Egypt and elsewhere in other countries known to be endemic for schistosomiasis (El-Malatawy et al. 1992; Barakat et al. 2000; Raja'a et al. 2000; Gryseels et al. 2006; Mazigo et al. 2012). Predisposition to infection according to age and gender is due to socioeconomic behavioral, ecological, and biological factors which influence the interaction between human and animal hosts and life cycle stages of the parasite (Mazigo et al. 2012).

Observation of SRP data according to governorates demonstrated that Kafr El-Sheikh Governorate had a high prevalence of *S. mansoni*. The study sample from Kafr El-Sheikh included 18,168 individuals from 44 villages and ezbas. Individual analysis of data from Kafr El-Sheikh provides a better understanding

Fig. 6 Implantation of rice in an Egyptian village. Science Photo Library #E768/0409

of the epidemiology in hot transmission areas. At baseline, in 1990, the overall prevalence was 39.3 ± 3.3 %, and the intensity of infection was 72.9 ± 7.3 epg. The risk of infection was higher among males (OR = 1.4). Infection peaked at 55.4 ± 3.2 % at 16 years of age, and intensity of infection peaked at 81.5 ± 12.1 in the age group 10–14 years. Frequent water contact activities increased the risk of infection (OR = 3). Morbidity data based on sonographic examination demonstrated splenomegaly in 55 % and PPF in 47.25 %, both increased with age and the latter increased with history of bathing in canal water (OR = 51); from the 44 villages *S. haematobium* was diagnosed in 41 samples only (Barakat et al. 2000).

Variability of prevalence and intensity of infection were evident among villages and ezbas indicating focal pattern of disease transmission. The lowest prevalence was 24.5 % and the highest was 68.9 %. Intensity of infection ranged from 37.8 epg to 129.7 epg. The study showed that prevalence and intensity of infection were positively correlated (Barakat et al. 2000).

The overall high prevalence of *S. mansoni* in Kafr El-Sheikh Governorates relative to other governorates in the Nile Delta might be due to extensive cultivation of rice in Kafr El-Sheikh. The season of rice cultivation occurs at the peak of snail shedding during the hot months of summer. It also requires frequent daily contact with water during the implantation period where all ages and both sexes participate in rice planting. This season is considered as a sociocultural event where villagers share work in cultivating their land (Fig. 6). The association of rice cultivation and high prevalence and intensity of infection of *S. mansoni* was recently reported from Côte d'Ivoire (Yapi et al. 2005). The clustering of high intensity of infection in a small segment of population follows the nonrandom distribution of eggs (Kirtorn and Hiagashi 1985). On the other hand, the observed low intensity of infection was explained by the availability of PZQ passive chemotherapy. Variability in force of infection between villages is due to variations in the socioeconomic standards and access to health services (Barakat et al. 2000).

In comparison to other studies in Kafr El-Sheikh Governorate in 1983 and 1990, a similar trend was observed for *S. haematobium* which continues to decline. However, the results for *S. mansoni* were variable, being 51 % in 1983 and 17 % in 1990. Comparisons of the three studies might be difficult because of differences in sample design (Kleinbaum et al. 1982). The high prevalence of the SRP study (39.2 %) as compared to the study done by Miller et al. 1978 (20 %) might be due to differences in the sensitivity of the diagnostic tool. In 1978 study, they used Merthiolate Iodine Formaldhyde Concentration technique (MIFC), while in SRP study, two Kato slides were examined from a single stool sample. The Kato Katz technique is more sensitive than MIFC technique due to examination of a large amount of stool (86 mg); also the Kato technique is a single-step technique without the potential of losing eggs as compared to MIFC which includes several steps such as sieving and centrifugation (Katz et al. 1972).

Influence of Aswan High Dam on the Epidemiology of Schistosomiasis in Egypt

The Aswan High Dam was constructed on the River Nile 7 km south of Aswan in 1967. Its impact on schistosomiasis transmission was controversial, while some authors underestimate the role of the High Dam (Miller et al. 1978); several studies have highlighted its role in changing the ecology of breeding habitat which favors flourishing of *B. alexandrina*, the snail intermediate host of *S. mansoni*. In summary these changes include (1) shift to perennial irrigation all through the Nile Delta and the Nile Valley, (2) changes in the water current velocity, and (3) absence of silt (Malek 1975; Abdel-Wahab et al. 1979; White 1988). It is noteworthy that reclamation of new areas in the desert and creation of the huge Lake Nasser led to mobilization of people from endemic areas thus creating new foci of transmission. In two new reclaimed areas in the desert close to Ismailia, prevalence of *S. mansoni* was 40 % and 49 % in 1992 (El-Sayed et al. 1995).

Despite the controversy in the analysis of epidemiological changes in human schistosomiasis after the construction of the High Dam, there has been a substantial decline of *S. haematobium* and an extension of *S. mansoni* to Upper Egypt. Although the density of *B. alexandrina* has increased, prevalence of *S. mansoni* has decreased due to ongoing control programs. The decline in *S. haematobium* all over Egypt is due to ecological changes created by the construction of the High Dam which interferes with the breeding of *B. truncatus* (Hotez et al. 2012). A cross-sectional study in Fayoum Governorate in 1991 showed that *S. mansoni* prevalence was 22.3 % while *S. haematobium* was 3.4 %; among *S. haematobium* infected cases there were only two children aged below 10 years. Reviewing records of Ministry of Health and Population (MOHP) demonstrated that *B. truncatus* has not been detected in local canals since 1986 and few uninfected snails were found between 1981 and 1985 (Abdel-Wahab et al. 1993). Therefore, these data indicate that interruption of *S. haematobium* transmission is due to unfavorable breeding habitat for the snail intermediate host.

Environmental and demographic changes associated with the development of water resources might facilitate spread of schistosomiasis (Patz et al. 2001; Steinmann et al. 2006). *S. mansoni* was introduced into Senegal and Mauritania after the construction of Senegal River Dam. Ten years after the construction of the Daima Barage Dam in 1985, prevalence of *S. mansoni* ranged from 4.4 % to 43.65 % in the Delta (Picquet et al. 1996). In Ethiopia, introduction of large scheme irrigation projects resulted in the rapid increase of *S. mansoni* prevalence reaching up to 82 % four decades after the start of the project (Kloos et al. 1988; Simonsen et al. 1990). Similarly, construction of For Kossou and Toabo Dams in Côte d'Ivoire increased the prevalence of *S. haematobium* from 14 % to 53 %, while *S. mansoni* remained stable (De Clercq et al. 1999). It seems that changes in transmission of specific Schistosome species is linked to changes in the ecology of the breeding habitat of the snail intermediate host created by hydrological changes which follow construction of dams and irrigation projects.

Control of Human Schistosomiasis in Egypt

Before 1984, control projects in Egypt were planned to interrupt transmission regardless of the force of infection in endemic communities. Molluscicíding and chemotherapy were the main components of all control projects. The largest of these was the Middle Upper control program which started in Middle Egypt in 1980 (Medhat et al. 1993; Talaat et al. 1999). The project was planned in three phase; (1) the intensive phase (wide application of mollusciciding and chemotherapy, (2) consolidation phase (focal mollusciciding and chemotherapy), and (3) maintenance phase. The impact of control activities was remarkable; prevalence of *S. haematobium* decreased from 30 % to 6.5 %. However, reinfection rates were high especially among young children (Kessler et al. 1987; Webbe and El-Hak 1990).

In 1990, the National Schistosomiasis Control Program (NSCP) adopted the strategy of morbidity control according to the new WHO strategy declared in 1980s (WHO 1985). The discovery of PZQ, as a safe drug, given in a single oral dose and effective for all human Schistosome species and the availability of the Kato technique as a sensitive diagnostic test easily processed under field conditions justified the shift to morbidity control. At the beginning of the program, selective chemotherapy was offered through passive and active case finding. In 1997, mass chemotherapy was offered to all children enrolled in schools (aged 6–18 years) and to all villages with prevalence of ≥20 %, in addition to focal mollusciciding, health education, and capacity building of personnel in rural health units. The threshold of mass chemotherapy was further reduced to ≥10 % in 1999, ≥5 % in 2000, ≥3.5 % in 2002, and to ≥3 % in 2003 (WHO 2011). These values are much lower than the recommended WHO regulations (Table 2). Since 1990, more than 50 million doses of PZQ were offered (Rollinson et al. 2012).

Table 2 WHO recommendation for preventive chemotherapy

Prevalence thresholds for schistosomiasis intervention
If prevalence of infection ≥50 % (high-risk community)
Treat all School-age children and other at risk groups once a year
If prevalence of infection ≥10 % and ≤50 % (moderate-risk)
Treat all School-age children and other at risk groups once every 2 years
If prevalence of infection <10 % (low-risk)
Treat all School-age children twice in childhood, and symptomatic cases in health facilities

Modified, with the permission of the publisher, from the "*Report of an Informal Consultation on Schistosomiasis Control*," World Health Organization 2011. (http://apps.who.int/iris/bitstream/10665/78066/1/9789241505017_eng.pdf)

In 1997 before the implementation of mass chemotherapy, an independent team evaluation for NSCP activities was conducted in Kafr El-Sheikh Governorate. A representative sample of 8000 individuals from four high prevalence villages (≥40 %) and four low prevalence villages (≤31 %) was investigated (Barakat et al. 1998). From 1991 to 1993, the SRP epidemiology team of Kafr El-Sheikh Governorate led the morbidity control by means of selective annual chemotherapy, while from 1993 to 1997, NSCP morbidity control measures solely were operating. A significant drop in prevalence and intensity of infection was achieved in all villages in 1993 after two rounds of selective chemotherapy offered by the epidemiology team. In 1997, the downward trend of prevalence was maintained in low prevalence villages; however, an upward increase in prevalence was observed for high prevalence villages, though still lower than the baseline prevalence in 1991 (Table 3) (Barakat et al. 1995, 1998). A very close prevalence was reported for Alexandria Governorate in 1998. A parasitological survey of stool from 3,281 individuals living in rural communities was undertaken to investigate the prevalence of schistosomiasis in this area. The results revealed that prevalence of *S. mansoni* accounted for 20.5 %, with low intensity of infection, and increased with age to reach a maximum of 40–46.3 % at 15–30 years of age. Intensity of infection followed the same pattern (Zaki et al. 2003). At this level of prevalence, selective chemotherapy offered to some of the infected population was not enough to maintain the downward trend in prevalence, indicating that mass treatment and percentage of chemotherapy coverage have to be considered (Wang et al. 2012). According to the World Health Assembly Resolution (WHA 54.19) held in May 2001, the minimum target is to cover 75 % of school children at risk by 2010 (WHO 2001).

The impact of mass chemotherapy adopted by the NSCP was substantial; overall prevalence of *S. mansoni* decreased from 16.4 % in 1988 to 4.2 % in 2000. Similarly, *S. haematobium* prevalence decreased from 11.9 % to 3 % during the same period (El-Khoby et al. 1998; Fenwick et al. 2003). Since 2003, a multisectoral approach involving mass and selective chemotherapy tailored to the force of infection, focal mollusciciding, health awareness and environmental sanitation, and surveillance was adopted, aiming to interrupt transmission. In villages with prevalence ≥3 %, mass chemotherapy, focal mollusciciding, potable water

Table 3 Prevalence of *S. mansoni* in high and low prevalence villages in Kafr El-Sheikh Governorate 1991–1997

	Prevalence of *S. mansoni*			
Village prevalence	1991 (baseline)	1992	1993	1997
Low	30.3	26.4	20.5	17.3
High	44.9	28.8	19.5	19.8

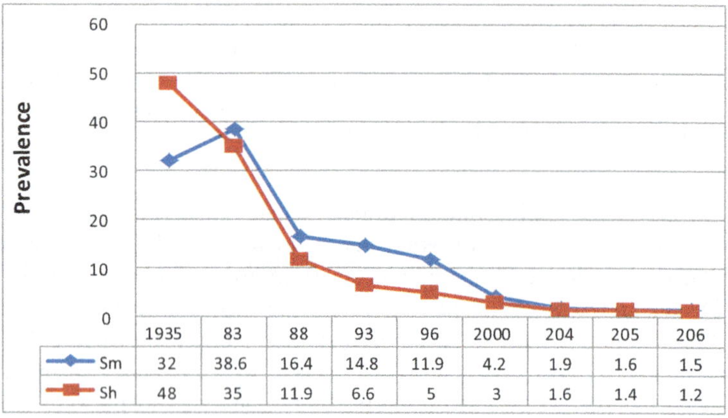

Fig. 7 Overall prevalence of schistosomiasis in Egypt, during the period 1935–2006, Modified, with the permission of the publisher, from the "*Report: Inter-country Meeting on Strategies to Eliminate Schistosomiasis from the Eastern Mediterranean Region*, World Health Organization 2007 (http://www.who.int/schistosomiasis/resources/EMRO_report_Schistosomiasis.pdf)

provisioning, and environmental sanitation were implemented. In villages with prevalence <3 %, the approach was limited to selective chemotherapy, focal mollusciciding, and environmental sanitation. In 2006, the overall prevalence of *S. mansoni* was 1.5 % and that of *S. haematobium* was 1.2 % (Fig. 7) (WHO 2007). In 2010, according to records of MoHP, there were only 20 villages in the whole country with a prevalence of more than 3.5 % and all were <10 % (WHO 2011).

Rollinson et al. (2012) emphasized that elimination must be seen as the extreme end of the control spectrum and not as a new goal by any means. The intersectoral approach between health, education, and agriculture guarantees the success of elimination programs (Holveck et al. 2007; Aagaard-Hansen et al. 2009). It is estimated that, in order to achieve elimination, the control program has to be sustained efficiently for 10–20 years (Curtale et al. 2010). However, morbidity sequelae may remain for decades. Endoscopic and histopathological study during the period from 2004 to 2009 covering 984 individuals aged 18–65 years presented with GIT manifestations demonstrated typical schistosomiasis colorectal lesions in 20.33 % of them (Gad et al. 2011). Serious complications such as hepatic decompensation, hypersplenism, and cor pulmonale are becoming rare. Therefore, in addition to field monitoring of infection and treatment outcomes, complete accurate longitudinal morbidity hospital-based data are paramount to document the success of control program.

Since the late 1980s, PZQ remains the mainstay in schistosomiasis control in all endemic countries, yet the drug is only effective against mature infection (Cioli and Pica-Mattoccia 2003). Drugs effective against premature infection such as artemether—an anti-malarial drug—will act in synergism with PZQ to interrupt transmission. Such a combination might be applicable in Egypt for two reasons; (1) difficulty in sustaining focal molluscicīding because of such a complicated irrigation scheme, and (2) Malaria is not a major health problem in Egypt. The impact of artemether has been investigated in a double blind randomized control trial involving 913 school-age children from hot transmission villages in Kafr El-Sheikh Governorate. At the end of the study, incidence of *S. mansoni* was 2.8 % among the artemether group compared to 6.5 % among the group that offered PZQ only (WHO-EM/TDR/007/E/12.04/2500, 2001-2002).

The success of the control program in Egypt so far is the outcome of sequential planned processes based on countrywide precise mapping of endemic communities, followed by integrated control strategies tailored to local situations and supported by political commitments in addition to mobilization of international funds. Currently, political instability in Egypt might hinder the sustainability of NSCP achievements; moreover, resurgence of high indices of transmission and a greater disease burden cannot be excluded.

Human Schistosomiasis in Morocco

History

Morocco is a Northern African country, bordering the North Atlantic Ocean and the Mediterranean Sea. In 2010, the population size was approximately 32 million. Urinary schistosomiasis is the only form of schistosomiasis in Morocco; it was introduced into Morocco from Egypt or Sub-Saharan Africa (WHO 1993). The first case of urinary schistosomiasis was reported from Marrakesh Province in 1914 (Doumenge et al. 1987).

Epidemiology

Foci of transmission were mainly limited to the southern part of the country especially in the oases along the pre-Saharan belt (Barneoud and Carrosse 1929; Barneoud 1932; Connet 1937; Gaud and Maurice 1946). However, in 1970, spread of infection to other areas in the central and northern parts of Morocco was associated with the establishment of irrigation projects in the late 1960s (Benmansour 1970; Laaziri and Benouna 1982; Khallaayoune et al. 1998a). Throughout Morocco transmission of schistosomiasis was maintained by the snail intermediate host *B. truncatus*, which is more abundant during summer leading to an increase in the intensity of transmission (Khallaayoune et al. 1998b).

Schistosomiasis is patchily distributed in Morocco. Before the implementation of control programs, the disease was endemic in 20 provinces and the rate of infection reached up to 50–60 % in the early 1980s (WHO 1987a, b). Following recreational and domestic activities, infection peaked among school-age children (Khallaayoune and Laamrani 1992; Watts et al. 1998).

Control

The National Program of Schistosomiasis Control (NPSC) started in the early 1970s following the spread of schistosomiasis to the central and northern parts of the country; it was fully implemented in 1982. Strategies of control included; (1) active and passive treatment of infected cases in schools and villages by mobile and local teams. Until 1986, infected cases were treated with metrofonate, thereafter PZQ was introduced, offered as a single oral dose of 40 mg/Kg, (2) Niclosamide focal molluscicıding, (3) health education programs in schools and villages, (4) intersectoral collaboration between health, education, and agriculture (Laamrani et al. 2000; Amarir et al. 2011).

From 1982 through 1997, the impact of the program was remarkable, overall prevalence decreased from 6.2 % to 0.3 %. Annual incidence was also reduced from 8.2/1000 to 1.3/1000. Moreover, there was a shift in the age specific infection rate among children aged 7–14 years; they represented 38 % of the total infected cases in 1996 as compared to 67 % in 1983 (WHO 1987a; Laamrani et al. 2000; Amarir et al. 2011).

The encouraging results of the NPSC lead to upgrading control strategies for elimination. The strategy of elimination was planned in stages with a target date for each province, the final to be reached in 2004 (Ministry of Health 1998). By 1999, the program was successful in 17 provinces (DELM 1999) and continuous decline of *S. haematobium* was evident from 1994 through 2006 (WHO 2007) (Fig. 8). In 2008, serological evidence of interrupted transmission was reported; 2,382 sera from children aged 1-16 years selected from the remaining endemic foci were tested to detect *S. haematobium* specific antibodies and all were negative (Amarir et al. 2011). Further longitudinal studies monitoring specific antibody titers can provide important information about the history of interruption of schistosomiasis transmission (Rollinson et al. 2012).

In conclusion, operational decentralization, integrated control strategies within the existing health services, and the intersectoral approach made achievement of the eradication goal feasible in Morocco (Rollinson et al. 2012). However, threats of resurgence cannot be ignored because the snail intermediate host is prevalent in the country. This necessitates carefully planned surveillance at the micro-level to maintain the success of the eradication program.

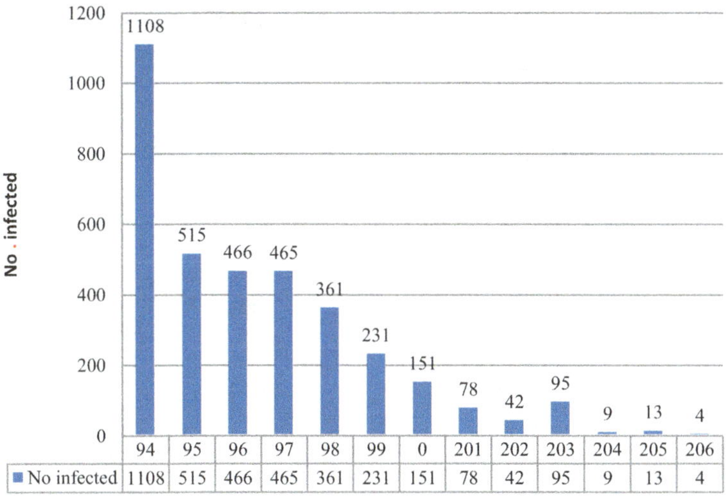

Fig. 8 Number of infected individuals with *S. haematobium* in Morocco during the period 1994–2006. Modified, with the permission of the publisher, from the "*Report: Inter-country Meeting on Strategies to Eliminate Schistosomiasis from the Eastern Mediterranean Region*, World Health Organization, 2007 (http://www.who.int/schistosomiasis/resources/EMRO_report_Schistosomiasis.pdf)

Human Schistosomiasis in Saudi Arabia

History

The Kingdom of Saudi Arabia (KSA) is one of the largest Arab countries of the Arabian Peninsula; it covers about 80 % of the Region. Geographically Saudi Arabia includes 6 regions; Eastern, Central, Northern, Northwest, Midwest, and Southwest regions; these regions are comprised of 13 provinces (Fig. 9). The population size amounts to 26 million. KSA has a high income economy based entirely on oil; however, agricultural economy has been expanding since the late 1970s due to development of new dams and irrigation projects. Schistosomiasis has been endemic in the Arabian Peninsula since the tenth century; Ibin Sina (Avi Cenna) described hematuria in his Medical texts (Alio 1967). From 1903 to 1961, schistosomiasis was reported from 12 provinces, the highest prevalence was found in the southwestern region. To date the eastern region is free from schistosomiasis (WHO 1987b; Al-Madani 1990).

Epidemiology

Both *S. haematobium* and *S. mansoni* are endemic in KSA, *S. mansoni* is mainly found in the high land of the western region, and *S. haematobium* is mainly reported

Fig. 9 Kingdom of Saudi Arabia map showing the 13 provinces. From mapsopensource.com (http://www.mapsopensource.com/saudi-arabia-map.html)

from Tabouk in the Northwest and from Baha and Mahael in the Low Land of the coastal plain in the Southwest region (Arfaa 1976; WHO 1987b; Ghandour et al. 1997, 1999; Shati 2009). Transmission of schistosomiasis is mediated by *B. pfeifferi* for *S. mansoni* and *Bulinus* species for *S. haematobium* (Magzoub and Kasim 1980; Ghandour et al. 1986, 1990); *B. truncatus*, *B. baccarii*, and *Bu. wrighti* have been reported as intermediate hosts for *S. haematobium* (Arfaa et al. 1989). Species specific gene markers have been used successfully to identify *Bu. truncatus* and *Bu. baccarii* collected from Asir, in the Southwestern region of KSA. Multiplex single step PCR helps to identify snails and to detect infection with *Schistosoma* species (Mostafa et al. 2012). Genetic variations of *Schistosoma* strains do exist at local, regional, and international levels, while Egyptian strains of *S. mansoni* are closely related to Saudi strains; Puerto Rican strains are clustered in different groups (Saoud 1966; Taylor and Nelson 1971; Voge and Mansour 1980; Fletcher et al. 1981, Jamjoom 2006).

The first comprehensive human and snail study was implemented by Alio 1967; the estimated prevalence of human schistosomiasis was 17 %; both *S. haematobium* and *S. mansoni* were endemic in the country except in the Eastern Region. Results of an extensive survey of the Ministry of Health in 1971 indicated that the disease is endemic in 12 regions at a prevalence ranging from 5 % to 20 % (Ashi et al. 1989). Arfaa in 1976 reported that transmission of schistosomiasis is only limited to a few foci in rural areas, and the disease is rare in large cities, e.g., Riyadh, Jeddah, Mecca, Taif, and Tabouk. He concluded that the snail habitats—which consist of wells, small canals, cisterns, small swamps, interrupted streams, and ponds—create a special type of transmission which can be defined as "oasis transmission," making control of the disease both simple and practical. However, the agricultural irrigation projects and construction of new dams have led to creation of permanent breeding habitats for the snail intermediate hosts over a wide region of KSA (Ghandour et al. 1986, 1990).

Recently, details of human infection were delineated in a study done in 2004; results showed that Saudis accounted for 61.2 % of total infected cases and infection peaked at 15–39 years, providing evidence about the endogenous source of infection. Infection by gender showed that males were four times more infected than females. Overall, *S. mansoni* represented 75 % of total infections. Prevalence was highest in most of southwestern region with a focus in Hail in the North. *S. haematobium* was limited to a few foci in the southwestern region in Jazan and Asir, both provinces bordering Yemen, where transmission of schistosomiasis is going on at an appreciable level. Ministry of Health statistical data in 2008 confirmed that Saudis are more infected than non-Saudis; the percentage of infection was 55.5 % and 45.5 % for Saudis and non-Saudis, respectively (Saudi Arabia Ministry of Health 2004, 2008). A morbidity study based on endoscopic and histopathology examination of samples collected from 2,458 individuals with GIT manifestations during the period from March 1979 through December 1988, revealed typical *S. mansoni* lesions in 8.8 %. Their ages ranged from 11 to 72 years. Another eight patients had schistosomal polyps (Mohamed et al. 1990). These values are lower than other countries in the Region where schistosomiasis is more prevalent (Gad et al. 2011).

Control

Primary healthcare centers played a key role in the implementation of control activities launched by the MOH in 1971. During 1973–1974, seven centers were designated to oversee the control activities. Mollusciciding, positive case finding, and treatment were effective in 1979.

During the period from 1990 to 1994, prevalence of *S. mansoni* was greatly reduced (Youssef et al. 1998); in 2003 prevalence had decreased down to 0.007 %. Thereafter, a comprehensive elimination program was launched; accordingly KSA was classified into three regions (1) schistosomiasis free areas, group A, (2) low endemic areas, group B, and (3) high endemic areas, group C. In addition to chemotherapy and mollusciciding, provision of potable water, environmental sanitation, and health education were applied (WHO 2007). Saudi Arabia is one of the ideal situations for effective application of snail control because the irrigation network is limited, and transmission foci are well defined (Arfaa 1976; Al-Madani 1990).

According to the annual MOH report, the outcome of the program was very efficacious; infection rates were 2.2, 2.9, and 2.78/100,000 in 2000, 2004, and 2008 respectively. Also, examination of 34,305 water bodies detected infected snails in

778 sites only (Saudi Arabia Ministry of Health 2004). However, the impact of the control program in Asir was not as effective as in other provinces (Fayed 1985).

Asir is situated in the southwestern part of KSA, bordering Yemen in a very limited area. Topography of Asir is an important factor contributing to difficulty in control of schistosomiasis. It consists of three regions: Asir Range comprising the mountainous area, Asir Plateau representing an upland that extends from Asir Range, the drainage of Asir Plateau is unsatisfactory and interrupted which favors flourishing of snail breeding places. The third zone is the coastal plain known as Tihama. The topography of the region helps in disseminating the snails from the high land of Asir Range to lower areas in Asir Plateau and the coastal plain of Tihama. Also, it is difficult to contact Bedouins living in this area due to their constant movements from place to place (Al-Madani 1991).

S. mansoni is mainly prevalent in Asir Range (9.1 %), while *S. haematobium* is prevalent in Asir Plateau and Tihama (7.5 %) (Fayed 1985). However the Annual Health report of the Ministry of Health recorded lower rates for *S. mansoni* (4.6 %) and *S. heamatobium* (2.2 %); still these rates are much higher than other provinces (Al-Madani 1991). Abiotic and biotic factors which might contribute to high transmission in Asir were investigated over an 8-year period (2000–2007). Inaccessibility to health services was one of the contributing factors in addition to hot humid summer climate, topography, and hydrographic changes; these factors influence breeding of snails (Shati 2009).

The contribution of schistosome genetic diversity to high prevalence in Asir needs to be investigated. Several studies in KSA explored genetic variability within the same strain collected from different regions (Shalaby et al. 2011; Mostafa et al. 2012). Genetic diversity might influence the force of infection, pathogenicity, immunogenicity, and the response to chemotherapy (Thiongo et al. 1997).

The role of animal reservoir hosts for *S. mansoni* in KSA should be considered. Infection was detected in Hamadryas baboons, which live in close proximity to humans in areas extending from the Yemen border to the Midwest Region (Nasher 1988; Yamane et al. 2003). Other factors that might hinder the success of control programs are: uncontrolled population movement from the highly endemic neighborhood in Yemen and the creation of multiple irrigation projects. However, schistosomiasis is mainly clustered in poor countries, while Saudi Arabia has a high income economy (one of the strongest in the Gulf region). It is worth mentioning that schistosomiasis has been eradicated from Japan despite widespread animal reservoir hosts contributing to over 80 % of transmission (Wang et al. 2005). Also, irrigation projects can be constructed to hinder breeding places and to limit human access for domestic activities; China is a good example where control of schistosomiasis is being maintained in spite of vast expansion of irrigation projects (Fenwick et al. 2006). It can be concluded that with comprehensive integrated control programs, elimination of schistosomiasis is likely to occur in high income economy.

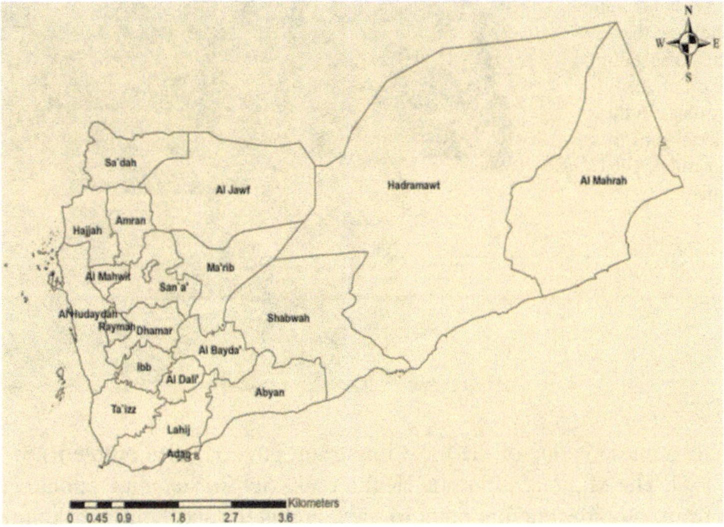

Fig. 10 Map of Yemen indicating the location of major cities and governorates Copyright: © Alyousefi et al. 2011. This is an open-access article, free of all copyright, and may be freely reproduced, distributed, transmitted, modified, built upon, or otherwise used by anyone for any lawful purpose. The work is made available under the Creative Commons CC0 public domain dedication

Human Schistosomiasis in Yemen

History

Yemen is a developing Middle Eastern country located at the southern part of the Arabian Peninsula with a total population of 25 million. This country depends totally on ground and rain water as a source of water. Recently, the country has fallen into a deep water crisis characterized by very rapid mining of groundwater, extreme water supply shortages in the major cities, and limited access of the population to safe drinking water. WHO reported that only 25 % of the population had easy access to safe water (WHO 2009; Alyousefi et al. 2011).

Being one of the poorest countries in the Middle East, 42 % of Yemen's total population is estimated to be under the national poverty line (World Bank 2004). The poverty ratio is higher in rural areas where 75 % of the population lives and only 25 % is covered with healthcare services compared to 80 % of urban areas (Alyousefi et al. 2011) (Fig. 10).

The economy of the country depends mainly on agriculture, most of the population being involved directly or indirectly in this field. New agricultural projects have been established and many of the previous ones extended, so the irrigation system is extensive (Nagi 2005).

Fig. 11 Artificial water pool of the type that contributes to sustaining transmission of schistosomiasis in the mountainous areas of Western Yemen. ©WHO/R. Ben Ismail

Schistosomiasis is one of the most important public health problems in Yemen (Nagi 2005). The Ministry of Public Health ranks *Schistosoma* infections second to malaria from a socioeconomic point of view and is the sixth major health problem in the country (Ministry of Public Health 1995). Yemen is the second in having the greatest number of schistosomiasis cases in the MENA Region (Hotez et al. 2012).

Epidemiology

Human schistosomiasis (both urinary and intestinal) are endemic in Yemen with three million people infected (Nagi 2000) and 600,000 suffering from clinical morbidity likely to result in death (Oshish et al. 2011; WHO 2012). Every year, around 2,000 people die of schistosomiasis, but this is not the exact figure as there are no reliable statistics in the country (IRIN Middle East 2008).

Transmission occurs in a large proportion of Yemen, with the mountainous areas in the western part of the country appearing to be the most severely affected (WHO 2012) (see Fig. 11).

Schistosomiasis was found to affect populations where substandard conditions of living are predominant, e.g., poor sanitation, insufficient safe water supply, and low standard of hygiene is practiced (Raja'a et al. 2001). Agricultural workers are at high risk of acquiring schistosomiasis infection because of their daily work in the fields and continuing contact with *Schistosoma*-infected water (Nagi 2005). Children who practice swimming are particularly at risk because of their prolonged and complete body exposure (Al-Shamiri et al. 2011), as well as their lower levels of acquired immunity (Woolhouse et al. 1991; Fulford et al. 1992). Children not enrolled in schools are thought to harbor particularly heavy infections (Nagi 2005; Hussein et al. 1996). Moreover, the erroneous habits of people as regards urination and defecation in canal water, bathing, washing utensils and clothes, walking barefoot during irrigation in agriculture, or fishing make them at risk of acquiring the infection (Sibomana 2009) (Fig. 11).

Despite the fact that safe, effective anti-*Schistosoma* treatment is at hand and considerable households have access to a safe water supply (51 %), the rate of *Schistosoma* infection is expected to increase (Raja'a et al. 2001). This is due to the suitable environment created by agricultural expansion linked with increased dam construction as a result of the water policy applied during the last decade of the millennium; combined, these changes have generated an optimal environment for both freshwater snails *Bi. arabica* and *Bu. truncatus*, which are the intermediate hosts for *S. mansoni* and *S. haematobium*, respectively (Steinmann et al. 2006). These snails were found to be highly susceptible to infection with local Yemeni strains of *S. mansoni* and *S. haematobium* under laboratory conditions (Nagi et al. 1999). Moreover, there is insufficiency, including inadequate household coverage with a safe water supply and lack of public health services, a high rate of illiteracy (50 %), and a lack of an indoor latrine in a considerable proportion of houses (37 %). Nevertheless, a comprehensive control policy is not effectively implemented due to costly requirements (Raja'a et al. 2001).

Schistosoma infection results in reduced school attendance via its major sequelae of anemia, growth stunting, and cognitive impairment (Engels et al. 2002) and is significantly associated with chronic abdominal pain, diarrhea, and malnutrition (King et al. 2005). The longer-term serious disease complications are organomegaly, particularly the liver and spleen, intestinal schistosomiasis, bladder cancer, and damage to the female genital tract from urinary schistosomiasis (Poggensee and Feldmeier 2001; King and Dangerfield-Cha 2008).

The distribution of *Schistosoma* infections across the country was a matter of research in several published papers. *Schistosoma* infections were reported from areas such as Sana'a and Saada (Yemen *Bilharzias* Control Project 1993), Marib (Nagi and Molan 1992), Taiz (Hazza et al. 1983), Ibb (Al-Haddad and Assabri 1998; Raja'a et al. 2000), Hajja (Azazy and Al-Dullaimi 1999), central highlands (Schaap et al. 1992), Aden and Yahr (Zain 1998).

Nagi et al. 1999 conducted an epidemiological survey during 1992. A total of 2,902 students of 13 schools in Sana'a Governorate, 800 students of three schools in Saada Governorate, and 2802 students of 14 schools in Hajja Governorate were included. The infectivity rate of *S. mansoni* was (7.6 %–18.8 %–76.3 %) and of *S. haematobium* was (52.2 %–10.1 %–49.0 %) in Hajja–Sana'a and Saada respectively. Males had a higher infection rate than females and *S. mansoni* was more prevalent than *S. haematobium*, which is similar to the findings of Hazza et al. (1983) and Farag (1985). Bloody stools were found in 61.7 % of *S. mansoni* cases and abdominal pain and/or diarrhea in 50 %. Heamaturia was seen in 62.5 % of *S. haematobium* cases while painful micturition in 33.3 %. Such findings have often been reported (Gilles 1982; Warren 1984).

Raja'a et al. (2000) conducted an epidemiological comparative survey aimed at determination of prevalence and focal distribution of *Schistosoma* infection and intestinal parasites to provide a reference for evaluating the need for community intervention. A total number of 230 children (5–18 years) from 7 villages that lie on the Assahul Valley of Ibb Governorate in Yemen were included. The *Schistosoma* infection rate was 37 % (S. *mansoni* 35 % and *S. haematobium* 5 %). Significant associations were found between *Schistosoma* infections with residence near the valley, male gender, and frequent water contact activities. The detected rates of infection (37 %) exceeded the estimations reported for the whole country (6 %) (Farag 1985), Mahweet (27 %) (Raja'a et al. 2001), and Hajja (21 %) (Azazy and Al-Dullaimi 1999). Exceedingly high rates (37 % for *S. haematobium* and 64 % for *S. mansoni*) were reported from Taiz (Hazza et al. 1983).

In year 2003, 515 school children from five schools in Abyan and Taiz Governorates in Yemen were screened by Ahmed (2009) for microhematuria using reagent strip method and for the presence of *S. haematobium* ova by filtration method as well as carrying out a questionnaire for hematuria. The prevalence of the infection as determined by filtration, questionnaire, and reagent strip method was 21.4 %, 22.2 % and 30.9 % respectively. The author concluded that the reagent strip method is practical, cheap, fast, and easy to perform in primary healthcare setting and method for screening and monitoring *S. haematobium* infection. Its performance can be enhanced when used in combination with questionnaire without additional costs.

Al-Shamiri et al. (2011) examined 1,406 stool samples and 1,484 urine samples of school children (5–16 years) from 32 basic schools in five districts in five schistosomiasis endemic areas (Al-Dhabab, Hedran, Warazan, Al-Barhand, Al-Shmaytin) during the period from June 2007 to March 2009. The overall prevalence was 20.8 % for *S. mansoni* and 7.4 % for *S. haematobium*. *S. mansoni* was more prevalent than *S. haematobium*, which was attributed to the distribution of *Biomphalaria* snails which is more abundant than that of *Bulinus* snails. Riadh Ben-Ismail, WHO's regional adviser (IRIN Middle East 2008), mentioned that intestinal schistosomiasis differs from that of the urinary variety; it might not fully respond to the drugs. Moreover, some animals (e.g., monkeys) carry intestinal schistosomiasis and can transmit it back to humans.

Control

Intestinal parasitic infections had received attention in Yemen as early as 1950s; most of these efforts were to combat schistosomiasis (Kuntz et al. 1953; Hazza et al. 1983). Preventive chemotherapy with the safe, effective, and cheap drug PZQ is the cornerstone of the WHO recommended approach against schistosomiasis, which aims to keep parasite numbers suppressed in order to avoid clinical manifestations of infection (Oshish et al. 2011). Many schistosomiasis control programs have distributed drugs using a school-based platform to reduce infection levels and related morbidity (Fenwick et al. 2009). This approach has considerable logistical and epidemiological advantages such as the ability to build on the school infrastructure in order to ensure high treatment coverage, the use of teachers to administer drugs and record compliance, and the fact that school-age children typically harbor the heaviest burden of infection, as well as subsequently making the greatest contribution to transmission (Ritcher 2003). Children are also easy to approach with a health education program, and they are easy to reach physically for chemotherapy. Moreover, children represent the future of developing countries (Nagi 2005). It has been postulated that by treating children at least three times during their school-age years, severe morbidity can be avoided in later life (Ritcher 2003).

Raja'a et al. 2001 studied the prevalence, intensity, and incidence of schistosomiasis among school children in a previously ignored area (Al Mahweet) in Yemen. A total of 897 pupils aged 5–18 years (453 from Al Mahweet town and 444 from rural surrounding areas) participated in the study. It was found that annual intervention with chemotherapy only, neglecting the other components of the comprehensive control program, is satisfactory decreasing the infection rate of *S. mansoni* by 62.5 %. However, the authors recommended that the time interval for retreatment of cases with *S. haematobium* should be shortened or combined with the other control measures. They also recommended mass treatment for boys of rural Al-Mahweet who visit the water source at least once per week.

Nagi (2005) conducted an intervention study targeting the community and school children in Khamir, located 90 km north of Sana'a for controlling schistosomiasis using chemotherapy and health education. Community and school baseline survey included 913 individuals of 100 houses and 323 children randomly selected from 14 schools in Khamir. The prevalence of *S. haematobium* infection 14 months post-intervention fell from 58.9 % to 5.8 % and the frequency of heavy infection from 40 % to 18.9 %. Health education sessions resulted in significant decrease in the frequency of contact with water sources and greater adherence to preventive measures. The author concluded that an integrated community and school-based program combining chemotherapy and health education can be effective for control of *S. haematobium* in endemic areas.

Subnational control has been ongoing since 2006 in Yemen via the distribution of PZQ against schistosomiasis and albendazole (ALB) against soil-transmitted helminthes using school-based treatment. In 2008, the Yemen National Schistosomiasis Control Program (NSCP)—the first of its kind in Yemen—was instigated with the aim of controlling schistosomiasis nationwide (Oshish et al. 2011).

With support from a separate World Bank grant, the NSCP utilized a school-based distribution system to treat school-age children (6–18 years) reaching enrolled and non-enrolled children in those areas with the highest endemicity of schistosomiasis across the country. The campaign was implemented in four phases during 2008 (first from 10 to 14 March, second 24 to 27 March, third 5 to 8 April, and the fourth phase 18 to 21 October) targeting 2,583,309 children from 5,495 schools in 107 districts in 14 provinces (69 % at primary schools and 31 % non-enrolled). PZQ and ALB were administered by 4,426 trained teachers and 3,034 health workers (IRIN Middle East 2008). This resulted in a very high coverage for enrolled school children (94 %) but a lower figure for non-enrolled school-age children (68 %) (Oshish et al. 2011).

To improve this and to expand treatment to adults in high and medium infected areas, a pilot program ran from 27 to 30 December 2009 in 10 high endemicity districts in the three governorates of Sana'a, Dhamar, and Hajjah using a combination of PZQ (against schistosomiasis; 40 mg/kg administered using a dose pole) and

ALB (against soil-transmitted helminthes; a single 400 mg tab/person). It was based on two complementary treatment approaches, campaign-based preventive chemotherapy and routine preventive chemotherapy. The campaign-based preventive chemotherapy constituted the active phase of the campaign. PZQ and ALB were distributed using fixed (schools and health facilities) and temporary sites (mobile teams). In 2009 the school-based teams did not treat non-enrolled children as this approach in 2008 yielded disappointing results. The treatment target was 300 persons/day per team for a period of 4–5 days. Health facilities were implemented at the same time as the neighboring schools in order to reach adults and non-enrolled school-age children. Leaflets—radio messages—announcements via public address systems attached to the vehicles of the village leaders and distribution teams were used to attract participants. Treatment target was 300 persons/day/team (Oshish et al 2011)

Temporary sites such as community leader houses, mosques, market places were used by mobile teams in order to reach as many of target population as possible; in most areas 300 persons per day per team was the target. Routine preventive chemotherapy was developed whereby sufficient drugs were retained within the routine health system of the Ministry of Public Health and Population (MOPHP) (health facilities, hospitals, dispensaries) in order to passively treat cases among individuals not targeted by the campaign-based preventive chemotherapy, those who missed treatment, or those who suffered reinfection. Health education and social mobilization were identified as important in ensuring community acceptance, high treatment coverage, and therefore success of the control program. The new approach achieved coverage of 90.1 % of non-enrolled children: a 40 % increase compared with the same districts in 2008, and coverage of 97.9 % of enrolled children, a 2 % increase compared to 2008. Coverage of females was 81.8 % and of adults in general was 73.9 % (Oshish et al. 2011).

The result of this pilot program was helpful informing the design and direction of the national program (2010–2015). The NSP aims to eliminate schistosomiasis-related morbidity in Yemen via repeated periodic chemotherapeutic treatment with PZQ (40 mg/kg) to all those who require it (Table 4), and the dissemination of relevant health education messages, over the 6-year life span of the program. This is supplemented with treatment with ALB to treat common soil-transmitted helminthiasis (Oshish et al. 2011).

The first nationwide control occurred in the operational year 2010 (July 2010–May 2011) with ambitious plans to treat approximately 10 million people across the country over three campaigns (Oshish 2011). The annual anti-*Schistosoma* campaign was launched on 10 March 2013, the first phase of the second operational year, 2012–2013. It will cover 160 districts in 12 governorates and target about 45 % of Yemen's 24 million population aged 6 years and older. This will be followed up with a second phase of the campaign that will target two million people in the remaining nine endemic governorates in mid-April 2013 (Weldon 2013)

Table 4 Treatment approach in areas of variable endemicity for the main treatment campaigns

Area	Criteria[a]	School-based treatment	Community-based treatment
High prevalence area	>40 % of either infection	Years 1, 2, 3, 4, 5, 6	1, 2, 4
Medium prevalence area	10–40 % of either infection	Years 1, 2, 4, 6	Year 1
Low prevalence area	<10 % of either infection	Years 2, 5	ND
Suspect area	Currently unmapped areas	Years 1, 2, 4, 6	Year 1

The districts (of which there are 333 in Yemen) is the unit of implementation
ND not done
[a]These cutoff points do not coincide with the WHO recommendations of >50 %, 10–50 %, and <10 % respectively (Steinmann et al. 2006);they have been altered in order to expand treatment to more needy people and to further decrease prevalence and intensity of infection. Year 1–6 refers to 2010–2015 respectively. Reproduced, with the permission of the publisher, from Oshish et al. (2011)

Yemen is the only country in the Middle East not to have eliminated the disease (IRIN Middle East 2008). The significant World Bank funds—being allocated to the 6-year control program in Yemen—will continue provided the political situation there returns to a level that allows drug distribution (Hotez et al. 2012)

Acknowledgments The authors are grateful to Prof. Dr. Soraya Sharaf, Parasitology Department, National Liver Institute (NLI), Menoufeya University, Egypt, who edited the manuscript.

References

Aagaard-Hansen J, Mwanga JR, Bruun B (2009) Social science perspectives on schistosomiasis control in Africa: past trends and future directions. Parasitology 136:1747–1758

Abdel-Wahab MF, Strickland GT, El-Sahly A, El-Kady N, Zakaria S, Ahmed L (1979) Changing pattern of schistosomiasis in Egypt, 1935-1979. Lancet 2:242–244

Abdel-Wahab MF, Yosery A, Narooz S, Esmat G, el-Hak S, Nasif S, Strickland GT (1993) Is *S. mansoni* replacing *S. haematobium* in the Fayoum? Am J Trop Med Hyg 49:697–700

Abdel-Wahab MF, Esmat G, Medhat E, Narooz S, Ramzy I, El-Boraey Y, Strickland GT (2000) The epidemiology of schistosomiasis in Egypt: Menoufeya Governorate. Am J Trop Med Hyg 62:25–34

Ahmed BA (2009) Two practical and cost effective methods for urinary schistosomiasis screening in Yemeni school children. Iran J Public Health 38:78–83

Al-Haddad AM, Assabri AM (1998) Health impact of uncompleted sewerage project in Maitam valley, Ibb town-Republic of Yemen. Yemen Med J 1:68–76

Alio IS (1967) Epidemiology of schistosomiasis in Saudi Arabia, with an emphasis on geographic distribution patterns. Report to the Arabian American Oil Company, Dharan

Al-Madani AA (1990) Schistosomiasis control in Saudi Arabia with special reference to the period 1983-1988. Public Health 104:261–266

Al-Madani AA (1991) Problems in the control of schistosomiasis in Asir Province Saudi Arabia. J Community Health 16:143–149

Al-Shamiri AH, Al-Taj MA, Ahmed AS (2011) Prevalence and co-infections of schistosomiasis/hepatitis B and C viruses among school children in an endemic area in Taiz, Yemen. Asian Pac J Trop Med 4:404–408

Alyousefi NA, Mahdy MK, Mahmud R, Lin YAL (2011) Factors associated with high prevalence of intestinal protozoan infections among patients in Sana'a City, Yemen. PLoS One 6:1–7

Amarir F, El-Mansouri B, Fellah H, Sebti F, Mohammed L, Handali S, Wilkins P, El- Idrissi AL, Sadak A, Rhajaoui M (2011) National serologic survey of *haematobium* schistosomiasis in Morocco: evidence for elimination. Am J Trop Med Hyg 84:15–19

Arfaa F (1976) Studies on schistosomiasis in Saudi Arabia. Am J Trop Med Hyg 5:295–298

Arfaa F, Mahboubi E, Al-Jeffri M, Selim A, Russell G (1989) The potential role of various species of intermediate hosts of *S. haematobium* in Saudi Arabia. Trans R Soc Trop Med Hyg 83:216–218

Ashi J, Arfaa F, Jeffri M, Suwairy M (1989) Progress achieved in the control of schistosomiasis in Saudi Arabia. J Trop Med Hyg 92:27–31

Assis AM, Barreto ML, Prado MS, Reis MG, Parraga IM, Blanton RE (1998) *S. mansoni* infection and nutritional status in school children: a randomized, double-blind trial in north eastern Brazil. Am J Clin Nutr 168:1247–1253

Azazy AA, Al-Dullaimi SS (1999) Prevalence of intestinal parasites in pupils of an elementary school in Hajja town, Yemen. Yemen Med J 3:66–68

Azim M (1935) The epidemiology of schistosomiasis in Egypt. J Egypt Med Assoc 18:215–230

Badr M (1981) Schistosomiasis in ancient Egypt. In: El-Bolkainy MN, Chu E (eds) Detection of bladder cancer associated with schistosomiasis, National Cancer Institute. Al-Ahram Press, Cairo, p 212

Barakat R (2012) Epidemiology of schistosomiasis in Egypt: travel through time. J Adv Res. http://dx.doi.org/10.1016/j.jare.2012.07.003

Barakat R, Farghaly A, el-Masry AG, el-Sayed MK, Hussein MH, Miller FD (1995) *S. mansoni* in the Nile Delta, Egypt: a large scale epidemiological study in Kafr El-Sheikh Governorate. Trop Geogr Med 47:259–265

Barakat R, Farghaly A, El-Morshedy H, Hassan M, de Miller W (1998) Impact of national schistosomiasis conorate, control program in Kafr El-Sheikh Governorate, Nile Delta, Egypt: an independent evaluation. J Egypt Public Health Assoc 73:737–753

Barakat R, Farghaly A, El-Masry AG, El-Sayed MK, Hussein MH (2000) The epidemiology of schistosomiasis in Egypt: patterns *of S. mansoni* infection and morbidity in Kafr El-Sheikh. Am J Trop Med Hyg 62:21–27

Barneoud J (1932) Bilharziose dans la Vallèe de Draa. Bull Inst d'Hygiène (Maroc) 2:80–88

Barneoud J, Carrosse J (1929) Bilharziose vèsicale à Marakech. Bull Inst Pasteur d'lger 7:51–79

Benmansour N (1970) Etude èpidèmiologique de la bilharziose vèsicale effectueè au Maroc entre 1966 et 1970. J Mèdical Maroc 6:237–243

Bilharz T (1853) Further observations concerning Distomum *haematobium* in the portal vein of man and its relationship to certain pathological formations with brief notes by Seibald. Z Wiss Zool 4:72

Chitsulo L, Engels D, Montresor A, Savioli L (2000) The global status of schistosomiasis and its control. Acta Trop 77:41–51

Cioli D, Pica-Mattoccia L (2003) Praziquantel. Parasitol Res 90:S3–S9

Cline BL, Richards FO, el-Alamy MA, el-Hak S, Ruiz-Tiben E, Hughes JM, McNeeley DF (1989) Nile Delta schistosomiasis survey: 48 years after Scott. Am J Trop Med Hyg 41:56–62

Connet M (1937) Existance d'un foyer de bilharziose v, Rabat vèsicale à Akka (dècouvert pour la première fois). Maroc Mèdical 7:365–366

Coutinho HM, Acosta LP, McGarvey ST, Jarilla B, Jiz M, Pablo A, Su L, Manalo DL, Olveda RM, Kurtis JD, Friedman JF (2006) Nutritional status improves after treatment of *S. japonicum*-infected children and adolescents. J Nutr 136:183–188

Curtale F, Mohamed MY, Youssef ZM (2010) Comprehensive primary health care, a viable strategy for the elimination of schistosomiasis. Trans R Soc Trop Med Hyg 104:70–72

De Clercq D, Vercruysse J, Picquet M, Shaw DJ, Diop M, Ly A, Gryseels B (1999) The epidemiology of a recent focus of mixed *S. haematobium* and *S. mansoni* infections around the 'Lac de Guiers' in the Senegal River Basin, Senegal. Trop Med Int Health 4:544–550

Deelder AM, Miller RL, de Jonge N, Krijger FW (1990) Detection of Schistosome antigen in mummies. Lancet 335:724–725

DELM (1999) Donnèe èpidèmiologiques des maladies sous surveillance, Bilan annèe 1999. Bull Epidemiol Hebd 8

Doumenge JP, Mott KE, Cheung C (1987) Atlas of the global distribution of schistosomiasis. Presses Universitairres de Bordeaux, Talence

Ebbel B (1937) The Ebers Papyrus, the Greatest Egyptian Medical Document. Oxford University Press, London, p 135

El-Enien M, Orieby A, Shawky E, Saad A, Shokrani N, El Fattah M, et al (eds) (1993) Schistosomiasis in Egypt: prevalence, intensity and morbidity in El Minya Governorate. International conference on schistosomiasis, Cairo

El-Hawey AM, Amr MM, Abdel-Rahman AH, El-Ibiary SA, Agina AM, Abdel-Hafez MA, Waheeb AA (2000) The epidemiology of schistosomiasis in Egypt: Gharbia Governorate. Am J Trop Med Hyg 62(2S):42–48

El-Khoby T, Galal N, Fenwick A (1998) The USAID/Government of Egypt's Schistosomiasis Research Project (SRP). Parasitol Today 14:92–96

El-Khoby T, Galal N, Fenwick A, Barakat R, El-Hawey A, Nooman Z, Habib M, Abdel-Wahab F, Gabr NS, Hammam HM, Hussein MH, Mikhail NN, Cline BL, Strickland GT (2000a) The epidemiology of schistosomiasis in Egypt: summary findings in nine governorates. Am J Trop Med Hyg 62:88–99

El-Khoby T, Hussein MH, Galal N, Miller FD (2000b) Epidemiology 1, 2, 3: origins, objectives, organization and implementation. Am J Trop Med Hyg 62:2–7

El-Malatawy A, El-Habashy A, Lechine N, Dixon H, Davis A, Mott KE (1992) Selective population chemotherapy among school children in Beheira governorate: the UNICEF/Arab Republic of Egypt/WHO Schistosomiasis Control Project. Bull World Health Organ 70:47–56

El-Sayed HF, Rizkalla NH, Mehanna S, Abaza SM, Winch PJ (1995) Prevalence and epidemiology of *S. mansoni* and *S. haematobium* infections in two areas of Egypt recently reclaimed from the desert. Am J Trop Med Hyg 52:194–198

El-Zawahry M (1964) A health survey in Egyptian Nubia, Part 1: objectives and design of survey, and epidemiological features of parasitosis. J Egypt Public Health Assoc 39:313–340

Engels D, Chitsula L, Montresor A, Savioli L (2002) The global epidemiological situation of schistosomiasis and new approaches to control and research. Acta Trop 82:139–146

Ezeamama AE, Friedman JF, Acosta LP, Bellinger DC, Langdon GC, Manalo DL, Olveda RM, Kurtis JD, McGarvey ST (2005) Helminth infection and cognitive impairment among Filipino children. Am J Trop Med Hyg 72:540–548

Ezeamama AE, McGarvey ST, Hogan J, Lapane KL, Bellinger DC, Acosta LP, Leenstra T, Olveda RM, Kurtis JD, Friedman JF (2012) Treatment for *S. japonicum*, reduction of intestinal parasite load, and cognitive test score improvements in school-aged children. PLoS Negl Trop Dis 6:1634

Farag HF (1985) Intestinal parasitosis in the population of Yemen Arab Republic. Trop Geogr Med 37:29–31

Fayed MA (1985) Schistosomiasis in Asir district in southern region of Saudi Arabia. J Egypt Soc Parasitol 15:289–292

Fedewa SA, Soliman AS, Ismail K, Hablas A, Seifeldin IA, Ramadan M, Omar HG, Nriagu J, Wilson ML (2009) Incidence analyses of bladder cancer in the Nile Delta Region of Egypt. Cancer Epidemiol 33:176–181

Felix AS, Soliman AS, Khaled H, Zaghloul MS, Banerjee M, El-Baradie M, El-Kalawy M, Abd-Elsayed AA, Ismail K, Hablas A, Seifeldin IA, Ramadan M, Wilson ML (2008) The changing patterns of bladder cancer in Egypt over the past 26 years. Cancer Causes Control 19:421–429

Fenwick A, Savioli L, Engels D, Robert Bergquist N, Todd MH (2003) Drugs for the control of parasitic diseases: current status and development in schistosomiasis. Trends Parasitol 19:509–515

Fenwick A, Rollinson D, Southgate V (2006) Implementation of human schistosomiasis control: challenges and prospects. Adv Parasitol 61:567–622

Fenwick A, Webster JP, Bosque-Oliva E, Blair L, Fleming FM, Zhang Y, Garba A, Stothard JR, Gabrielli AF, Clements AC, Kabatereine NB, Toure S, Dembele R, Nyandindi U, Mwansa J, Koukounari A (2009) The Schistosomiasis Control Initiative (SCI): rationale, development and implementation from 2002-2008. Parasitology 136(13):1719–1730

Fletcher M, LoVerde PT, Woodruff DS (1981) Genetic variation in *S. mansoni*: enzyme polymorphisms in population from Africa, Southwest Asia, South America and the West Indies. Am J Trop Med Hyg 30:406–421

Fulford AJ, Butterworth AE, Sturrock RF, Ouma JH (1992) On the use of age-intensity data to detect immunity to parasitic infections with special reference *to S. mansoni* in Kenya. Parasitology 105:219–227

Gad YZ, Ahmad NA, El-Desoky I, Arafa MM, Farag RE (2011) Clorectal schistosomiasis: Is it still endemic in delta Egypt, early in the third millennium? Trop Parasitol 1:108–110

Gaud J, Maurice A (1946) Foyers de bilharziose vèsicale dans le Souss. Bull Inst d'Hygiène (Maroc) 6:61–62

Ghandour AM, Al-Robai AA, El-Gohary M (1986) An ecological study on some aspects of schistosomiasis in mid-western region of Saudi Arabia. Arab Gulf J Sci Res 4:203–219

Ghandour AM, Al-Ghamdi HS, Al-Robai AA (1990) A review of snail intermediate hosts of schistosomiasis in Saudi Arabia. J Med Appl Malacol 2:79–91

Ghandour AM, Tricker K, Doenhoff MJ, Al-Robai AA, Banaja AA (1997) An enzyme-linked immunosorbent assay using *S. mansoni* purified egg antigen for the diagnosis of schistosomiasis in Saudi Arabia. Trans R Soc Trop Med Hyg 91:287–289

Ghandour AM, Zahid NZ, Banaja AA (1999) Epidemiological study on the transmission of schistosomiasis in Saudi Arabia (Western region). Ann Trop Med Parasitol 93:193–195

Gilles HM (1982) Infection with *S. haematobium*. In: Jordan P, Webbe G (eds) Schistosomiasis: epidemiology, treatment and control. William Heinemann Medical Books Ltd, London, pp 79–104

Gryseels B, Polman K, Clerinx J, Kestens L (2006) Human schistosomiasis. Lancet 368:1106–1117

Habib M, Abdel-Aziz F, Gamil F, Cline BL (2000) The epidemiology of schistosomiasis in Egypt: Qalubeia Governorate. Am J Trop Med Hyg 62:49–54

Hammam HM, Allam FA, Moftah FM, Abdel-Aty MA, Hany AH, Abdel-Motagaly KF, Nafeh MA, Khalifa R, Mikhail NN, Talaat M, Hussein MH, Strickland GT (2000a) The epidemiology of schistosomiasis in Egypt: Assiut Governorate. Am J Trop Med Hyg 62:73–79

Hammam HM, Zarzour AH, Moftah FM, Abdel-Aty MA, Hany AH, El-Kady AY, Nasr AM, Abd-El-Samie A, Qayed MH, Mikhail NN, Talaat M, Hussein MH (2000b) The epidemiology of schistosomiasis in Egypt: Qena Governorate. Am J Trop Med Hyg 62:80–87

Hazza YA, Arfa'a F, Haggar M (1983) Studies in schistosomiasis in Taiz province. Yemen Arab Republic. Am J Trop Med Hyg 32:1023–1028

Holveck JC, Ehrenberg JP, Ault SK, Rojas R, Vasquez J, Cerqueira MT, Ippolito-Shepherd J, Genovese MA, Periago MR (2007) Prevention, control, and elimination of neglected diseases in the Americas: pathways to integrated, inter-programatic, intersectoral action for health and development. BMC Public Health 17:6, http://www.biomedcentral.com/1471-2458/7/6

Hotez PJ, Savioli L, Fenwick A (2012) Neglected tropical diseases of the Middle East and North Africa: review of their prevalence, distribution and opportunities of control. PLoS Negl Trop Dis 6:1475. doi:10.1371/journal.pntd.0001475, http://www.plosntds.org

Hussein MH, Talaat M, El-Saied MK, El- Badawi A, Evans DB (1996) Who misses out two with school-based health programs? A study of schistosomiasis control in Egypt. Trans R Soc Trop Med Hyg 90:362–365

International Association for Medical Assistance to Travelers IAMAT (2012) World schistosomiasis risk chart. http://www.iamat.org.

Jamjoom MB (2006) Molecular identification of some *Schistosoma mansoni* isolates in Saudi Arabia. World J Med Sci 1:102–107

Jukes MC, Nokes CA, Alcock KJ, Lambo JK, Kihamia C, Ngorosho N, Mbise A, Lorri W, Yona E, Mwanri L, Baddeley AD, Hall A, Bundy DA, Partnership for Child Development (2002) Heavy schistosomiasis associated with poor short-term memory and slower reaction times in Tanzanian school children. Trop Med Int Health 7:104–117

Katz N, Chaves A, Pellegrino J (1972) A simple device for quantitative stool thick-smear technique in schistosomiasis *mansoni*. Rev Inst Med Trop Sao Paulo 14:397–400

Kessler P, Southgate B, Klumpp R, Mahmoud M, Restrand L, Saleh L (1987) Report of an independent evaluation mission on the National *Bilharzia* Control Program, Egypt, 1985 (abridged version). Trans R Soc Trop Med Hyg 81:1–57

Khalil M, Azim M (1938) Further observations on the introduction of infection with *S. haematobium* infection. J Egypt Med Assoc 21:95–101

Khallaayoune K, Laamrani H (1992) Seasonal patterns in the transmission of *S. haematobium* in Attaouia, Morocco. J Helminthol 166:89–95

Khallaayoune K, Laamrani H, Madsen H (1998a) Distribution of *Bulinus truncatus*, the intermediate host of *S. haematobium*, in an irrigation system in Morocco. J Freshw Ecol 13:129–133

Khallaayoune K, Madsen H, Laamrani H (1998b) Evaluation of three methods to control *Bulinus truncatus*, the intermediate host of *S. haematobium* in an irrigation scheme, Tessaout-Amont, Morocco. Acta Trop 69:51–63

King CH, Dangerfield-Cha M (2008) The unacknowledged impact of chronic schistosomiasis. Chronic Illn 4:65–79

King CH, Dickman K, Tisch DJ (2005) Reassessment of the cost of chronic helmentic infection: a meta-analysis of disability—related outcomes in endemic schistosomiasis. Lancet 365:1561–1569

Kirtorn UD, Hiagashi GI (1985) *S. haematobium* in Upper Egypt: analysis of dispersion patterns. Am J Trop Med Hyg 34:331–340

Kleinbaum DG, Kupper LL, Morgenstern H (1982) Epidemiologic research: principles and quantitative methods. Lifetime Learning Publications, Belmont, MA

Kloos H, Lo CT, Birrie H, Ayele T, Tedla S, Tsegay F (1988) Schistosomiasis in Ethiopia. Soc Sci Med 26:803–827

Kuntz R, Malakatis G, Lawless D, Strome C (1953) Medical mission to the Yemen, Southwest Arabia, 1951.II.A cursory survey of the intestinal protozoa and helminth parasites in the people of the Yemen. Am J Trop Med Hyg 2:13

Laamrani H, Mahjour J, Madsen H, Khallaayoune K, Gryseels B (2000) *S. haematobium* in Morocco: moving from control to elimination. Parasitol Today 16:257–260

Laaziri M, Benouna M (1982) Guide de la lutte contre la bilharziose, direction des affaires techniques. Ministère de la Santè Publique, Maroc

Latham MC, Stephenson LS, Kurz KM, Kinoti SN (1990) Metrifonate or praziquantel treatment improves physical fitness and appetite of Kenyan schoolboys with *S. haematobium* and hookworm infections. Am J Trop Med Hyg 43:170–179

Leiper R (1915) Report on the results of the *Bilharzia* mission in Egypt: Part I. Transmission. J R Army Med Corps 25:1–55

Lotfy M (2009) Human schistosomiasis in Egypt: Historical review, assessment of the current picture and prediction of the future trends. J Med Res Inst 30:1–7

Magzoub M, Kasim AA (1980) Schistosomiasis in Saudi Arabia. Ann Trop Med Parasitol 74:511–513

Malek E (1975) Effect of the Aswan High Dam on prevalence of schistosomiasis in Egypt. Trop Geogr Med 27:359–364

Mazigo HD, Nuwaha F, Kinung'hi SM, Morona D, Pinot de Moira A, Wilson S, Heukelbach J, Dunne DW (2012) Epidemiology and control of human schistosomiasis in Tanzania. Parasit Vectors 5:247

Medhat A, Abdel-Aty MA, Nafeh M, Hammam H, Abdel-Samia A, Strickland GT (1993) Foci of *S. mansoni* in Assiut province in Middle Egypt. Trans R Soc Trop Med Hyg 87:404–405

Michelson MK, Azziz FA, Gamil FM, Wahid AA, Richards FO, Juranek DD, Habib MA, Spencer HC (1993) Recent trends in the prevalence and distribution of schistosomiasis in the Nile Delta Region. Am J Trop Med Hyg 49:76–87

IRIN Middle East (2008) Yemen: new campaign targets three million with *Bilharzias*. http://www.irinnews.org/report/77209

Miller F, Hussein M, Mancy K, Hilbert M (1978) Aspects of environmental health impacts of the Aswan High Dam on rural population in Egypt. Prog Water Technol Ltd 11:173–180

Miller F, Hussein M, Mancy KH, Hilbert MS, Monto AS, Barakat RM (1981) An epidemiological study of *S. haematobium* and *S. mansoni* in thirty-five Egyptian villages. Trop Geogr Med 33:355–365

Miller RL, Armelagos GJ, Ikram S, De Jonge N, Krijger FW, Deelder AM (1992) Palaeo-epidemiology of *Schistosoma* infection in mummies. BMJ 304:555–556

Ministry of Health (1998) Etat d'avancementbdes programs de lute contre maladies parasitaires (annèes 1996-1997); Direction de l'èpidèmiologie et de Lutte Contre les Maladies. Service des Maladies Parasitaires Ministère de la Santè Publique, Rabat, Morocco

Ministry of Public Health (1995) Year plan of health development 1996-2000. Sana'a, MoPD, p 29

Mohamed AR, Al-Karawi M, Yasawy MI (1990) *Schistosoma*l colonic disease. Gut 131:439–442

Mostafa OM, Bin Dajem SM, Al-Qahtani A, Ibrahim EH, Al-Quraishy SA (2012) Developing, species-specific primers to identify *Bulinus truncatus* and *Bulinus baccari*, the intermediate hosts of *S. haematobium* in Saudi Arabia. Gene 499:256–261

Nagi MA (2000) The present status of schistosomiasis and intestinal helminthes in Yemen. In: Report on the regional workshop on the integrated control of soil-transmitted helminthes and schistosomiasis, Cairo, Egypt 16–18 October. WHOEM/CTD/017/E/I. World Health Organization Regional Office for the Eastern Mediterranean, Cairo

Nagi MA (2005) Evaluation of a program for control of *S. haematobium* infection in Yemen. East Mediterr Health J 11(5–6):977–987

Nagi MA, Molan AL (1992) Schistosomiasis among school children in Marib province of Republic of Yemen. Int Med J: 3212

Nagi MA, Kuma A, Mubarak JS, Bamashmoos SA (1999) Epidemiological, clinical and hematological profile of schistosomiasis in Yemen. East Mediterr Health J 5(1):177–180

Nash TE, Cheever AW, Ottesen EA, Cook JA (1982) Schistosome infections in humans: perspectives and recent findings. Ann Intern Med 97:740–745

Nasher AK (1988) Zoonotic parasite infections of the Arabian sacred baboon Papio hamadryas arabicus Thomas in Asir Province, Saudi Arabia. Ann Parasitol Hum Comp 63:448–454

Nazel MW, el-Morshedy H, Farghaly A, Shatat H, Barakat R (1999) *S. mansoni* infection and cognitive functions of primary school children, in Kafr El-Sheikh, Egypt. J Egypt Public Health Assoc 74:97–119

Nooman ZM, Hasan AH, Waheeb Y, Mishriky AM, Ragheb M, Abu-Saif AN, Abaza SM, Serwah AA, El-Gohary A, Saad A, El-Sayed M, Fouad M (2000) The epidemiology of schistosomiasis in Egypt: Ismailia Governorate. Am J Trop Med Hyg 62:35–41

Oshish A, Al-Kohlani A, Hamed A, Kamel N, Al-Soofi A, Farouk H, Ben-Ismail R, Gabrielli AF, Fenwick A, French MD (2011) Towards nationwide control of schistosomiasis in Yemen: pilot project to expand treatment to the whole community. Trans R Soc Trop Med Hyg 105:617–627

Patz JA, Graczyk TK, Geller N, Vittor AY (2001) Effects of environmental changes on emerging parasitic diseases. Int J Parasitol 30:1395–1405

Picquet M, Ernould JC, Vercruysse J, Southgate VR, Mbaye A, Sambou B, Niang M, Rollinson D (1996) The epidemiology of human schistosomiasis in the Senegal river basin. Trans R Soc Trop Med Hyg 90:340–346

Poggensee G, Feldmeier H (2001) Female genital schistosomiasis: facts and hypothesis. Acta Trop 79(3):193–210

Population Reference Bureau (n.d.) Data Finder. http://www.prb.org/DataFinder.aspx. Accessed 23 June 2013

Raja'a YA, Assiragi HM, Abu-Luhom AA, Mohammed AB, Albahr MH, Ashaddadi MA, Al-Muflihi AN (2000) Schistosome infection rate in relation to environmental factors in school children. Saudi Med J 21:635–638

Raja'a YA, Sulaiman SM, Mubarak JS, El-Bakri MM, Al-Adimi WH, El-Nabihi MT, El-Dhobri MA, Raja'a JA (2001) Some aspects in the control of schistosomiasis and soil-transmitted helminthosis in Yemeni children. Saudi Med J 22(5):428–432

Ritcher J (2003) The impact of chemotherapy on morbidity due to schistosomiasis. Acta Trop 86:161–183

Rollinson D, Knopp S, Levitz S, Stothard JR, Tchuenté LA, Garba A, Mohammed KA, Schur N, Person B, Colley DG, Utzinger J (2012) Time to set the agenda for schistosomiasis elimination. Acta Trop 128:423–40, http://dx.doi.org/10.1016/j.actatropica.2012.04.013

Ruffer MA (1910) Note on the presence of *Bilharzia haematobia* in Egyptian Mummies of the XXth Dynasty [1250-1000 BC]. BMJ 1:16

Salem S, Mitchell RE, El-Alim El-Dorey A, Smith JA, Barocas DA (2011) Successful control of schistosomiasis and the changing epidemiology of bladder cancer in Egypt. BJU Int 107 (2):206–211

Saoud MF (1966) The infectivity and pathogenicity of geographical strains of *S. mansoni*. Trans R Soc Trop Med Hyg 60:583–600

Saudi Arabia Ministry of Health Statistic Book (2004) http://www.moh.gov.sa/statistics/1425/default.html

Saudi Arabia Ministry of Health Statistic Book (2008) http://www.moh.gov.sa/statistics/1429/default.html

Schaap HB, Den Dulk MO, Polderman AM (1992) Schistosomiasis in the Yemen Arab Republic; prevalence of *S. mansoni* and *S. haematobium* infection among school children in the central highlands and their relation to altitude. Trop Geogr Med 44:19–22

Scott J (1937) The incidence and distribution of the human schistosomiasis in Egypt. Am J Hyg 25:566–614

Shalaby I, Gherbawy Y, Banaja A (2011) Genetic diversity among *S. mansoni* population in the western region of Saudi Arabia. Trop Biomed 28:90–101

Shati AA (2009) Factors affecting the prevalence of human schistosomiasis in Aseer region, Saudi Arabia. J Biol Sci 9:815–819

Shokeir AA, Hussein MI (1999) The urology of pharaonic Egypt. BJU Int 84:755–761

Sibomana L (2009) Association of schistosomiasis with socio-demographic status measures in sub-Saharan Africa. MSc thesis, University of Pittsburgh, Pittsburgh, PA

Simonsen PE, Nega A, Furu P (1990) Intestinal schistosomiasis among children in a labor village of Wonji sugar estate, Ethiopia. East Afr Med 67:532–538

Steinmann P, Keiser J, Bos R, Tanner M, Utzinger J (2006) Schistosomiasis and water resources development: systematic review, meta-analysis, and estimates of people at risk. Lancet Infect Dis 6:411–425

Talaat M, El-Ayyat A, Sayed HA, Miller FD (1999) Emergence of *S. mansoni* infection in Upper Egypt: the Giza Governorate. Am J Trop Med Hyg 60(5):822–6

Taylor MG, Nelson GS (1971) A comparison of the susceptibility to niridazole of two geographical strains of *S. mansoni* in mice with a note on the susceptibility of *S. mattheei*. Trans R Soc Trop Med Hyg 65(2):169–74

Thiongo FW, Madsen H, Ouma JH, Andreassen J, Christensen NO (1997) Host-parasite relationship in infections with two Kenyan isolates of *S. mansoni* in NMRI mice. J Parasitol 83:330–332

Voge M, Mansour NS (1980) An unusual structural feature of the Egyptian strain of *S. mansoni*. J Parasitol 66:862–863

Wang TP, Vang Johansen M, Zhang SQ, Wang FF, Wu WD, Zhang GH, Pan XP, Ju Y, Ornbjerg N (2005) Transmission of *S. japonicum* by humans and domestic animals in the Yangtze River valley, Anhui province China. Acta Trop 96:198–204

Wang X, Gurarie D, Mungai PL, Muchiri EM, Kitron U, King CH (2012) Projecting the long-term impact of school- or community-based mass-treatment interventions for control of *Schistosoma* infection. PLoS Negl Trop Dis 6(11):e1903. doi:10.1371/journal.pntd.0001903

Warren KS (1984) Schistosomiasis. In: Weatheral DJ, Ledingham JGG, Warrel DA (eds) Oxford textbook of medicine. Oxford University Press, Oxford

Watts S, Khallaayoune K, Bensefia R, Laamrani H, Gryseels B (1998) The study of human behavior and schistosomiasis transmission in an irrigated area in Morocco. Soc Sci Med 46:755–65

Webbe G, El-Hak S (1990) Progress in the control of schistosomiasis in Egypt 1985–1988. Trans R Soc Trop Med Hyg 84:394–400

Weldon A (2013) Yemen starts national campaign to fight schistosomiasis. www3.imperial.ac.uk/newssummary/news_11-3-2013-14-35-27

White G (1988) The environmental effect of the High Dam at Aswan. Environment 30:5–39

Wilson MS, Mentink-Kane MM, Pesce JT, Ramalingam TR, Thompson R, Wynn TA (2007) Immunopathology of schistosomiasis. Immunol Cell Biol 85:148–54

Woolhouse ME, Taylor P, Matanhire D, Chandiwana SK (1991) Acquired immunity and epidemiology *of S. haematobium*. Nature 351:757–759

World Bank (2004) Towards a general typology of small-scale fisheries in Yemen: Social assessment and development prospects. Republic of Yemen Inset 2: Yemeni fisheries at a glance (Sect. 4.5 Towards a general typology of small-scale fisheries in Yemen, Sect. 4.6 Current trends, Sect. 5.3 Towards an operational framework for a fisheries project, Sect. 6 General conclusion). International Development Agency (World Bank), IFAD. http://siteresources.worldbank.org/INTYEMEN/Data%20and%20Reference/20477216/FisheriesFAO.pdf

World Bank (n.d.) Data: Middle East and North Africa. http://data.worldbank.org/region/MNA. Accessed 22 June 2013

World Bank (n.d.) Poverty. http://data.worldbank.org/topic/poverty. Accessed 22 June 2013

World Health Organization (1985) The control of schistosomiasis. Geneva [contract no.: 728]

World Health Organization (1987a) Atlas of the global distribution of schistosomiasis. CEGET-CNRS/OMSWHO, Rabat, Morocco

World Health Organization (1987b) Atlas of the global distribution of schistosomiasis. CEGET-CNRS/OMSWHO, Saudi Arabia

World Health Organization (1993) The control of schistosomiasis. WHO Tech Rep 830

World Health Organization (2001) World Health Assembly Resolution 54.19. Schistosomiasis and soil transmitted helminthic infection. http://www.who.int/neglected_diseases/mediacentre/WHA_54.19_Eng.pdf.pdf. Accessed 26 June 2013

World Health Organization (2002) EM/TDR/007/E/12.04/2500. Evaluation of the prophylactic effect of artemether on human schistosomiasis: a randomized controlled trial. Small Grants Scheme (SGS) No. 39, Final report summaries 2001-2002

World Health Organization (2007) EMRO report of an inter-country meeting on strategies to eliminate schistosomiasis from the Eastern Mediterranean Region, Muscat, Oman, 6–8 November

World Health Organization (2009) Yemen: coverage with primary health care indicator. World Health Organization 2 World Bank (2004) Yemen Republic at a glance 3. Kuntz R, Malakatis G Lawless D. http://www.plosone.org/article/fetchObjectAttachment?uri=info%3

World Health Organization (2011) Informal consultation on schistosomiasis control. WHO, Geneva

World Health Organization (2012) Schistosomiasis and soil-transmitted helminthiasis: more than 9 million people treated in Yemen, 5 October/Geneva

World Health Organization, weekly epidemiological report (2012) Schistosomiasis: population requiring preventive chemotherapy and number of people treated in 2010. http://www.who.int/wer/

Wright W (1973) Geographical distribution of Schistosomes and their intermediate host. In: Ansari N (ed) Epidemiology and control of schistosomiasis (*bilharziasis*). University Park Press, Baltimore, pp 42–48

Wynn TA, Thompson RW, Cheever AW, Mentink-Kane MM (2004) Immunopathogenesis of schistosomiasis. Immunol Rev 201:156–167

Yamane A, Shotake T, Mori A, Boug AI, Iwamoto T (2003) Extra-unit paternity of hamadryas baboons (Papio hamadryas) in Saudi Arabia. Ethol Ecol Evol 15:379–87

Yapi YG, Briët OJ, Diabate S, Vounatsou P, Akodo E, Tanner M, Teuscher T (2005) Rice irrigation and schistosomiasis in savannah and forest areas of Côte d'Ivoir. Acta Trop 93:201–211

Yemen Bilharzia Control Project. Annual report MoPH, Sana'a (1993) 3. Nagi MA, Molan AL. Schistosomiasis among school children. http://www.smj.org.sa/DetailArticle.asp?ArticleId=255

Youssef AR, Cannon JM, Al-Juburi AZ, Cockett AT (1998) Schistosomiasis in Saudi Arabia, Egypt, and Iraq. Urology 51:170–174

Zain GH (1998) Prevalence of intestinal parasitic infections among school children in Yemen. [Dissertation] Khartoum: U of K, p 73. http://www.smj.org.sa/DetailArticle.asp?ArticleId=255

Zaki A, Bassili A, Amin G, Aref T, Kandil M, Abou-Basha LM (2003) Morbidity of schistosomiasis *mansoni* in rural Alexandria, Egypt. J Egypt Soc Parasitol 33:695–710

Fasciolosis in the MENA Region

Mohammad Bagher Rokni, Wael M. Lotfy, Kayhan Ashrafi, and K. Darwin Murrell

Contents

M.B. Rokni (✉)
Department of Medical Parasitology and Mycology, School of Public Health, Tehran University of Medical Sciences, P.O. Box 14155-6446, Tehran, Iran
e-mail: roknimoh@tums.ac.ir

W.M. Lotfy
Department of Parasitology, Medical Research Institute, Alexandria University, Alexandria, Egypt

K. Ashrafi
Research Center for Fascioliasis and Parasitic Diseases, Gilan University of Medical Sciences, Rasht, Iran

Department of Medical Microbiology, School of Medicine, Gilan University of Medical Sciences, Rasht, Iran

K.D. Murrell
Faculty of Health and Medical Sciences, Department of Veterinary Disease Biology, University of Copenhagen, Dyrlaegevej 100, 2, 1870 Frederiksberg, Denmark

M.A. McDowell and S. Rafati (eds.), *Neglected Tropical Diseases - Middle East and North Africa*, Neglected Tropical Diseases, DOI 10.1007/978-3-7091-1613-5_3,

Abstract Fasciolosis, a helminthic parasitic disease caused by liver fluke *Fasciola* spp. is considered an emerging and reemerging parasitic disease of global significance. In the MENA region, which includes more than 20 countries of the Middle East and North Africa, human fasciolosis is most important in Iran and Egypt with more than 30,000 cases reported. Human and animal fasciolosis has been reported from all countries of the region but to a lesser extent. Important factors in this region that affect the prevalence of infectious diseases such as fasciolosis are the political turmoil and government instability that impede proper public health management. In this chapter, the authors describe the status, basic biology, epidemiology and clinical disease of fasciolosis in the region.

Keywords Fasciolosis • Fasciola hepatica • Fasciola gigantica • Epidemiology • Middel East • North Africa

Background

Fasciola hepatica Linnaeus, 1758 and *Fasciola gigantica* Cobbold, 1855 (Platyhelminthes: Trematoda: Digenea) are the causative agents of fasciolosis in domestic animals and humans.

Fasciolosis is a disease caused by two trematode species of the genus *Fasciola* including *F. hepatica* and *F. gigantica*. Estimates of human infection in Asia and Africa vary from 2.4 to 17 million people (Mas-Coma 2005). People at risk are estimated at 91.1×10^6 (Keiser and Utzinger 2005). In 2005, it was estimated that about 56.2 million people were infected with food-borne trematodes, 7.9 million had severe sequelae, and the global burden was 665,352 DALYs, of which the most important share was attributed to fasciolosis (Fürst et al. 2012) (Table 1).

The MENA Region

According to World Bank, the MENA region includes countries of the Middle East and North Africa: Middle East: Afghanistan, Bahrain, Iran, Iraq, Israel, Jordan,

Table 1 Parasite-specific and region-specific modeled point estimates for fasciolosis in 2005, based on GBD 2010 study regions (Fürst et al. 2012)

	Total number of infected	Number of heavy infections	Number of deaths	YLD	YLL	DALYs
MENA region	1,119,812	133,268	0	17,275	0	17,275
Global	2,646,515	299,510	0	35,206	0	35,206

YLD years lived with disability, *YLL* years of life lost, *DALY* disability-adjusted life

Kuwait, Lebanon, Oman, Pakistan, Qatar, Saudi Arabia, Syria, United Arab Emirates, West Bank and Gaza; North Africa: Algeria, Djibouti, Egypt, Libya, Malta, Morocco, Somalia, Sudan, Tunisia, and Yemen. Almost 400 million people, approximately 5 % of the world's population, live in the MENA region, led by Egypt (80 million), Iran (75 million), Algeria (36 million), and Morocco and Iraq (31–32 million each) as the most populated countries (Hotez et al. 2012).

Methodology

For this review, publication databases related to fasciolosis in the MENA countries were created. Search engines utilized were PUBMED, SCIRUS, GOOGLE SCHOLAR, CABI, and IMEMR. Database searches were not restricted by date. In addition to full text papers, abstracts were also reviewed. The panel of key words utilized was "Fasciolosis," "Fascioliasis," "*Fasciola hepatica*," "*Fasciola gigantica*," "*Fasciola*." In addition to papers in English, papers published in other languages such as French, Persian, and Arabic

General Features of Fasciola *and Fasciolosis*

Life Cycle

Fasciola's life cycle encompasses two stages: the adult sexual stage and the larval asexual phase. The adult parasite normally resides in biliary ducts and gall bladder of the definite host but at times may invade ectopic sites such as eye, skin, spleen, etc. Eggs released from the parasite escape the host's body via the feces. After spending 9–15 days in the water, the undifferentiated ovum in the egg develops into a miracidium which hatches out of the egg and searches for a snail intermediate host belonging to the family Lymnaeidae, including species of the *Galba/Fossaria* group for *F. hepatica* and species of *Radix* for *F. gigantica*. In a suitable snail, the larval parasite undergoes several development stages (i.e., sporocyst, followed by redia) in which considerable asexual multiplication occurs. Finally, tailed cercariae, measuring 200–300 μm in length, emerge from the snail and swim through the water until locating suitable vegetation on which it encysts, becoming an infective metacercaria. Metacercariae may also be suspended in the water column as well and be infective if swallowed. After ingestion by the definite host, either by cyst-contaminated vegetation or water, the metacercaria excysts in the duodenum, releasing the juvenile stage, which burrows through the gut mucosa and migrates to the liver parenchyma. After 3–4 months, the juvenile flukes move into the bile ducts where they mature sexually and initiate egg production. The adults lay on average between 8,000 and 25,000 eggs per day and the average life span of the parasite is estimated at 9–13 years.

Systematics and Geographical Distribution

Fasciola hepatica is distributed worldwide, while *F. gigantica* occurs mainly in tropical regions of Africa, South and East Asia, and the Middle East. Humans and at least 46 other species of both domestic and wild mammals are reported as final hosts, either naturally or experimentally.

Previous studies in Africa have demonstrated that *F. hepatica* mainly occurs in the North, in Algeria, Tunisia, and Morocco and *F. gigantica* in the Southern African countries of Burkina Faso, Senegal, Kenya, Zambia, and Mali (Amor et al. 2011a). Both *F. hepatica* and *F. gigantica* have been reported from animal infections in Yemen (Farag 1985, 1998). Studies dealing with the genetic characterization of *F. hepatica* from North Africa are few and most species differentiations are based on morphological characteristics. In Asia and Africa, *F. hepatica* and *F. gigantica* appear to be sympatric (Mas-Coma et al. 2005), and this makes it difficult to identify morphologically each species (Alasaad et al. 2007).

According to Amor et al., some recovered specimens identified morphologically as *Fasciola* sp. from *Equus caballus* from the city of Tunis (north of Tunisia) were genetically confirmed by sequences of the 1st (ITS-1), the 5.8S and 2nd (ITS-2) Internal Transcribed Spacers (ITS) of nuclear ribosomal DNA (rDNA) (Amor et al. 2011b). From North Africa (Tunisia and Algeria), the liver flukes from sheep, cattle, and horse represent the single species *F. hepatica*, and two haplotypes (FhITS-H1 and FhITS-H2) were defined showing the same substitution C/T in position 859 in the ITS-2 sequences (Farjallah et al. 2009). The FhITS-H1 haplotype revealed a widespread distribution, indicating that this is the main haplotype involved in the spread of *F. hepatica* from Tunisia and Algeria (89.23 %) (Farjallah et al. 2009). The second haplotype of *F. hepatica* (FhITS-H2) appears to be less common, being reported as 10.76 % from Tunisia and Algeria (Farjallah et al. 2009).

Detection and Diagnosis

All diagnostic methods including clinical findings, biochemical and haematological data and imaging, as well as immunological techniques, and direct detection of eggs in stools have their drawbacks and inherent constraints (Mansour et al. 1983; Shehab et al. 1999). Definitive diagnosis is usually based on high antibody or antigen titer by serology (Shehab et al. 1999). Parasitological diagnosis of chronic fasciolosis is based on the identification of the parasite eggs in faeces or material obtained by duodenal or biliary drainage.

The diagnosis of fasciolosis is based primarily on fecal exams as the gold standard but the sensitivity in this method is not high. So in the MENA region a compendium of diagnostic methods including stool exam, serology, surgery, endoscopy, and imaging have been utilized for more reliable diagnosis (Rahimi et al. 2011; Rokni et al. 2002, 2006).

In Pakistan, Shahzad et al. concluded that PCR assay of fecal specimens is a more sensitive technique than microscopic methods for diagnosing fasciolosis. This conclusion was based on a finding that the percentage of positive samples by PCR for *F. hepatica* was higher for sheep and goat (4 % and 6.75 %) fecal and bile samples compared to the standard microscopic method (2.75 % and 5.25 %, respectively) (Shahzad et al. 2012).

In a study from Iraq, the diagnosis of fasciolosis included a compendium of hepatobiliary surgeries and endoscopic retrograde cholangiopancreatography (ERCP) which allowed direct visualization of the flukes discovered during a radiological examination for stones (Hawramy et al. 2012). The authors stated that precise serology is needed in Iraq to "solve many of the misdiagnoses problems and avoiding unnecessary surgeries as drug treatments completely cure the disease" (Hawramy et al. 2012).

Clinical Manifestation

Nearly all reported clinical manifestations have been observed in the MENA region. Overall, fasciolosis has two important phases in the final host: migration and final adult residency or acute and chronic phases, respectively. The first phase exhibits signs and symptoms that include epigastric pain, fatigue, fever, right quadrant abdominal pain, indigestion, weight loss, and malaise. Extrahepatic abnormalities such as pulmonary infiltrates, pleuropericarditis, meningitis, and lymphadenopathy result as well.

During the chronic period, numerous manifestations may occur: hepatomegaly, obstructive jaundice due to biliary obstruction, cholecystitis, cholelithiasis, fibrosis, cirrhosis, liver abscesses, hyperplasia of the ductal epithelium, and subcapsular hemorrhages. Anemia, of normocytic normochromic nature, can also be expected. The presence of small, multiple intrahepatic stones, along with cholangitis and cholelithiasis constitute the features of human chronic phase. In a recent report, abdominal pain was the most common manifestation as 88.8 % followed by malaise (61.1 %) (Hawramy et al. 2012).

No cases of death has been reported from fasciolosis in the region so far, and normally the rate of death in this disease is very rare (Mas-Coma and Bargues 1997).

Ectopic cases of fasciolosis have been occasionally reported from the MENA region (Fattah et al. 1964; Rokni 2008). Nearly all parts of the body have been reported as having ectopic fasciolosis, including gastrointestinal tract, subcutaneous tissue, heart, blood vessels, the lung and pleural cavity, brain, orbit, abdominal wall, appendix, pancreas, spleen, inguinal nodes, cervical node, skeletal muscle, and epididymis. It is difficult to diagnose ectopic cases of fasciolosis by routine methods; consequently, ectopic fasciolosis is probably very underreported in the region (Mas-Coma and Bargues 1997).

Treatment

For more than two decades, triclabendazole (TCBZ) has remained the drug of choice for the treatment of fascioliasis in humans and animals. It displays high efficacy against both immature and adult flukes. The drug was originally manufactured by Novartis under the commercial names Fasinex (liquid form for animals) and Egaten. Currently, Egaten is the only human form of TCBZ and it is not available commercially. The drug is distributed free of charge through the national control programme of the MoHP. The World Health Organization (WHO) offers Egaten to endemic countries up on an annual government request.

Based on the recommendation of the WHO, the drug of choice globally is TCBZ, including the MENA. However, due to some restrictions other fasciolocides have been used as well. In Iraq, five human cases with fasciolosis were treated with albendazole because TCBZ was not available, and apparently, the result was positive for all cases but one (Hassan et al. 2013; Hawramy et al. 2012). In another study, only three patients could provide TCBZ from out of Iraq because this drug is not available there but eight patients received albendazole 400 mg t.d.s. for 1 week postoperatively (Hawramy et al. 2012).

In Saudi Arabia, nine patients with fascioliosis were successfully treated with mirazid; two capsules (600 mg) an hour before breakfast for 6 consecutive days (El-Mathal and Fouad 2005). In Morocco, anthelmintics are routinely used by local farmers for treatment of animal fasciolosis rather than for prophylaxis. This results to limited success in the context of controlling the disease (Khallaayoune et al. 1991).

Economic Impact of Fasciolosis

In Saudi Arabia, 8.6 % of livers from slaughtered cattle were condemned due to fasciolosis and the economic importance of such infections in terms of lost meat and offal was estimated as US$20,000 annually (Degheidy and Al-Malki 2012).

Camel is the most important animal in Saudi Arabia in terms of meat and milk for many people, therefore, a prevalence of 4.22–15 % with *F. gigantica* imposes a remarkable loss on the country's food production (Banaja and Ghandour 1994).

The sheep industry in northern Tunisia also suffers severely from ovine fasciolosis. Akkari et al. (2011) have report a high prevalence of fasciolosis in flock lambs and ewes, 60 % and 65 %, respectively. An even higher infection rate was recorded in tracer lambs (>70 %) at postmortem. Considering Tunisian sheep population is about 6.5 million, (the most dense sheep population in North Africa), the economic losses due to fasciolosis is considerable, because sheep contribute 50 % of the national red meat production (Rekik and Ben Hammouda 2000).

In Iran, the average percentage of liver condemnations due to *Fasciola* spp. over a 3-year period was 2.12 %. Over all, the average annual cost from condemned livers was 8.2 million USD with *Fasciola* spp. constituting one of the main causes (Jahed Khaniki et al. 2012).

Losses due to animal fasciolosis in Egypt were estimated as US$27,176,765 annually (Soliman 2008).

Epidemiology

Sources of Infection

Plants

The main source of infection is ingestion of metacercaria encysted on water plants such as watercress (*Nasturtium officinale*). In Iraq 14 patients (77.7 %) infected with fasciolosis had a history of ingestion of raw watercress purchased in markets (Hawramy et al. 2012). In Tunisia, the plants, *Apium nodiflorum*, *Oxalis cernua*, and *Sonchus maritimus* are considered the main source for animal infection, while *A. nodiflorum* is involved in human infection (Hammami et al. 2007).

Water

Metacercaria of *Fasciola* spp. may also remain suspended in water and contaminate water used for cooking and drinking. This risk was found to be significantly reduced in Egypt by the introduction of "Washing Units," in which water from a canal was directed though a swimming pool filter; the prevalence of fasciolosis in humans in one village was reduced from 18 % to 2 % within a few years, confirming that drinking water can be a significant infection risk for cooking and washing vegetables (Curtale 2008). Further evidence of the importance of contaminated water is a study using PCR to detect *F. hepatica* DNA in samples of water in the Nowshehra district of Khyber Pakhtunkhwa Province of Pakistan. The results indicated that

29/300 (9.66 %) of samples tested were positive; the prevalence of *F. hepatica* DNA was 10 % in tube well water, 8 % in open well water, 1.66 % in tap water, and 16 % in drainage water (Khan et al. 2012). Similarly, washing of kitchen utensils is another potentially important risk for transmission of metacercariae. According to a World Bank report because of the level of poverty is high in the MENA region, many people are unable to obtain safe water for washing of dishes and drinking (World Bank 2011).

In some parts of the region people often drink beverages made of local plants, e.g., stalks of cane in Pakistan, carrot in Iran, etc. These plants have a high possibility of contamination with metacercaria due to irrigation by contaminated waters.

Raw Liver

Although the potential of ingestion of raw liver as a source of infection has been proposed (Taira et al. 1997), the habit of eating raw or semicooked liver in MENA region is scarce, so the potential from this source is low.

Chewing Khat

Chewing the leaves of khat (*Catha edulis* Forsk.) may be a potential source of infection (Cats et al. 2000). This habit is prevalent in East African and Middle Eastern Countries (Al-Motarreb et al. 2010). Cats et al. (2000) reported a case of fasciolosis in a 36-year old Somalian man that putatively acquired the infection from khat, although the diagnosis was made only on the basis of serology. However, a similar case attributed to chewing khat occurred in a Yemenite woman residing in the UK (Doherty et al. 1995). Because khat shrubs are often cultivated in areas where the grazing of sheep on irrigated pastures occurs and its leaves when harvested are wrapped and kept are kept damp during transport, opportunity for contamination with metacercariae exists (Al-Motarreb et al. 2010).

Host Gender

Globally it is estimated that of 758,660 male and 768,630 female examined for fasciolosis, 0.202 % and 0.207 % were infected, so sex ration for M/F was determined as 0.979 (Fürst et al. 2012).

Review of literature shows that in Egypt and Iran, females are infected more often than males. In Egypt, prevalence, but not intensities, appeared to be significantly higher in females, i.e., 8.8 % versus 14.8 % for M/F, which was significant (Esteban et al. 2003). Results from another study in Egypt indicated women to be at more risk than men (a prevalence of 10.3 % in females and 4.4 % in males) (Farag et al. 1979). Fasciolosis was also higher in females (7.69 %) than in males (5.85 %)

in another study in Egypt (Samaha 1989). In Egypt, rural girls are often involved in household and farm work and more exposed to sources of infection than boys. The lower school attendance for girls in rural areas appears to also be an important factor in increasing the risk of infection (Soliman 2008). In the endemic region of northern Iran, studies during the first reported outbreak showed a statistically significant difference between the two genders with higher rates in females ($P < 0.05$) (Forghan-parast et al. 1993). However, in other studies there since, the prevalence was higher in females than males but this was not statistically significant (Ashrafi et al. 2012; Moghaddam et al. 2004a; Sarkari et al. 2012).

In rural communities of both countries, females involved in agricultural tasks, meal and salad preparation, and washing activities probably have a higher risk of infection, but these risk factors need confirmation.

Control and Preventive Measures

The most critical obstacle to the establishment control programs many MENA countries is the political turmoil that frequently occurs in this region. Since 2011 onwards, governmental shifts, in addition to frequent civil agitations and demonstrations, public authorities are often inhibited from implementing control programs, not only for fasciolosis but other infectious diseases as well. Because of the poverty in this region such political turmoil extracts a severe toll: in 2005 it was estimated that 3.6 % of the MENA population lives below the World Bank poverty figure of US$1.25 per day, while 16.9 % lives below US$2 per day (Hotez et al. 2012).

One of the most difficult tasks in a control effort is persuading people not to consume certain high risk, traditional foods. It would be useful to teach people the five keys to safer food, provided by WHO (WHO/SDE/PHE/FOS/01.1. Food Safety. World Health Organization), as follows: keep clean; separate raw and cooked; cook thoroughly; keep food at safe temperatures; use safe water and raw materials.

Major Features of Human and Animal Fasciolosis in Different MENA Countries

Although human and animal fasciolosis occur throughout the Mena region, the zoonosis is most prominent in Iran and Egypt (Hotez et al. 2012; Mas-Coma 2005). For this reason, fasciolosis in these two countries are treated in greater detail below then are the Other Mena region countries.

Iran

History

During the last few decades, the public health importance of human fasciolosis has significantly increased (Mas-Coma and Bargues 1997). Up until 1989, human fasciolosis was reported only sporadically in Iran. The first documented case of the disease was an ectopic case in the thyroid (Adl and Sedigh 1956); subsequently a few case reports of hepatobiliary fasciolosis were reported. This status changed in 1989 when the world's largest ever outbreak occurred in Gilan Province, northern Iran, affecting more than 10,000 people (Table 2) (Assmar et al. 1991; Massoud 1989). A second large outbreak occurred 10 years later affecting an estimated 10,000–15,000 people (Forghan-Parast and Ashrafi (2001) (Table 2). Several hundred additional human cases were reported before and after the second outbreak, establishing that Gilan Province is a highly endemic region for human fasciolosis, particularly in Bandar-Anzali City. Due to its unique epidemiological characteristics a specific pattern termed the "Caspian Pattern" of transmission has been proposed for Caspian Sea areas; this area is defined as a hypoendemic area with large scale epidemics sometimes affecting more than 10,000 people (Mas-Coma 2007). Recent surveys reveal prevalence of 0.4 and 1.2 % using coprological and serological methods, respectively, confirming a hypoendemic situation (Ashrafi et al. 2012). World Health Organization (WHO) has also included Iran among six countries which are known to have a serious fasciolosis problem (WHO 2010). In addition to the outbreak data (Table 2), prevalence studies confirm the high rates of infection in Gilan province (Table 3).

The large outbreaks and prevalence data from these endemic areas have greatly increased the awareness of fasciolosis among Iranian physicians and parasitologists, and this has undoubtedly contributed to the increase in reports of cases from various localities all over the country, including ectopic and normal residence cases (Alavi-Naini et al. 2013; Aminian et al. 2012; Mohammadi-Ghalehbin et al. 2012; Rokni 2008).

In the outbreaks in Gilan and Kermanshah the highest number of infected individuals was seen in the 10–29 year and in 10–19 year age groups, respectively. In comparison, in non-epidemic situations, the numbers of infected cases were higher in older (>20 year) age groups (Ashrafi et al. 2012; Moghaddam et al. 2004a).

Animal Fasciolosis

Animal fasciolosis has been reported from different areas of Iran. Mahami-Oskouei et al. (2012) reported an overall 1.10 % prevalence infection in sheep and cattle slaughtered in six different provinces of Iran. Khorasan Razavi and Fars provinces had the highest (14.54 ± 3.16) and lowest (7.75 ± 0.79) prevalence of infection,

Table 2 Fasciolosis outbreaks occurring in Iran

Date	Place (province)	No. of cases	Sex ratio F/M	Age (years)	Diagnostic methods	References
1989–1990	Gilan	10,000	–	–	Coprology Serology	Assmar et al. (1991), Massoud (1989)
2000	Kermanshah	17	53/47	Mean 21.6	Serology (ELISA and CCIE)	Hatami et al. (2000)
2000–2001	Gilan	10,000–15,000	–	–	Coprology Serology	Forghan-Parast and Ashrafi (2001)
1999–2002	Mazandaran	107	19/17 (data only for 36 cases)	–	Coprology (health centers' data)	Moghaddam et al. (2004a)

Table 3 Prevalence surveys conducted in Iran from 1991 to 2012

Date	Location (province)	Prevalence (number tested)	Age (range years)	Male:Female infections	Diagnostic method	References
1991	Gilan	50 % (452)	10–19	73/153	Serology (ELISA)	Assmar et al. (1991)
2012	Gilan Province	1.2 % (1,283)	41–50	8/7	Serology	Ashrafi et al. (2012)
2012	Gilan Province	0.4 % (1,283)	51–60	3/2	Coprology	Ashrafi et al. (2012)
2011	Kohgyluyeh and Boyreahmad Province	1.8 % (1,000)	41–50	9/9	Serology (ELISA)	Sarkari et al. (2012)
2012	Ilam	0.7 % (600)	30–50	1/3	Serology (ELISA)	Abdi et al. (2013)
2012	Ardabil	1.9 % (458)	40–49	4/5	Serology (ELISA)	Asadian et al. (2013)

respectively. Moghaddam et al. (2004b) reported 7.3 and 25.4 % overall prevalence of fasciolosis in sheep and cattle, respectively from Mazandaran Province, using coprological surveys. In a study in Ilam Province animal fasciolosis was demonstrated in 53 % of slaughtered cattle, 36.5 % of sheep, 10.5 % of goats, and overall prevalence as 1 % (Abdi et al. 2013). Moshfea et al. (2003) reported fasciolosis in slaughtered animals in Yasuj, as 12.5 % in cattle, 11.75 % in sheep, and 7.16 % in goats. In northern parts of Iran, Eslami et al. (2009), reported fecal samples of 32 % of sheep, 32.1 % of cattle, 17 % of buffaloes, and 50 % of horses infected with *Fasciola* egg.

Etiology

In Iran, the distribution of *F. gigantica* and *F. hepatica* overlaps in almost all areas and both species may be obtained from a single definitive host. The fluke species involved in human fasciolosis is not clarified taxonomically. Iranian *F. hepatica*-like specimens are larger than *F. hepatica* from other regions, as is the *F. gigantica*-like specimens recovered in Iran (longer and narrower than "classical" *F. gigantica* but with a smaller body area). In the Iranian fasciolid populations, intermediate forms are also present (Ashrafi et al. 2006b). To overcome hybridization effects, PCR-restriction fragment length polymorphism () and random-amplified Polymorphic DNA (p)-derived sequences have been utilized to distinguish between the two species. ITS1 and ITS2 sequences from rDNA have proved to be reliable genetic markers for identification (Rokni et al. 2010b). Rokni et al. showed the variability of *F. hepatica* isolates in Iran, using RAPD markers. No intraspecies variation has been seen in the Iranian *F. hepatica* isolates at ITS1 rRNA gene loci, indicating the highly conserved nature of this region (Rokni et al. 2010a).

Snail Intermediate Hosts

Several species of *Lymnaea* including *Galba truncatula* and *L. gedrosiana* have been confirmed to be important the in transmission of *F. hepatica* and *F. gigantic*, respectively (Ashrafi 2004; Ashrafi et al. 2007; Mansoorian 2000).

Sources of Infection

In Iran, *Nasturtium* spp. *Eryngium caucasicum* and *Mentha* spp. are important sources (Ashrafi et al. 2006a). Two very important sources in the endemic regions of northern Iran are green salt (Local name: Dalar) and elaborated olive (Local name: Zeitoon parvardeh). Green salt is prepared by a mixture of ground aquatic plants such as *Mentha pulegium* (local name Khlivash) as well as *Mentha piperita* (Bineh) and 30–40 % salt; the final pH is 5.0. It is usually eaten with cucumber, prunes, yogurt, etc. Zeitoone parvardeh, an appetizer, is a mixture of stone-free olive, ground aquatic plants mostly, *Eryngium coucasicum* ("Choochagh"), walnuts, various spices, garlic, and sour-pomegranate juice. Metacercariae maintained 2 weeks in Zeitoon-parvardeh or Dalar had survival rates of 66.6 % and 57.8 %, respectively (Ashrafi et al. 2006a).

Clinical Manifestations

During the first reported outbreak of human fasciolosis, the most common clinical manifestations were weight loss (88 %) followed by epigastric pain (87 %) (Yadegari et al. 1991). In a small outbreak in Kermanshah Province, western Iran, the main clinical manifestations were weight loss (47 %), epigastric pain (41 %), abdominal pain (29 %), hepato-splenomegaly (29 %), right upper quadrant pain (24 %), and right hypeochondria tenderness, as well as sweating (24 %) (Hatami et al. 2000). In new outbreaks reported in Yasuj, central Iran, abdominal pain, allergic manifestations and headache were observed (Sarkari et al. 2012).

Treatment

In Gilan Province, bithionol (40 mg/kg for 15 days) was used for treatment of 31 patients during an outbreak with 66–69 % effectiveness, although 60 % of the patients were hospitalized due to severe drug side effects (Sarshad et al. 1990). Because of the high incidence of human fasciolosis in Iran, the first clinical trial of TCBZ was carried out, using the WHO protocol for the veterinary formulation of TCBZ (Yadegari et al. 1999). In this randomized clinical trial, which lasted 6 years the efficacy of TCBZ was 94 %, with good tolerance and minimum side effects. Afterwards, TCBZ was recommended as the drug of choice for treatment of human fasciolosis by the WHO. The results of another randomized trial, employing a single, double, and triple dose of a human formulation of TCBZ supported the safety and efficacy of 10 mg/kg of the drug for 1–3 days (Talaie et al. 2004). An attempt to treat with praziquantel (70 mg/kg) in 100 cases yielded only a 2 % cure rate, demonstrating the ineffectiveness of this drug for treatment of human fasciolosis (Yadegari et al. 1991). At present, treatment of human fasciolosis in Iran uses Egaten (human pharmaceutical preparation of TCBZ) which is donated by WHO.

Prevention

Based on the cultural and epidemiological characteristics of the region and WHO proposal, the following control measures were implemented:

1. Treatment of infected people with TCBZ;
2. Establishment of effective veterinary public health measures including treatment of livestock with TCBZ;

3. Increase awareness of people living in endemic areas about the danger of eating raw, uncooked aquatic and semi-aquatic plants.

Egypt

History

Egypt has been plagued by the global emergence of human fasciolosis since the late 1970s (WHO 1995). The highest prevalence of human fasciolosis in the Middle East and Africa has been reported from this country (Amor et al. 2011a; Hotez et al. 2012). Remains of *Fasciola* flukes have been detected in ancient Egyptian mummies (Curry et al. 1979; David 1997; Tapp 1986), confirming that human fasciolosis has existed in Egypt since the Pharaonic times (Amor et al. 2011a).

Etiology

Fasciola gigantica has been present in the country for millennia (Amor et al. 2011a). *Fasciola hepatica*, however, apparently was not present in the Nile Delta till the end of the nineteenth century (Looss 1896). In 1957, Halawani and Gindy (1957) reported that *F. gigantica* was the common liver fluke in Egypt, while *F. hepatica* was found only in the oases. However, Kuntz et al. (1958) reported in 1958 that during examination of numerous wild and domestic animals in the Nile Delta they found both *F. hepatica* and *F. gigantica* to be common. Most probably, *F. hepatica* has been introduced to the Nile Delta in herds imported from Europe sometime during the twentieth century (Nagaty 1942; Selim et al. 1970; Soliman and Farid 1960). The species most incriminated in human fasciolosis in the Nile Delta and Valley is *F. hepatica* (Ghavami et al. 2009). However, human infection with *F. gigantica* in endemic foci in Egypt cannot be excluded (Hammond 1974; Ragab and Farag 1978). In contrast to Egypt, *F. gigantica* is thought to be the more common species infecting humans in northern Iran (Amor et al. 2011a), and Vietnam (Abdouslam et al. 2003). By using a computer image analysis system (CIAS) to characterize the *Fasciola* eggs shed by humans, it was found that the size of eggs from Egypt corresponds to the *F. hepatica* morph, while the size of eggs from Vietnam corresponds to the *F. gigantica* morph (Amor et al. 2011a).

Prevalence of Human Fasciolosis

Altogether, the population at risk in Egypt is considered to be 27 million (Lotfy and Hillyer 2003) and the prevalence ranging from 2 to 11 % (Soliman 2008). Before 1978 only sporadic cases of human fasciolosis were reported from the country (Abdouslam et al. 2003; Chester 1928; Kuntz et al. 1958). Apparently, human fasciolosis emerged after recent wide spread irrigation systems built in Egypt (Amor et al. 2011a). Aswan Low and High Dams reduced the current of the Nile flow, which resulted in flare-up of the freshwater snail populations and the associated helminthic infections (Amor et al. 2011a). The great increase in irrigation canals associated with the dams also greatly increased snail habitat, which also contributed to an increase in schistosomiasis.

In 1978, Ragab and Farag diagnosed five cases of human infection. By tracing the place of origin of the patients, the authors detected a focus of human fasciolosis in the Abis area located south of Alexandria; the prevalence of human fasciolosis was determined to 7.3 %. They concluded that the problem of human fasciolosis in Egypt needed more effort to assess its magnitude and to avoid further spreading of the disease.

Mansour et al. (1983) conducted further studies in the same village, and based on stool exams they reported that of 30 patients, 20 had *F. hepatica* eggs, 8 had *F. gigantica* eggs, and 2 had eggs of both species (Mansour et al. 1983). Of course based on the current knowledge, the differentiation between different species of *Fasciola* based on egg detection is very problematical (Valero et al. 2001). In 1995, the WHO estimated the overall prevalence in the Nile Delta at 3 %, with at least 830,000 people infected and 27.7 million people being at risk of infection (WHO 1995).

Age-Related Prevalence

In 2002, in a study in four newly identified endemic foci in Behera, Egypt, the highest prevalence and intensity of infection (5.7–11.9 %) was highest in the 9–11 years age group. The results of this study emphasized that primary schoolchildren are at high risk of contracting the infection and should be considered the main target for control measures (Curtale et al. 2003). However, according to Mas-Coma et al., there is a wide range of age among patients with fasciolosis in Egypt (Mas-Coma 2005).

Animal Fasciolosis

The prevalence of fasciolosis in Egypt in cattle and buffalo was 3–59.5 % and in sheep as 11–53 % (Soliman 2008). Another study on livestock conducted in Upper Egypt showed an overall fasciolosis prevalence of 30.3 %, including 28.6 % in cows, 33.7 % in buffaloes, and 17.2 % in sheep (Hussein and Khalifa 2010).

Snail Hosts

Radix natalensis (synonym *R. caillaudi*, Lymnaeidae) is the main intermediate host for *F. gigantica* in Egypt and Africa (Ahmed and Ramzy 1999; El-Dafrawy 2002; Hussein and Khalifa 2008). Natural infections of this snail species with *Fasciola* spp. have been reported (El-Shazly et al. 2002, 2012). Experimental infection of *R. natalensis* with *F. hepatica* was confirmed. Thus, *R. natalensis* can be considered a potential intermediate host of *F. hepatica* in Egypt (Amer et al. 2011; Ghavami et al. 2009).

Galba truncatula (synonym *Lymnaea truncatula*), the preferential intermediate host of *F. hepatica* in temperate regions, was found naturally infected in Eygpt with *F. gigantica* (Aissi et al. 2009) and *Fasciola* spp. (El-Shazly et al. 2002, 2012). Successful experimental infections with both species of *Fasciola* were obtained under the laboratory conditions (Amor et al. 2011a).

Pseudosuccinea columella (synonym *Lymnaea columella*, Lymnaeidae) was found infected with *F. gigantica* under the field conditions (Ahmed and Ramzy 1999). *Lymnaea stagnalis* was found naturally infected with *Fasciola* spp. (El-Shazly et al. 2012).

Biomphalaria alexandrina (Planorbidae) was reported to be naturally infected with *F. gigantica* (Amor et al. 2011a) or *Fasciola* spp. (El-Shazly et al. 2002). Yet, this snail was not susceptible to infection with *F. gigantica* under experimental conditions (Abd El Bagi et al. 2004).

Diagnosis

In Egypt, different stool examination techniques have been evaluated and used for the diagnosis of human fasciolosis. Based on the results of the different studies it may be concluded that at least in mesoendemic foci, like those in Egypt, the Kato-Katz technique has enough sensitivity to detect and efficiency to quantify *F. hepatica* eggs in human feces. The technique's simplicity and ease of performance can help to ensure its application in epidemiological surveys (Ahmed et al. 1994; Amor et al. 2011a). Ultrasonographic changes are of great help in diagnosis of fasciolosis, especially when stool examinations are repeatedly negative for *Fasciola* eggs (Ahmed et al. 1994). This technique coupled to stool examination may be used to assess complete cure in drug trails (Amor et al. 2011a). Liver biopsy may accidentally reveal the presence *Fasciola* eggs (El-Shabrawi et al. 1997).

Source of Infection

Many species of green leafy vegetables are eaten fresh as salads, including arugula (*Eruca sativa*), lettuce (*Lactuca sativa*), and kurrat (*Allium kurrat*). Although not classified as aquatic plants like watercress, they are grown adjacent to freshwater canals and are frequently irrigated. After harvesting, they are also prepared for marketing by washing in the nearby canals The processes of irrigation and washing exposes them to the *Fasciola* cercariae, which can encysts and become infective within a few hours (Amor et al. 2011a).

Clinical Manifestations

Although rashes and urticaria are considered common symptoms of the acute stage of infection in other epidemic foci (Abdouslam et al. 2003), they are not very specific in Egyptian patients (Curtale et al. 2003). Fever, which is usually considered the first symptom of *Fasciola* invasion (Abdouslam et al. 2003), did not appear more frequently in Egyptian cases compared with uninfected persons. The high prevalence of fever causing diseases, affecting the rural population in the Nile Delta, might have reduced the specificity of this symptom (Curtale et al. 2003). During the acute stage of the disease, ultrasonography revealed hypoechoic areas inside the liver, sometimes expanding into the peritoneal cavity, and splenomegaly are typically observed in Egypt (El-Shabrawi et al. 1997; Fawzy et al. 1992).

In Egyptian studies, the parasite may be recovered from ectopic lesions (Rashed et al. 2010). Rarely, worms may be retrieved from patients suspected of fasciolosis, or patients with cholelithiasis or obstructive jaundice of unknown cause by ERCP (Al Qurashi et al. 2012). In addition, motile parasites (echogenic masses) and associated pathology in the gall bladder or the biliary tree may be identified by abdominal ultrasonography (El-Shabrawi et al. 1997; El-Shazly et al. 2001; Fawzy et al. 1992). During the chronic stage, abnormalities essentially affect the biliary tract (El-Shabrawi et al. 1997; Fawzy et al. 1992). In this stage, imaging abnormalities including hepatomegaly, splenomegaly, periportal fibrosis, thickened gall bladder wall, dilated common bile duct, parasites in gall bladder and common bile duct, cholelithiasis, biliary duct stones, cystic liver lesions, and focal lesions in the liver and ascites were reported (El-Shazly et al. 2001). Excluding viral liver infections, human fasciolosis is a significant cause of cholestasis in endemic areas of Egypt ($P < 0.05$) (El-Shazly et al. 2005). It was reported that the major complications of the disease in humans were bleeding, biliary cirrhosis, and ectopic lesions (Ragab and Farag 1978). Documented uncommon complications due chronic infection include gall bladder rupture accompanied with development of a liver abscess and acalcular cholecystitis with empyema of the gall bladder (Abou Basha et al. 1989).

Different types of anaemia were encountered in infected persons (Bassiouny et al. 1991; El-Khashab et al. 1993; El-Shazly et al. 1991, 2005; Amor et al. 2011a). The type of anemia that develops could be considered a biomarker of the chronicity period of the infection (Amor et al. 2011a). It changes from normocytic to macrocytic in the early chronic period and to microcytic in the advanced chronic; it is changed from normochromic in the early chronic period to hypochromic in the advanced chronic period

Prevention

In 2007, the WHO reported that Egypt is the only country that has implemented an organized control program against human fasciolosis, which began in 1996 in six endemic districts in Egypt's Behera Governorate. Concurrently, school surveys have been conducted in all the Delta governorates and in some Upper Egypt governorates (Hawramy et al. 2012). In 1998, the Egyptian Ministry of Health and Population (MOHP) launched the first school-based intervention to control the disease. Because of a lack of precedents, the MOHP designed and implemented an innovative selective treatment strategy, based on their experience from other helminth control programmes. In this selective treatment approach, chemotherapy (TCBZ) was targeted to specific high-risk age groups and villages; these were identified by a baseline surveys that allowed targeting of screening and selective treatment of schoolchildren within those districts. During the period from 1998 to 2002, the program screened almost 36,000 schoolchildren in the six districts and treated 1,280 cases. Prevalence of the disease was reduced from 5.6 to 1.2 %. The selective chemotherapy approach was feasible and appropriate in addressing a low prevalence infection and more cost-effective than mass treatment (Curtale et al. 2005).

Chemotherapy

The use of TCBZ for the treatment of human fasciolosis in Egypt is effective (Amor et al. 2011a). Reasonable success (50 % cure rate) has been observed in the treatment of acute fasciolosis with an initial course of parenteral dehydroemetine in Egyptian children (Farid et al. 1988, 1990). Egypt was the first country in the world to register TCBZ for human use. In 2011, the WHO reported that TCBZ is registered for human use in only four countries which are Egypt in 1997, Ecuador in 2001, Venezuela in 2001, and France in 2002 (Hawramy et al. 2012). Reliance on a single drug has raised considerable concern that tolerance or even resistance to the drug might develop. Internationally, a number of TCBZ -resistant *F. hepatica* infections in animals have been confirmed which indicates that resistance does exist (Fairweather 2011). The presence of TCBZ resistance has not been confirmed in Egypt.

Egyptian clinical trials have also shown that praziquantel failed to cure human fasciolosis (Farag et al. 1986; Farid et al. 1986, 1988), even when used in doses 25 times the recommended dose for schistosomiasis (Ahmed et al. 1994). Bithionol has been used in treatment of human fasciolosis in Egypt (Abou Basha et al. 1989; Amor et al. 2011a; Farid et al. 1988, 1990) but it showed low efficacy against cases of acute infection and few cases of chronic infection. Mirazid is relatively new drug manufactured by an Egyptian company (Pharco Pharmaceuticals). It was presented to the market as a schistosomicidal and fasciolicidal drug. It is prepared from myrrh, either Arabian or Somali, which is an oleo gum resin obtained from the stem of thorny trees (*Commiphora molmol*) and probably other related species of the Bursearaceae (Greene 1993). There has been considerable debate over the efficacy and effectiveness of Mirazid in treatment of fasciolosis, with most investigators claiming it is ineffective (Botros et al. 2009; Osman et al. 2010). However, some Egyptian physicians still recommend the drug, motivated, perhaps because it is a natural plant product. In a single clinical trial in Egypt, artemether showed no or only little effect against human fasciolosis and hence could not represent an alternative to TCBZ (Keiser et al. 2011).

Other MENA Countries

There are some case reports of human fasciolosis from Lebanon (Birjawi et al. 2002), although hepatobiliary parasitic diseases are rare in Lebanon natives. The disease was identified by sonography. In another study from 1997 to 1998 and 2007 to 2008 in Lebanon, only one case was reported by stool exam (Araj et al. 2011).

In Tripoli, Libya, animal fasciolosis has been reported as 4 % of local bred sheep and 18 % of imported sheep. Besides, 0 % of local bred cattle and 65 % of imported cattle were infected (Ben Amer and Ahmed 1980). There are reports of human and animal fasciolosis from Algeria, Morocco as well (Khallaayoune 1995).

Unfortunately, the amounts of documents are so scarce, which restrains depicting any clear feature of the disease in the related country.

Algeria

Four human cases of fasciolosis have been reported from Algeria (Zait and Hamrioui 2005) (Table 2). A prevalence ranging from 6.3 % to 27.3 % has been reported in cattle (Mekroud et al. 2004).

Iraq

The first case of fasciolosis in Iraq, an ectopic infection in the eye due to *F. gigantica*, was reported in 1964 (Fattah et al. 1964). In 2004, a case of biliary fasciolosis was identified (Hawrami 2004), followed by a case with an adult *F. hepatica* in the gallbladder in 2010 (Hawrami 2010). Two previous cases were diagnosed during different surgeries (Hawrami 2010).

Five cases of eosinophilic granulomatous abscesses in Sulaimaniyah (Eastern Kurdish region of Northern Iraq) were attributed to *F. hepatica* (Hassan et al. 2013). The diagnosis was based on serological methods (ELISA test) only and no larvae or eggs of *Fasciola* were found by histopathology. Cases of fasciolosis in cattle (27 %) and lymnaeid snails (4 %) already have been reported in that area (Hawramy et al. 2012). In another study, Hawramy et al. (put in date) reported 18 cases of fasciolosis in Sulaimani based on medical records of the patients from 1997 to 2012

Table 4 Human fasciolosis in specific countries of the MENA region (excluding Egypt and Iran)

Country	No. of cases	Sex ratio F/M	Age (range years)	Diagnostic methods	References
Algeria	4	3/1	10–42	Parasitology/IEP	Zait and Hamrioui (2005)
Iraq	18	15/3	25–82	Hospital-based documents	Hawramy et al. (2012)
Iraq	10	–	–	Surgery-medical inspection	Hassan et al. (2013), Hawramy et al. (2012)
Israel	2	–	–	Case report (one imported case in 65-year-old female from Afghanistan)	Dan et al. (1981)
Pakistan	21	0.30/0.28	<20	Parasitology	Qureshi et al. (2005)
Pakistan	14	5.08/4.4	<30	IHA	Qureshi and Tanveer (2009)
Saudi Arabia	2 5	– –	– –	Parasitology Serology	Degheidy and Al-Malki (2012)
Tunisia	2	–	19; 41	Serology	Hammami et al. (2007)
Yemen	185	–	–	Parasitology/biliary drainage	Farag (1985)
Kuwait	1	F	–	Endoscopy	Al-Mekhaizeem et al. (2004)

(Table 4). The authors emphasized that fasciolosis is an emerging, and underestimated disease in the region.

Fasciolosis in cattle due to *F. gigantica* has been reported 50 % in Babylon Province. The differentiation with *F. hepatica* was based on morphological characters (Al Qurashi et al. 2012).

Israel

Fasciola hepatica was extracted from the common bile duct during a cholecystectomy operation in a 65-year-old woman from Israel (Dan et al. 1981). The patient had been raising sheep in Afghanistan until 1970, when she immigrated to Israel. This cases plus another one in July 1980 composes the documented cases of fasciolosis in Israel (Dan et al. 1981).

Jordan

No reports of human fasciolosis have appeared form Jordan. However, a prevalence of 3.2 % was found in sheep imported from Romania (Maraqa et al. 2005).

Kuwait

Kuwait is regarded a non-endemic area. But because of the many immigrants workers originating from endemic countries it is possible that human fasciolosis may occur at least occasionally. A case of human fasciolosis has been reported in a 47-year-old woman with obstructive jaundice secondary to biliary obstruction (Al-Mekhaizeem et al. 2004). The patient had a history of travel to Egypt.

Pakistan

The 0.31 % prevalence of human fasciolosis has been reported in Lahore, based on fecal exams performed in 2003–2005 (Qureshi et al. 2005) (Table 4). The highest infection rate in the context of season was in summer (0.42 %) and the lowest in spring and autumn (0.17 %), while the highest prevalence was in August and January (0.6 %) and the lowest in March (0.0 %). July is the optimal month for the emergence of cercariae from snails in Pakistan. All these areas are rural and sanitation conditions are not good. Apparently, raw vegetables, especially lettuce, are washed with contaminated water and consumed in the area. The authors have not stated how many times they have collected stools from the cases, but considering that out of 7,200 fecal samples only 21 samples were positive for fasciolosis, and regarding the low sensitivity of coprological methods for fasciolosis diagnosis, more positive cases were expected. This result was recognized by the authors and they have advised for serological verification to clarify the issue. In another study, using **indirect hemagglutination antibody** (IHA), human fasciolosis prevalence was 4.67 % in Punjab, Pakistan (Qureshi and Tanveer 2009) (Table 4). The highest rate of infection was seen in 11–20-year-old group. Females were more susceptible than males but of no significant difference.

There are many reports of animal fasciolosis in Pakistan. In Punjab, around Lahore, *F. hepatica* prevalence in sheep was 14.67 % (Ijaz et al. 2009; Lashari and Tasawar 2011) and 21.41 % in southern Punjab (Lashari and Tasawar 2011), while in goats *F. hepatica* infection was 28.75 % Multan (Tasawar et al. 2007) and 7.58 % in Toba Tek Singh (Ayaz et al. 2013). In the Pothwar region, the overall seroprevalence of *F. hepatica* in sheep and goats using ELISA was 39.2 % and 4.08 %, respectively, while the prevalence based on fecal egg count (FEC) was 28.4 % and 5.1 %, respectively (Afshan et al. 2013). Bovine fasciolosis prevalence is reported to be 25.46 % in Punjab, Pakistan. The occurrence of *F. gigantica* (22.40 %) was higher ($P < 0.05$) than *F. hepatica* (3.06 %) (Khan et al. 2009). Whereas it was reported 14.71 % earlier (Maqbool et al. 2002). The probable reasons affecting this increase are stated as follows:

(1) Socioeconomic status of the farmers to treat the nuisance, (2) development of resistance due to improper use of fasciolicides, and (3) lack of regular evaluation of local available drugs (Khan et al. 2009). Both species of *Fasciola* have been reported as the agent of fasciolosis in Pakistan (Khan et al. 2009).

Saudi Arabia

Seven human fasciolosis cases are reported from Saudi Arabia (Degheidy and Al-Malki 2012) (Table 4). Another study reports nine cases of fasciolosis among male immigrant manual workers (El-Mathal and Fouad 2005). The diagnosis was confirmed by observing eggs in fecal samples after sedimentation and Kato-Katz techniques. Animal fasciolosis has also been reported among imported and locally reared sheep (El-Mathal and Fouad 2005).

In Saudi Arabia, the most common clinical manifestations observed were abdominal distension, flatulence, tender right-upper quadrant and easy fatigability, and only occasionally the tinge of jaundice. Symptoms were right upper quadrant pains, colicky abdominal pains and vomiting, epi-gastric pain, mild fever, and tympanitic abdomen. Anemia and eosinophilia were also observed (El-Mathal and Fouad 2005).

Sudan

Based on 1998–2007 slaughter house records from Sudan the overall prevalence rate of *F. gigantica* infection was higher in cattle (6.05 %) compared to sheep (2.37 %) (Babiker et al. 2013). The prevalence rate was higher in the wet season compared to the dry and cold season, but the difference was not significant. More than 90 % of domestic livestock are owned by nomads. Sudan is considered a major exporter of meat and leather, which comprises about 20 % of the country foreign trade (Babiker et al. 2013). *Fasciola gigantica* infection is a cause of great economic loss due to total or partial condemnation of livers. A rate of 12.5 % positivity for *F. gigantica* eggs was reported in goats in central Sudan (Koko et al. 2003). The authors concluded that the prevalence of fasciolosis is significantly related to inner irrigation location of animals rather than animal grazing in the area. This is probably because *Lymnae natalensis*, the intermediate host of *F. gigantica* in Gezira, Sudan, are common in stagnant canal water used for irrigation. No reports of human fasciolosis in Sudan were found in the literature review.

Tunisia

In Tunisia, human fasciolosis is not common although animal infection is frequent in the north and the southwest areas of Tunisia. However, human fasciolosis prevalence is reported to be 6.6 % in southwest Tunisia (Hammami et al. 2007). Most patients are natives of the north or the southwest Tunisian oases. Gafsa oases constitute a new location for the development of fasciolosis in the southwest of Tunisia. Animal fasciolosis has been reported as 14.3 % in cattle, 35–55 % in sheep, and 68 % in goats (Hamed et al. 2009; Hammami et al. 2007). *Fasciola hepatica* has been found on *Bulinus truncatus*, *Galba truncatula*, and *Lymnaea truncatula* from North and Southwest of Tunisia (Hammami and Ayadi 2008). The prevalence of the infection of the intermediate host *G. truncatula* was 19.2 %.

In Tunisia, control measures are critical. Strategic flukicide treatments with TCBZ, aimed at reducing mollusk infestation and subsequent pasture contamination by metacercariae, during two main periods, September and February, is of high importance (Hammami et al. 2007).

Yemen

Human fasciolosis was identified in 0.5 % of 37,000 people during the period 1980–1982; *Fasciola* ova were found in 185 cases (Farag 1985). Three additional cases have been reported previously (Chen and Mott 1990).

Concluding Remarks and Future Priorities

The transmission dynamics of human fasciolosis in any particular area is strongly affected by the climate, farming practices, human dietary habits, wandering animals, and diversity as well as multiplicity of intermediate hosts. All these factors are seen in the MENA region to varying degrees. Although, the region is generally an arid desert area, irrigated agriculture is a common practice in some countries. Livestock husbandry is also an integral part of farming in agricultural communities in the area. Such animals act as the source of infection for human pollution. In addition, climate and farming practices that favor the survival of the snail intermediate hosts are often common throughout the region. It is not surprising that Egypt and Iran, where there are the highest prevalence rates of fasciolosis in human and animals occur, are among the largest producers of agricultural commodities in the region. This review of the status of fasciolosis indicates that fasciolosis has an emerging and reemerging nature in Iran and Egypt. The fact that early reports on human fasciolosis in the region failed to stimulate greater interest in this zoonosis was the failure to recognize that the occurrence of only a single case undoubtedly was associated with a much greater outbreak and should have been considered as an indicator an unrecognized problem, perhaps newly emerging.

In Egypt several factors contribute to the high rate of fasciolosis. The climate and ubiquitous canals favor *Lymnaea* snails and allow their egg masses to persist year round. Transmission is favored also by continued exposure of the animals (reservoir hosts) to encysted metacercariae through grazing on the plant covered banks of the canals; infection control in these hosts is nearly non-existent (Soliman 2008). Another factor is the common practice of people in these areas to wash their vegetables in the contaminated canals, which may become contaminated and then consumed or marketed. Other significant factors include the poor sanitary habits and low health awareness of the at risk populations.

In northern endemic areas of Iran, numerous irrigation canals, agricultural crop traditions (mainly rice), temperatures higher than 20 °C, high rainfall (1,300–1,800 mm annually), and short dry seasons favor fasciolosis transmission and lymnaeid populations (Mas-Coma 2005). Fuentes et al. (1999) have demonstrated by mathematical modeling the impact of annual rainfall in determining the level of transmission in endemic areas.

With regard to the control of the disease, a key obstacle to establishing and maintaining effective programs in this region is political turmoil, which has serious impacts on national public health management. One consequence of this risk is the increase in number of people at risk for fasciolosis. Other significant factors are environmental changes, alteration in human behavior, increase in urbanization, migration, dam building, and expansion of irrigation.

The policy on monitoring fasciolosis should embrace the following key points:

Declaration of all cases of fasciolosis should be statutory requirement;
Egaten® (Human form of TCBZ) should be provided free of charge for all patients;
All notes stated in the text should be considered and implemented.

However, mentioned data in this chapter merely show the tip of the clinical iceberg. Considering that there is gross underreporting of the issue of fasciolosis in the region, following points should be regarded on setting stage for monitoring the disease and establishing the priorities of the research:

- Calculation of the Disability Adjusted Life Years DALYs for this disease in the region
- Verifying all blind spots on differences between human and animal fascioliasis
- Verifying the spectrum of species infecting human patients (*hepatica* or *gigantica*)
- Verifying the nature of seasonal distribution of fasciolosis
- Establishing a simple molecular method to differentiate *F. hepatica* and *F. gigantica* available and applicable in all countries of the region
- Establishing a serological test to follow up patients after treatment
- Clarifying all details on genotyping different species of *Fasciola* in the region and specify the intermediate forms, if any

- Conducting surveys on clarifying the prevalence and intensity of the disease in different countries of the region
- Verifying the coinfection between fasciolosis and parasitic diseases
- Verifying the relationships between the intermediate hosts and the transmission pattern
- Mapping the disease using GIS, modern diagnostic tests, and epidemiology criteria
- Clarification the relation between the fasciolosis clinical manifestations and their appearance or disappearance
- Determining the exact sources of infection for each country

Acknowledgments The authors declare that there is no conflict of interest.

References

Abd El Bagi ME, Sammak BM, Mohamed AE, Al Karawi MA, Al Shahed M, Al Thagafi MA (2004) Gastrointestinal parasite infestation. Eur Radiol 14(Suppl 3):E116–E131

Abdi J, Naserifar R, Rostami Nejad M, Mansouri V (2013) New features of fascioliasis in human and animal infections in Ilam province, Western Iran. Gastroenterol Hepatol Bed Bench 6: 152–155

Abdouslam OE, Al-Bassam LS, Al-Izzi SA, Azwai SM (2003) Prevalence of external and internal parasites in llamas (Lama glama) at Surman park in Libya. J Camel Pract Res 10:61–65

Abou Basha L, Fadali G, Nour B, Abdalla M (1989) Uncommon complications of human fascioliasis in Alexandria. J Egypt Soc Parasitol 19:743

Adl Y, Sedigh M (1956) A case of Dwu in thyroid gland. Name Mahaneh Daneshkadeh Pezeshki 8:13

Afshan K, Qayyum M, Rizvi SSR, Mukhtar M, Mushtaq M, Miller JE (2013) Serological and coprological comparison for rapid diagnosis of *Fasciola hepatica* infection in small ruminants from sub-tropical area of Pakistan. Small Rumin Res 113:267–272

Ahmed AH, Ramzy RM (1999) Infection of two lymnaeid snails with Fasciola gigantica in Giza, a field study. J Egypt Soc Parasitol 29:687–696

Ahmed L, Salama Z, El-Kady N, Farag A, Salama H, Ibraheem M, Mahmoud S (1994) Human fascioliasis. A field study in Tamya Centre Fayoum Governorate. J Egypt Med Assoc 77: 287–96

Aissi M, Harhoura K, Gaid S, Hamrioui B (2009) Preliminary study on the prevalence of fasciolosis due to Fasciola hepatica in some cattle farms in north-central Algeria (the Mitidja). Bull Soc Pathol Exot 102:177–178

Akkari H, Gharbi M, Darghouth M (2011) Infestation of tracer lambs by Fasciola hepatica in Tunisia: determining periods for strategic anthelmintic treatments. Rev Sci Tech 30:917

Al Qurashi H, Masoodi I, Al Sofiyani M, Al Musharaf H, Shaqhan M, All GN (2012) Biliary fascioliasis-an uncommon cause of recurrent biliary colics: report of a case and brief review. Ger Med Sci 10:Doc10. doi:10.3205/000161

Alasaad S, Huang CQ, Li QY, Granados JE, García-Romero C, Pérez JM, Zhu XQ (2007) Characterization of Fasciola samples from different host species and geographical localities in Spain by sequences of internal transcribed spacers of rDNA. Parasitol Res 101:1245–1250

Alavi-Naini R, Sharifi-Mood B, Khazaei A-R, Khorgami P (2013) A case of Fascioliasis from south-east of Iran. Zahedan J Res Med Sci 3:29–30

Al-Mekhaizeem K, Al-Mukhaizeem F, Habib MA (2004) Fasciola hepatica infestation presenting as biliary obstruction 11 years after open cholecystectomy and CBD exploration. Kuwait Med J 36:293–295

Al-Motarreb A, Al-Habori M, Broadley KJ (2010) Khat chewing, cardiovascular diseases and other internal medical problems: the current situation and directions for future research. J Ethnopharmacol 132:540–548

Amer S, Dar Y, Ichikawa M, Fukuda Y, Tada C, Itagaki T, Nakai Y (2011) Identification of Fasciola species isolated from Egypt based on sequence analysis of genomic (ITS1 and ITS2) and mitochondrial (NDI and COI) gene markers. Parasitol Int 60:5–12

Aminian K, Rezayat KA, Shafaghi A, Tanhaeevash R (2012) Living Fasciola hepatica in biliary tree: a case report. Ann Hepatol 11:395

Amor N, Farjallah S, Salem M, Lamine DM, Merella P, Said K, Ben Slimane B (2011a) Molecular characterization of Fasciola gigantica from Mauritania based on mitochondrial and nuclear ribosomal DNA sequences. Exp Parasitol 129:127–136

Amor N, Halajian A, Merella P, Farjallah S, Said K, Ben B (2011b) Genetic characterization of Fasciola spp. from Tonekabon city (northern Iran) based on the ribosomal internal transcribed spacer regions. Pak J Zool 43:1061–1067

Araj G, Musharrafieh U, Haydar A, Ghawi A, Itani R, Saliba R (2011) Trends and prevalence of intestinal parasites at a tertiary care center in Lebanon over a decade. J Med Liban 59:143–8

Asadian S, Mohebali M, Mahmoudi M, Kia EB, Heidari Z, Asgari M, Aryaeipour M, Moradi S, Rokni MB (2013) Seroprevalence of human fascioliasis in Meshkin-Shahr District, Ardabil Province, Northwestern Iran in 2012. Iran J Parasitol 8:516–521

Ashrafi K (2004) A survey on human and animal fascioliasis and genotypic and phenotypic characteristics of fasciolids and their relationship with lymnaeid snails in Gilan province, northern Iran PhD, Tehran University of Medical Sciences, Iran

Ashrafi K, Valero MA, Massoud J, Sobhani A, Solaymani-Mohammadi S, Conde P, Khoubbane M, Bargues MD, Mas-Coma S (2006a) Plant-borne human contamination by fascioliasis. Am J Trop Med Hyg 75:295–302

Ashrafi K, Valero MA, Panova M, Periago MV, Massoud J, Mas-Coma S (2006b) Phenotypic analysis of adults of *Fasciola hepatica*, *Fasciola gigantica* and intermediate forms from the endemic region of Gilan, Iran. Parasitol Int 55:249–260

Ashrafi K, Massoud J, Naieni KH, Jo-Afshani MA, Mahmoodi M, Ebadati N, Rezvani SM, Artigas P, Bargues MD, Mas-Coma S (2007) Nuclear ribosomal DNA ITS-2 sequence characterization of *Fasciola hepatica* and *Galba truncatula*. Iran J Public Health 36:42–49

Ashrafi K, Saadat F, Rahmati B (2012) Evaluating the level of endemicity of human fascioliasis in rural communities of Bandar-e-Anzali district; an important step for implementation of disease control programs. First international and 8th national congress of parasitology and parasitic diseases in Iran, Kerman, Iran, p 121

Assmar M, Milaninia A, Amir-Khani A, Yadegari D, Forghan-Parast K, Nahravanian H, Piazak N, Esmayli A, Hovanesian A, Valadkhani Z (1991) Seroepidemiological investigation of fascioliasis in northern Iran. Med J Islam Repub Iran 5:23–27

Ayaz MM, Raza MA, Murtaza S, Akhtar S (2013) Epidemiological survey of helminths of goats in southern Punjab, Pakistan. Trop Biomed 30:62–71

Babiker E, Elmansory YHA, Elsadig A, Majid AM (2013) Prevalence of fasciolosis in cattle and sheep slaughtered for human consumption in the White Nile state, Sudan: a retrospective. Univ Khartoum J Vet Med Anim Prod 2:105–119

Banaja A, Ghandour A (1994) A review of parasites of camels (Came/us dromedarius) in Saudi Arabia. J King Abdulaziz Univ Sci 6:75–86

Bassiouny HK, Soliman NK, El-Daly SM, Badr NM (1991) Human fascioliasis in Egypt: effect of infection and efficacy of bithionol treatment. J Trop Med Hyg 94:333–337

Ben Amer K, Ahmed Z (1980) Studies on the prevalence of liver flukes with special reference to Dicrocoelium dendriticum (Rudolphi, 1891) Looss, 1899 in sheep and cattle from Tripoli region, Libya. Riv Parassitol 41:15–18

Birjawi G, Sharara A, Al-Awar G, Tawil A, Moukaddam H, Khouzami R, Haddad M (2002) Biliary fascioliasis: case report and review of the literature. J Med Liban 50:60
Botros SS, El-Lakkany NM, Badawy AA, Mahmoud SS, Ebeid FA, Fenwick A (2009) Mirazid shows insignificant activity against ovine fascioliasis. Ann Trop Med Parasitol 103:605–616. doi:10.1179/000349809X12459740922453
Cats A, Scholten P, Meuwissen SGM, Kuipers EJ (2000) Acute *Fasciola hepatica* infection attributed to chewing khat. Gut 47:584–585. doi:10.1136/gut.47.4.584
Chen M, Mott K (1990) Progress in assessment of morbidity due to Fasciola hepatica infection: a review of recent literature. Trop Dis Bull 87:R1–R38
Chester WL (1928) *Fasciola hepatica*: a fatal case in a European soldier. J R Army Med Corps 51:463
Curry A, Anfield C, Tapp E (1979) Electron microscopy of the Manchester mummies. In: David AR (ed) Manchester museum mummies project, multidisciplinary research on ancient Egyptian mummified remains. Manchester Museum, Manchester, pp 103–111
Curtale F (2008) The use of community washing units to reduce transmission of liver flukes and other water borne parasites in Egypt. 3rd K.E. Mott Symposium on Flukes and Schistosomes, 10th European Multicolloquium of Parasitology (EMOP 10), Paris
Curtale F, Hassanein YA, El Wakeel A, Mas-Coma S, Montresore A (2003) Distribution of human fascioliasis by age and gender among rural population in the Nile Delta, Egypt. J Trop Pediatr 49:264–268
Curtale F, Hassanein YA, Savioli L (2005) Control of human fascioliasis by selective chemotherapy: design, cost and effect of the first public health, school-based intervention implemented in endemic areas of the Nile Delta, Egypt. Trans R Soc Trop Med Hyg 99: 599–609
Dan M, Lichtenstein D, Lavochkin J, Stavorowsky M, Jedwab M, Shibolet S (1981) Human fascioliasis in Israel. An imported case. Isr J Med Sci 17:430–432
David A (1997) Disease in Egyptian mummies; the contribution of new technologies. Lancet 349:1760–1763
Degheidy NS, Al-Malki JS (2012) Epidemiological studies of fasciolosis in human and animals at Taif, Saudi Arabia. World Appl Sci J 19:1099–1104
Doherty JF, Price N, Moody AH, Wright SG, Glynn MJ (1995) Fascioliasis due to imported khat. Lancet 345:462
El-Dafrawy SM (2002) Morphological and ecological studies on *Lymnaea natalensis* the snail vector of *Fasciola gigantica* in Egypt. J Egypt Soc Parasitol 32:447–456
El-Khashab MN, Mourad AA, El-Ghoneimy S, Mahmoud AM, Nasr SM (1993) Anaemia of human hepatic fascioliasis. Zag J Med Physiol 2:237
El-Mathal EM, Fouad MA (2005) Human fascioliasis among immigrant workers in Saudi Arabia. J Egypt Soc Parasitol 35:1199–1207
El-Shabrawi M, El-Karaksy H, Okasha S, El-Hennawy A (1997) Human fascioliasis: clinical features and diagnostic difficulties in Egyptian children. J Trop Pediatr 43:162–166
El-Shazly AE, Handousa AE, Youssef ME, Rizk H, Hammouda MM (1991) Human fascioliasis: a parasitic health problem in Dakahlia Governorate, Egypt. J Egypt Soc Parasitol 21:553–559
El-Shazly AM, Soliman M, Gabr A, Haseeb AN, Morsy AT, Arafa MA, Morsy TA (2001) Clinico-epidemiological study of human fascioliasis in an endemic focus in Dakahlia Governorate, Egypt. J Egypt Soc Parasitol 31:725–736
El-Shazly AM, Helmy MM, Haridy FM, El-Sharkawy EM, Morsy TA (2002) *Fasciola* immature stages sought in *Lymnaea* species and *Biomphalaria* species in the water bodies of Dakahlia Governorate. J Egypt Soc Parasitol 32:109–118
El-Shazly AM, El-Nahas HA, Abdel-Mageed AA, El Beshbishi SN, Azab MS, Abou El Hasan M, Arafa WA, Morsy TA (2005) Human fascioliasis and anemia in Dakahlia governorate, Egypt. J Egypt Soc Parasitol 35:421–432
El-Shazly AM, Nabih N, Salem DA, Mohamed MZ (2012) Snail populations in Dakahlia Governorate, Egypt, with special reference to lymnaeids. Egypt J Biol 14:45–49

Eslami A, Hosseini SH, Meshgi B (2009) Animal fasciolosis in north of Iran. Iran J Public Health 38:132–135
Esteban JG, Gonzales C, Curtale F, Munoz-Antoli C, Valero MA, Bargues MD, el-Sayed M, el-Wakeel AA, Abdel-Wahab Y, Monteresor A (2003) Hyperendemic fascioliasis associated with schistosomiasis in villages in the Nile Delta of Egypt. Am J Trop Med Hyg 69:429–437
Fairweather I (2011) Raising the bar on reporting 'triclabendazole resistance'. Vet Rec 168: 514–515. doi:10.1136/vr.d2867
Farag H (1985) Intestinal parasitosis in the population of the Yemen Arab Republic. Trop Geogr Med 37:29
Farag HF (1998) Human fascioliasis in some countries of the Eastern Mediterranean Region. East Mediterr Health J 4:156–160
Farag H, Barakat R, Ragab M, Omar E (1979) A focus of human fascioliasis in the Nile Delta, Egypt. J Trop Med Hyg 82:188
Farag HF, Regab M, Salem A, Sadek N (1986) A short note on praziquantel in human fascioliasis. J Trop Med Hyg 89:79–80
Farid Z, Trabolsi B, Boctor F, Hafez A (1986) Unsuccessful use of praziquantel to treat acute fascioliasis in children. J Infect Dis 154:920–921
Farid Z, Kamal M, Woody j (1988) Treatment of acute toxaemic fascioliasis. Trans R Soc Trop Med Hyg 82:299
Farid Z, Mansour N, Kamal M, Kamal K, Safwat Y, Woody JN (1990) The treatment of acute Fasciola hepatica infection in children. Trop Geogr Med 42:95–96
Farjallah S, Sanna D, Amor N, Mehel BB, Piras MC, Merella P, Casu M, Curini-Galletti M, Said K, Garippa G (2009) Genetic characterization of Fasciola hepatica from Tunisia and Algeria based on mitochondrial and nuclear DNA sequences. Parasitol Res 105:1617–1621
Fattah F, Babero B, Karaghouli A, Shaheen A (1964) The zoonosis of animal parasites in Iraq X. A confirmed case of human ectopic fascioliasis. Am J Trop Med Hyg 13:291–294
Fawzy RK, Salem AE, Osman MM (1992) Ultrasonographic findings in the gallbladder in human fascioliasis. J Egypt Soc Parasitol 22:827–831
Forghan-Parast K, Ashrafi K (2001) Comparison of the formalin-Ether and Kato-Katz in the parasitological diagnosis of human fascioliasis. J Med Fac Gilan Univ Med Sci Iran 9:1–6
Forghan-parast K, Yadegari D, Asmar M (1993) Clinical epidemiology of human fascioliasis in Gilan. J Guilan Univ Med Sci 6:4–11
Fuentes M, Valero M, Bargues M, Esteban J, Angles R, Mas-Coma S (1999) Analysis of climatic data and forecast indices for human fascioliasis at very high altitude. Ann Trop Med Parasitol 93:835–850
Fürst T, Keiser J, Utzinger J (2012) Global burden of human food-borne trematodiasis: a systematic review and meta-analysis. Lancet Infect Dis 12:210–221
Ghavami MB, Rahimi P, Haniloo A, Mosavinasab SN (2009) Genotypic and phenotypic analysis of fasciola isolates. Iran J Parasitol 4:61–70
Greene DA (1993) Gold, frankincense, myrrh, and medicine. N C Med J 54:620–622
Halawani AA, Gindy M (1957) A short account on the economic importance of *Fasciola gigantica*, its effect on Livestock and snail vector. J Egypt Med Assoc 40:547
Hamed N, Hammami H, Khaled S, Rondelaud D, Ayadi A (2009) Natural infection of Fasciola hepatica (Trematoda: Fasciolidae) in Bulinus truncatus (Gastropoda: Planorbidae) in northern Tunisia. J Helminthol 83:271–273
Hammami H, Ayadi A (2008) Molluscicidal and antiparasitic activity of Solanum nigrum villosum against Galba truncatula infected or uninfected with Fasciola hepatica. J Helminthol 82: 235–240
Hammami H, Hamed N, Ayadi A (2007) Epidemiological studies on Fasciola hepatica in Gafsa oases (South West of Tunisia). Parasite 14:261–264
Hammond JA (1974) Human infection with the liver fluke *Fasciola gigantica*. Trans R Soc Trop Med Hyg 68:253–254

Hassan HA, Majid RA, Rashid NG, Nuradeen BE, Abdulkarim QH, Hawramy TA, Rashid RM, Farris AB, Guarner J, Hughson MD (2013) Eosinophilic granulomatous gastrointestinal and hepatic abscesses attributable to basidiobolomycosis and fasciolias: a simultaneous emergence in Iraqi Kurdistan. BMC Infect Dis 13:91. doi:10.1186/1471-2334-13-91

Hatami H, Asmar M, Massoud J, Aryanifar S, Mansori F, Fatemi S, Shahrezaei A, Rezaei R, Namdaritabar H (2000) Report of the first outbreak of human fasciolosis in Kermanshah Provinve. Moddares J 3:79–87

Hawrami T (2004) Liver fluke causing Dilatation of Common Bile Duct. Basrah J Surg 10: 103–104

Hawrami T (2010) Case report: *Fasciola hepatica* worm swims in the Gallbladder. Zanco J Med Sci 14:1–4

Hawramy TAH, Saeed KA, Qaradaghy SHS, Karboli TA, Nore BF, Bayati NHA (2012) Sporadic incidence of Fascioliasis detected during Hepatobiliary procedures: a study of 18 patients from Sulaimaniyah governorate. BMC Res Notes 5

Hotez PJ, Savioli L, Fenwick A (2012) Neglected tropical diseases of the Middle East and North Africa: review of their prevalence, distribution, and opportunities for control. PLoS Negl Trop Dis 6:e1475

Hussein AN, Khalifa RM (2008) Experimental infections with *Fasciola* in snails, mice and rabbits. Parasitol Res 102:1165–1170. doi:10.1007/s00436-008-0888-5

Hussein A-NA, Khalifa R (2010) Fascioliasis prevalences among animals and human in Upper Egypt. J King Saud Univ Sci 22:15–19

Ijaz M, Khan M, Avais M, Ashraf K, Ali M, Khan M (2009) Infection rate and chemotherapy of various helminthes in diarrhoeic sheep in and around Lahore. J Anim Plant Sci 19:13–16

Jahed Khaniki G, Kia E, Raei M (2012) Liver condemnation and economic losses due to parasitic infections in slaughtered animals in Iran. J Parasit Dis: 1–5. doi:10.1007/s12639-012-0172-6

Keiser J, Utzinger J (2005) Emerging foodborne trematodiasis. Emerg Infect Dis 11:1507

Keiser J, Sayed H, El-Ghanam M, Sabry H, Anani S, El-Wakeel A, Hatz C, Utzinger J, El-Din SS, El-Maadawy W, Botros S (2011) Efficacy and safety of artemether in the treatment of chronic fascioliasis in Egypt: exploratory phase-2 trials. PLoS Negl Trop Dis 5:e1285. doi:10.1371/journal.pntd.0001285

Khallaayoune K (1995) Liver fluke in Morocco: when to prevent it. Magazine de l'Actualite de l'Elevage de la Sante Animale et de l'Hygiene 3:7

Khallaayoune KH, Stromberg BE, Dakkak A, Malone JB (1991) Seasonal dynamics of *Fasciola hepatica* burdens in grazing timahdit sheep in Morocco. Int J Parasitol 21:307–314

Khan MK, Sajid MS, Khan MN, Iqbal Z, Iqbal MU (2009) Bovine fasciolosis: prevalence, effects of treatment on productivity and cost benefit analysis in five districts of Punjab, Pakistan. Res Vet Sci 87:70–75

Khan I, Khan AM, Ayaz K, Anees M, Khan SA (2012) Molecular detection of Fasciola hepatica in water sources of District Nowshehra Khyber Pakhtunkhwa Pakistan. Int J Adv Res Technol 1(7):106–117, 106–117

Koko W, Abdalla H, Galal M (2003) Caprine fascioliasis in the Gezira State, Central Sudan. J Anim Vet Adv 2:396–399

Kuntz RE, Lawless DK, Langbehn HR, Malakatis GM (1958) Intestinal protozoa and helminths in the peoples of Egypt living in different type localities. Am J Trop Med Hyg 7:630–639

Lashari MH, Tasawar Z (2011) Prevalence of some gastrointestinal parasites in sheep in southern Punjab, Pakistan. Pak Vet J 31:295–298

Looss A (1896) [Recherches sur la faune de l'Égypte]. Première Partie. Mémoires de l'Institut Égyptien 3:1–252

Lotfy WM, Hillyer GV (2003) Fasciola species in Egypt. Exp Pathol Parasitol 6:9–22

Mahami-Oskouei M, Forouzandeh-Moghadam M, Rokni MB (2012) Prevalence and severity of animal fasciolosis in six provinces of Iran. Feyz J Kashan Univ Med Sci 16:254–260

Mansoorian A (2000) Some freshwater snails from Northern Iran. Iran J Public Health 29

Mansour NS, Yousef FG, Mikhail EM, Boctor FN (1983) Use of partially purified *F. gigantica* worm antigen in the serological diagnosis of human fascioliasis in Egypt. Am J Trop Med Hyg 32:550–554

Maqbool A, Hayat CS, Akhtar T, Hashmi HA (2002) Epidemiology of fasciolosis in buffaloes under different managemental conditions. Vet Arh 72:221–228

Maraqa A, Amr Z, Rifai L, Al-Melhim W (2005) An abattoir survey of liver and lung helminthic infections in local and imported sheep in Jordan. Turk J Vet Anim Sci 29:1–2

Mas-Coma S (2005) Epidemiology of fascioliasis in human endemic areas. J Helminthol 79: 207–216

Mas-Coma S (2007) Parasitic diseases, global change and the developing world: the example of emerging fascioliasis. Sci Parasitol 1:10–21

Mas-Coma S, Bargues M (1997) Human liver flukes: a review. Res Rev Parasitol 57:145–218

Mas-Coma S, Bargues MD, Valero MA (2005) Fascioliasis and other plant-borne trematode zoonoses. Int J Parasitol 35:1255–1278

Massoud J (1989) Fasciolosis outbreak in man and drug test (Triclabandazole) in Caspian littoral, northern part of Iran. Bull Soc France Parasitol 8:438

Mekroud A, Benakhla A, Vignoles P, Rondelaud D, Dreyfuss G (2004) Preliminary studies on the prevalences of natural fasciolosis in cattle, sheep, and the host snail (Galba truncatula) in north-eastern Algeria. Parasitol Res 92:502–505

Moghaddam AS, Massoud J, Mahmoodi M, Mahvi AH, Periago MV, Artigas P, Fuentes MV, Bargues MD, Mas-Coma S (2004a) Human and animal fascioliasis in Mazandaran province, northern Iran. Parasitol Res 94:61–69. doi:10.1007/s00436-004-1169-6

Moghaddam AS, Massoud JF, Mahmood M, Messaoud K, Artigas P, Periago MV, Fuentes MV, Bargues MD, Mas-Coma S (2004b) Distributional outline of lymnaeid snails (Gastropoda) in the fascioliasis endemic area of Mazandaran, Iran. Acta Parasitol 49:145–152

Mohammadi-Ghalehbin B, Chinifroush-Asl MM, Ramzi F (2012) Extra-hepatic fascioliasis with peritoneal malignancy tumor feature. J Parasit Dis 36:78–80

Moshfe A, Bagheri M, Mohebi Nobandeghany Z (2003) Prevalence of Fasciola hepatica in slaughtered livestock in Yasuj's slaughterhouse 1381–82. Armaghan Danesh (Journal of Yasuj University of Medical Sciences) 8:25–33

Nagaty HF (1942) On some parasites collected in Egypt from food producing mammals. J Egypt Med Assoc 25:110–111

Osman MM, El-Taweel HA, Shehab AY, Farag HF (2010) Ineffectiveness of myrrh-derivative Mirazid against schistosomiasis and fascioliasis in humans. East Mediterr Health J 16:932–936

Qureshi A, Tanveer A (2009) Seroprevalence of fasciolosis in buffaloes and humans in some areas of Punjab, Pakistan. Pak J Sci 61:91–96

Qureshi AW, Tanveer A, Waheed S, Qureshi A, Gill TJ, Ali SA (2005) Epidemiology of human fasciolosis in rural areas of Lahore, Pakistan. Punjab Univ J Zool 20:159–168

Ragab M, Farag HF (1978) On human fascioliasis in Egypt. J Egypt Med Assoc 61:773–780

Rahimi MT, Ashrafi K, Koosha S, Abdi J, Rokni MB (2011) Evaluation of fast-ELISA versus standard-ELISA to diagnose human fasciolosis. Arch Iran Med 14:18

Rashed AA, Khalil HH, Morsy AT (2010) Zoonotic ectopic fascioliasis: review and discussion. J Egypt Soc Parasitol 40:591–608

Rekik M, Ben Hammouda M (2000) A steering frame for the genetic improvement of sheep and goats in Tunisia. Options Méditerr Ser A 43:129–136

Rokni MB (2008) The present status of human helminthic diseases in Iran. Ann Trop Med Parasitol 102:283–295

Rokni MB, Massoud J, O'Neill SM, Parkinson M, Dalton JP (2002) Diagnosis of human fasciolosis in the Gilan province of Northern Iran: application of cathepsin L-ELISA. Diagn Microbiol Infect Dis 44:175–179

Rokni MB, Lesan S, Massoud J, Kia EB, Molawi G (2006) Comparative Evaluation of Fast Enzyme Linked Immunosorbent Assay (Fast-ELISA) and Standard-ELISA For The Diagnosis Of Human Hydatidosis. Iran J Public Health 35:29–32

Rokni MB, Mirhendi H, Behnia M, Jalalizand N (2010a) Molecular characterization of Fasciola hepatica isolates by RAPD-PCR and ribosomal ITS1 sequencing. Iran Red Crescent Med J 12:27–32

Rokni MB, Mirhendi H, Mizani A, Mohebali M, Sharbatkhori M, Kia EB, Abdoli H, Izadi S (2010b) Identification and differentiation of *Fasciola hepatica* and *Fasciola gigantica* using a simple PCR-restriction enzyme method. Exp Parasitol 124:209–213

Samaha H (1989) The zoonotic importance of fascioliasis in Abbis, Alexandria governorate [Egypt]. Assiut Vet Med J 21:118–122

Sarkari B, Ghobakhloo N, Moshfea A, Eilami O (2012) Seroprevalence of human fasciolosis in a new-emerging focus of fasciolosis in yasuj district, southwest of iran. Iran J Parasitol 7:15–20

Sarshad A, Yadegari D, Forghan-Parast K, Haji-Abdolbaghi M (1990) A report on treatment of 31 cases of human fascioliasis with bithionol. First national congress of parasitic diseases in Iran, pp 82–83

Selim MK, Refaii AH, El Amrousi S, Hosney Z (1970) Studies on the various parasites harbouring imported animals to U.A.R. with particular reference to their pathology. Vet Med J Giza 17:173–193

Shahzad W, Mehmood K, Munir R, Aslam W, Ijaz M, Ahmad R, Khan MS, Sabir AJ (2012) Prevalence and molecular diagnosis of fasciola hepatica in sheep and goats in different districts of Punjab, Pakistan. Pak Vet J 32:535–538

Shehab AY, Hassan EM, Basha LM, Omar EA, Helmy MH, El-Morshedy HN, Farag HF (1999) Detection of circulating E/S antigens in the sera of patients with fascioliasis by IELISA: a tool of serodiagnosis and assessment of cure. Trop Med Int Health 4:686–690

Soliman MF (2008) Epidemiological review of human and animal fascioliasis in Egypt. J Infect Dev Ctries 2:182–189

Soliman KN, Farid AS (1960) An outbreak of parasitic bronchopneumonia in a flock of imported sheep. In: Proceedings of the first annual veterinary congress, Cairo, Egypt, pp 269–75

Taira N, Yoshifuji H, Boray JC (1997) Zoonotic potential of infection with *Fasciola* spp. by consumption of freshly prepared raw liver containing immature flukes. Int J Parasitol 27: 775–779

Talaie H, Emami H, Yadegarinia D, Nava-Ocampo AA, Massoud J, Azmoudeh M, Mas-Coma S (2004) Randomized trial of a single, double and triple dose of 10 mg/kg of a human formulation of triclabendazole in patients with fascioliasis. Clin Exp Pharmacol Physiol 31:777–782

Tapp E (1986) Histology and histopathology of the Manchester mummies. In: David AR (ed) Science in egyptology. Proceedings of the 1979 and 1984 Symposia. Manchester University Press, Manchester, pp 347–50

Tasawar Z, Minir U, Hayat C, Lashari M (2007) The prevalence of Fasciola hepatica in goats around Multan. Pakistan Vet J 27(1):5–7

Valero MA, Darce NA, Panova M, Mas-Coma S (2001) Relationships between host species and morphometric patterns in *Fasciola hepatica* adults and eggs from the Northern Bolivian Altiplano hyper endemic region. Vet Parasitol 102:85–100

WHO (1995) Control of foodborne trematode infections. WHO technical report series no. 849. World Health Organization, Geneva, Switzerland

WHO (2010) The "neglected" neglected worms. Action against worms. December 2007, Issue 10

World Bank n.d. (2011) Poverty. http://data.worldbank.org/topic/poverty. Accessed 26 June 2011

Yadegari D, Forghanparast K, Assmar M (1991) Survey of praziquantel effect on fascioliasis. Med J Islam Repub Iran 5:43–44

Yadegari D, Talaie H, Massoud J (1999) Clinical trial of Triclabendazole on human fascioliasis: long term follow up. Med J Islam Repub Iran 13:89–91

Zait H, Hamrioui B (2005) [New cases of human fascioliasis in Algeria]. Med Trop (Mars) 65:395–396

Trachoma

Siamak Zarei-Ghanavati, Alireza Eslampoor, Mojtaba Abrishami, and Sophie X. Deng

Contents

S. Zarei-Ghanavati (✉)
Cornea Research Center, Mashhad University of Medical Sciences, Mashhad, Iran
e-mail: zareis@mums.ac.ir

A. Eslampoor
Eye Research Center, Mashhad University of Medical Sciences, Mashhad, Iran
e-mail: Eslampoura@mums.ac.ir

M. Abrishami
Eye Research Center, Farabi Eye Hospital, Tehran University of Medical Sciences, Tehran, Iran
e-mail: mojtaba_abrishami@yahoo.com

S.X. Deng
Jules Stein Eye Institute, UCLA, Los Angeles, CA, USA
e-mail: deng@jsei.ucla.edu

M.A. McDowell and S. Rafati (eds.), *Neglected Tropical Diseases - Middle East and North Africa*, Neglected Tropical Diseases, DOI 10.1007/978-3-7091-1613-5_4,

Abstract Trachoma is one of the oldest infectious diseases known to mankind and is the leading cause of infectious blindness in the world. It belongs to a group of diseases known as "Neglected Tropical Diseases (NTDs)".

Trachoma is caused by *Chlamydia trachomatis*. Ocular surface Chlamydia infection causes inflammation, which manifests as chronic conjunctivitis. Chronologically untreated infections as well as repeated infections induce a chronic inflammatory state resulting in conjunctival fibrosis and scarring. The progressive cicatrizing disease distorts the lids causing entropion and trichiasis. These result in corneal scar and irregularity which are the main causes of the blindness, secondary to trachoma.

Trachoma is the leading cause of preventable blindness, and the economic burden of trachoma on the lives of individuals, families, and communities is enormous. The World Health Organization (WHO) leads an international effort to eliminate trachoma as a blinding disease based on the WHO-developed strategy.

Keyword Trachoma • Chlamydia trachomatis • Bacterial conjunctivitis • Eye infections • Neglected tropical diseases

Trachoma is a new Latin word, which comes from Greek trachoma (τράχωμα) meaning "roughness." It was also called "granular conjunctivitis" and "Egyptian ophthalmia." Trachoma is one of the oldest infectious diseases known to mankind and it is the leading cause of infectious blindness in the world. It belongs to a group of diseases known as "**Neglected Tropical Diseases (NTDs)**" (Merriam Webster Dictionary http://www.merriam-webster.com/dictionary/trachoma)

History

Trachoma was well known in the last several decades and before that as an eye disease with visible symptoms and documented as "ophthalmias." History of this disease was constructed via modern data and ancient records. It began as early as 8000 B.C. (Boldt 1904).

Taborisky (1952) believed that Central Asia is most likely the place where trachoma originated by his research on main tribes from Central Asia, Mongols, and the Finns. MacCallan (1936) also agreed that the origins of trachoma were in Central Asia. However, as infectious diseases often emerge simultaneously in different areas, several other evidences suggest that Central Asia was one of the regions of origin (Boldt 1904).

Abnormal changes in the structure of Australian skeletons were the oldest possible signs founded. The geographic distribution of the skulls containing the lesion and the presence of different diseases in Australia's history were analyzed. Scientists found trachoma in same areas where the skulls with lesion were found. It was concluded that the most compatible disease must be trachoma. The skulls' age dated back to 8000 B.C. (Webb 1990)

Chinese therapies for trachoma date back to 2600 B.C. Therapies for trachoma are found in the Ebers papyrus from the sixteenth century B.C. The Hippocratic Corpus, the collection of work from Hippocrates' school of medicine (fifth century B.C.), contains sections about trachoma. Written documents from several physicians, including Celsus (first century A.D.), Discorides (40–91 A.D.), Galen (129–216 A.D.), Grassus (twelfth century A.D.), and Chauliac (1300–1368 A.D.), are also available (Karasch 1993; Tower 1963). The word "trachoma" was first used around 60 A.D. in Dioscorides's work Materia Medica (Albert and Edwards 1996). Moreover, several hints indicating trachoma in the plays by Aristophanes, including Plutus, are seen. In this play, during the era of the widespread trachoma in Greece, Plutus, god of riches, was blinded by Zeus (Tower 1963; Albert and Edwards 1996).

Between 1200 A.D. and 1700 A.D., although remained largely unnoticed, trachoma had been present in Europe. Military activities expanded trachoma quickly. In 1798 blinding eye diseases incapacitated thousands of Napoleon soldiers. "Egyptian military ophthalmia" infected three thousands of Napoleon's troops, blinding many. Military ophthalmia was not a single disease; it was actually a combination of several eye infections. Napoleonic Wars accelerated spread of trachoma across the Europe. Infected soldiers were recruited into the ranks of the Italian, Russian, and Belgian armies. The British soldiers were also infected with trachoma as a result of their military campaigns in Egypt (Larner 2004; Watts 2003; Thygeson 1962; Millar and Lane 1988).

Trachoma spread very quickly, because little was known about the treatment. Investigations by British scientists were initiated from 1810, and they proposed cleanliness, isolation, and improvement in living of British soldiers. Although they were unable to treat it, they now know how to prevent its spread (Albert and Edwards 1996).

Immigrants to the new land, America, at the end of the nineteenth century and beginning of the twentieth century became a source of infection. Due to wide transport of disease, legislations were passed by the U.S. Congress. In 1897, trachoma was the first disease classified as dangerous contagious disease by the U.S. government. Steamship began to screen for infected passengers because in the case of infection, they should reject the immigrant back to Europe (Albert and Edwards 1996). United States Public Health Service physicians across the country were employed to screen the immigrants for contagious infectious diseases, primarily for trachoma. The exam included eversion of the lids and examining the underside of the eyelids (Markel 2004; Allen and Semba 2002).

In those days, Trachoma Belt was defined in the USA and it was Virginia, Kentucky, Tennessee, Missouri, Arkansas, Alabama, and Oklahoma, the states where the trachoma rates were high (Allen and Semba 2002). Moreover, on some Native American reservations, the prevalence of trachoma reached 90 %, while 10 % of students of New York City public schools suffered trachoma (Markel 2004; Allen and Semba 2002).

Despite the attention given to identifying cases of trachoma, little was known about effective treatments. Although no effective remedy was available, many different treatments were being administered. Some treatments aimed to cover the symptoms. These symptoms covering treatments were used to trick the immigration inspectors (Allen and Semba 2002).

With immigrants and travelers moving around the world, trachoma control became something to address on a global level. Because of widespread distribution of the disease, many international organizations worked together to combat against trachoma. La Ligue Contre Le Trachome and the International Organization against Trachoma were founded before World War II. Additionally, the World Health Organization (WHO) immediately after foundation classified trachoma as a dangerous disease for the human beings and, in 1949, published a conjoined report by collecting all available statistics on trachoma (WHO 1949).

In the twentieth century, cytoplasmic inclusion bodies were found for the first time in samples from diseased apes and human beings. By continuing the researches, these inclusion bodies were found in female genital epithelium, male urethritis, and non-gonorrheal ophthalmia neonatorum. Finally in 1930 these inclusion bodies were accepted as the cause of trachoma. *Chlamydia trachomatis* was first cultured by T'ang and his colleagues from China in 1954. At first, because of small size and the inability to culture it except in living cells, the causing agent of trachoma was believed to be a virus (Thygeson 1962). By the 1970s, *Chlamydia trachomatis* DNA and RNA were isolated, and because of susceptibility to antibiotics, it was assumed to be a bacterium (Mabey and Bailey 1999).

In the mid-twentieth century, the breakthrough of antimicrobials aided the development of trachoma treatment. Primarily, in 1937 sulfanilamide was discovered to be effective for treating trachoma. Allergies to sulfa antibiotics and severe skin reactions forced scientists to find an alternative treatment. In the early 1950s, both topical and oral tetracyclines were investigated. Although systemic tetracyclines were effective, damaged teeth and slowed bone growth of young children were the most concerning side effects. Topical tetracyclines had fewest side effects and were chosen as the most effective therapy and as the treatment of choice until the late twentieth century. In 1990s oral azithromycin therapy became the preferred treatment for trachoma because it requires only one dose. With these new therapies available, trachoma became a more preventable epidemy (Taylor 2001; Munoz and West 1997; Asbury et al. 2000; Baneke 2012; Wright et al. 2008).

Pathogenesis

Chlamydiae are obligate intracellular bacteria. They include three species: *Chlamydia trachomatis*, *Chlamydia psittaci*, and *Chlamydia pneumoniae*. Trachoma is caused by serotypes A, B, and C of *Chlamydia trachomatis*. Different serotypes (which are determined by polymorphisms in a surface-exposed protein) predominate in different families and in different communities. The species *C. trachomatis* also causes genital infections (serotypes D–K) and lymphogranuloma venereum (serotypes L1–L3) (Schachter et al. 1988). Other serotypes except A–C (like D–K) rarely cause a chronic follicular conjunctivitis which is not distinguishable from trachoma. It manifests as follicular conjunctivitis with pannus and, at times, conjunctival scarring. It is called "paratrachoma." However, these serotypes do not involve in the pathogenesis of trachoma blindness (Solomon et al. 2004).

Ocular surface *Chlamydia* infection causes inflammation, which manifests as a predominantly chronic follicular conjunctivitis with lymphocytic, monocytic, and plasma cells and macrophages infiltrate. The follicles are germinal centers with islands of B cells and T cells surround the germinal centers. Conjunctival scarring is a result of the prolonged inflammation of recurrent and chronic conjunctival follicular reinfection (Solomon et al. 2004). Scarring is associated with disruption of the epithelial architecture due to atrophy of the conjunctival epithelium, loss of goblet cells, and replacement of the normal, loose, vascular subepithelial stroma with thick compact bands of type IV and type V collagen. As a result, new vertically orientated fibers are formed which are firmly attached to the tarsal plate. Conjunctival inflammation in the phase of scarring is associated with a T-cell infiltrate (Whittum-Hudson et al. 1986).

The human immune response to *C. trachomatis* is poorly understood. Although anti-chlamydial antibodies have been found in the tears and serum of patients with clinically active trachoma, humoral immune response seems to be minimally effective in trachoma. Cell-mediated response plays the most important role in the resolution of the infection. However, this may also play a major role in the pathogenesis of trachomatous scarring. Infection itself is usually confined to epithelial cells, but inflammatory cells penetrate deep into the substantia propria. Persistent or repeated inflammatory reactions to the infection cause the scarring complication of trachoma and blindness.

Clinical Manifestations

Trachoma is clinically subdivided into active (early) and cicatricial (late-stage) disease. Active infection is more commonly found in children. Symptoms during the active disease can vary from ocular irritation and a slight watery discharge in mild cases to photophobia and copious watering in more severe ones. However, it is

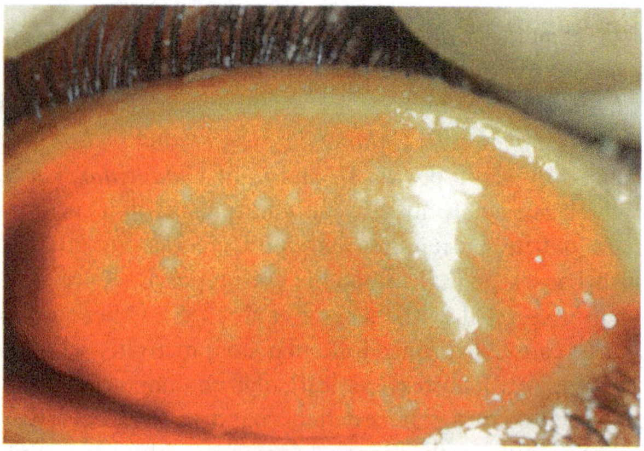

Fig. 1 Mixed follicular and papillary reaction (TF). Jack Kanski. Clinical Diagnosis in Ophthalmology, Elsevier Mosby Inc, USA, 2006. (ISBN 9780323037617)

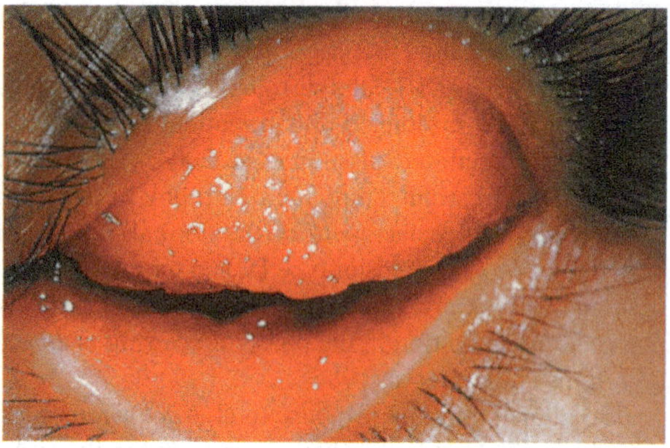

Fig. 2 Severe predominantly follicular conjunctivitis (TI). Jack Kanski. Clinical Diagnosis in Ophthalmology, Elsevier Mosby Inc, USA, 2006. (ISBN 9780323037617)

not uncommon to find asymptomatic individuals with significant conjunctival inflammation.

Initial signs include a mucopurulent conjunctivitis characterized by diffuse papillary hypertrophy and lymphoid follicles, particularly of the superior tarsal conjunctiva (Figs. 1 and 2). Follicles consist of lymphoid cell collections subjacent to the conjunctival epithelium. Their size can vary from 0.2 to 2 mm in diameter, yet only those above 0.5 mm are regarded significant in the WHO classification scheme. They may be absent in children under 2 years of age, in whom papillary reactions predominate. Infection with *C. trachomatis* concurrently occurs in other extraocular mucous membranes, commonly the nasopharynx.

Tender preauricular adenopathy could also be seen in acute phase. Follicles are of special significance in that their involution at limbal conjunctiva result in shallow depressions known as Herbert's pits (Fig. 3) which are pathognomonic for trachoma diagnosis. Also, tarsal follicles may become necrotic and heal, forming stellate (Fig. 4) or linear scars (Arlt's line, named after the Austrian ophthalmologist Carl Ferdinand Von Arlt) (Fig. 5). The cornea could be involved in various ways during

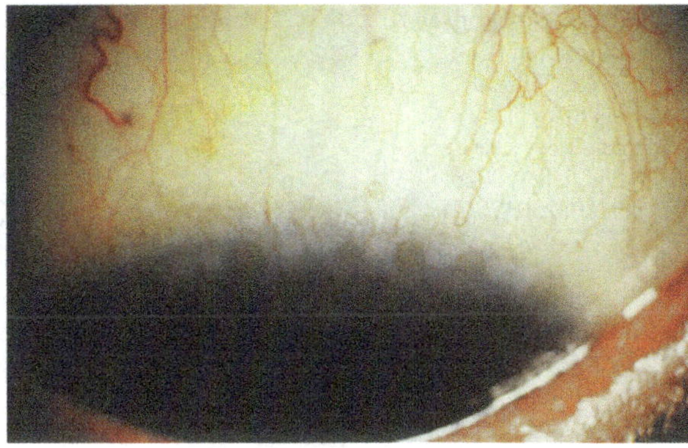

Fig. 3 Herbert's pits. Jack Kanski. Clinical Diagnosis in Ophthalmology, Elsevier Mosby Inc, USA, 2006. (ISBN 9780323037617)

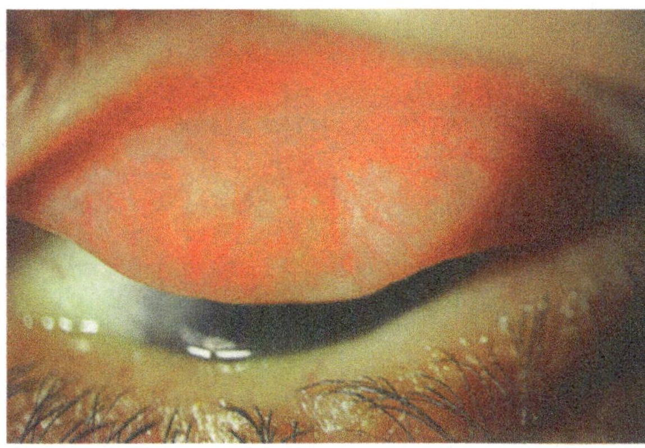

Fig. 4 Chronic conjunctival inflammation with linear and stellate scarring (TS). Jack Kanski. Clinical Diagnosis in Ophthalmology, Elsevier Mosby Inc, USA, 2006. (ISBN 9780323037617)

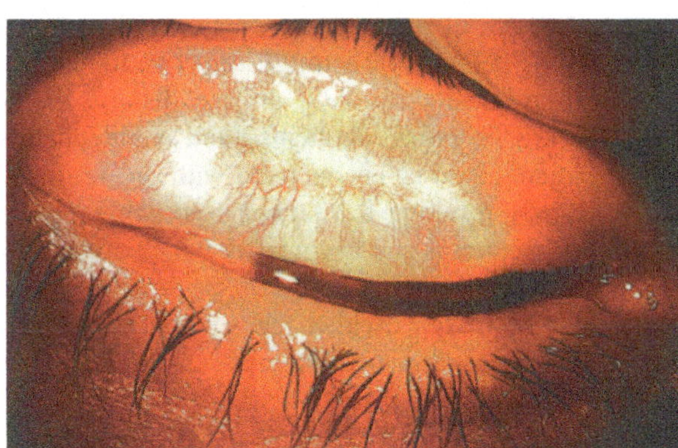

Fig. 5 More severe scarring (Arlt's line). Jack Kanski. Clinical Diagnosis in Ophthalmology, Elsevier Mosby Inc, USA, 2006. (ISBN 9780323037617)

active disease: superficial vascular pannus, punctate epithelial keratopathy, superficial infiltrates, swelling of the limbus, and development of limbal follicles (Fig. 6).

Chronologically untreated infections as well as repeated infections induce a chronic inflammatory state resulting in conjunctival fibrosis and scarring.

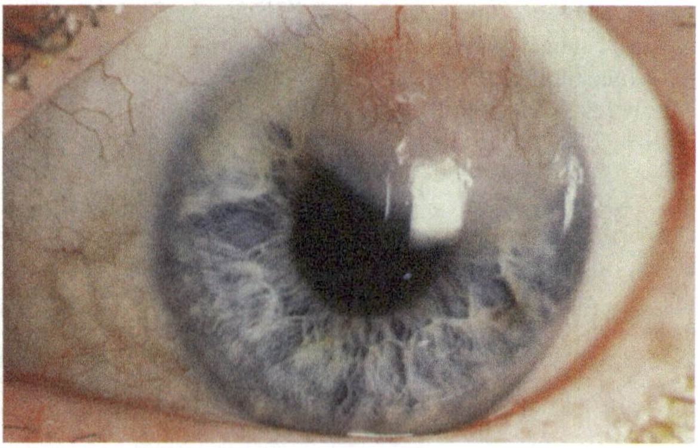

Fig. 6 Corneal opacity and pannus formation (CO). Jack Kanski. Clinical Diagnosis in Ophthalmology, Elsevier Mosby Inc, USA, 2006. (ISBN 9780323037617)

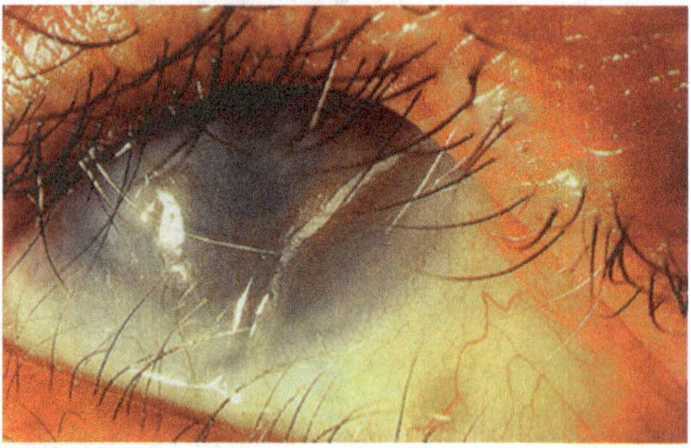

Fig. 7 Trichiasis, corneal scarring, and cicatricial entropion (TT). Jack Kanski. Clinical Diagnosis in Ophthalmology, Elsevier Mosby Inc, USA, 2006. (ISBN 9780323037617)

The final cicatricial stage of trachoma occurs no sooner than third decade of life, although in hyperendemic settings it may be found in children as well.

The progressive cicatrizing disease distorts the lids causing entropion and trichiasis (Fig. 7). Fornix shortening and symblepharon can be seen in late stages.

The patient could be encountered with severe forms of dry eye due to aqueous tear deficiency resulting from damage to goblet cells and lacrimal glands. These result in corneal scar and irregularity which are the main causes of the blindness secondary to trachoma.

In 1908 McCallan divided the clinical course of trachoma into four stages (MacCallan 1936). Staging system is summarized below:

- **Stage 1 (Incipient trachoma):** Hyperemia of palpebral conjunctiva and immature follicles (Fig. 1).
- **Stage 2 (Established trachoma):** Appearance of mature follicles and papillae with progressive corneal pannus (Fig. 2).
- **Stage 3 (Cicatrizing trachoma):** Scarring of palpebral conjunctiva with easily visible white bands (Arlt's line) (Figs. 4 and 5).
- **Stage 4 (Healed trachoma):** Inactive disease but symptomatic secondary to sequelae of cicatrization (Figs. 3, 6, and 7).

The WHO recommends a simplified grading system for trachoma as described below (Taylor 1987; Dawson et al. 1981):

- **Follicular trachoma (TF)** is the presence of 5 or more follicles (each at least 0.5 mm in diameter) on the central part of the upper tarsal conjunctiva. Follicular trachoma is an indicator of active disease. TF is most often found in 3-5 year-old children. Follicles are germinal centers that primarily consist of lymphocytes and monocytes (Fig. 1).
- **Trachomatous inflammation-intense (TI)** is the inflammatory thickening of the upper tarsal conjunctiva. This thickening has a velvety appearance obscuring most of the normal deep tarsal vessels. Intense inflammatory trachoma is also an indicator of active disease. Risk of significant conjunctival scarring and blinding disease increase after development of TI (Fig. 2).
- **Trachomatous scarring (TS)** is the presence of easily visible scars in the tarsal conjunctiva. TS is an indicator of past inflammatory disease and associated with the development of trichiasis and dry eye syndrome (Figs. 4 and 5).
- **Trichiasis (TT)** is defined as at least 1 eyelash rubs on the eyeball or evidence of recent removal of in-turned eyelashes. Trichiasis is the result of subconjunctival fibrosis over the tarsus and leads to Corneal scarring. If TT is corrected timely, vision can be restored (Fig. 7).
- **Corneal opacity (CO)** is defined as corneal opacity that is so dense that it blurs part of the pupillary margin at least. CO is the blindness stage of trachoma. Opacity includes pannus, epithelial vascularization, and infiltration (Fig. 6).

Diagnosis

Typical clinical manifestations in an endemic area seem enough for the diagnosis. However, laboratory tests are needed to confirm the diagnosis.

Intra-cytoplasmic inclusions were detected first by Giemsa stain of samples from *C. trachomatis* infection. Embryonated hens' eggs were primarily used to culture chlamydia species in 1957 (T'ang et al. 1957). Tissue-culture systems are also used for detection of trachoma. Although it is a specific test, its sensitivity is low. On the one hand they are difficult and time-consuming procedures; on the other hand they are the only available method to identify the live organism for evaluating the antimicrobial resistance (Cohen et al. 2005).

Monoclonal antibodies are used for cytological methods of evaluating chlamydial infection. Serological techniques with limited diagnostic value have been used for epidemiological researches (Stephens et al. 1982).

Nucleic acid amplification tests by Polymerase Chain Reaction (PCR) could provide greater precision in diagnosis. These tests are the current paradigm of laboratory diagnosis of trachoma (Mabey and Bailey 1999; Wright et al. 2008; Schachter et al. 1988) but have low correlation with clinical presentation (Burton et al. 2003; Lietman et al. 2000). Seventeen percent of patients with follicular conjunctivitis in hypo-endemic areas, 27 % in endemic areas, and 64 % in hyper-endemic areas were positive with these tests (Wright and Taylor 2005; Michel et al. 2006).

Chlamydial infection has a cycle. Latent disease is followed by active infection with a variable duration. Afterwards, recovery phase is seen when the organism has been cleared and clinical signs slowly resolve (Taylor et al. 1982). In endemic areas patients might be able to clear the organism rapidly due to their active immune system, but they may be reinfected before their clinical signs have resolved. Rapid cycling may be the cause of inability of tests to identify the organism in animal models after repeat inoculation (Taylor et al. 1984).

Simplified use of a system such as WHO grading system can mean that some individuals who are nucleic acid amplification test-positive but are not graded as having follicular disease actually do have trachoma (Dawson et al. 1981).

It could be concluded that discrepancy between nucleic acid amplification test results and clinical diagnosis may be due to several causes such as long recovery phase, incorrect grading of disease, masquerading diseases of trachoma (e.g., viral/allergic conjunctivitis), short incubation period of disease, contamination, and missed mild clinical signs.

In general, more severe the disease, more likely the results are positive for the PCR tests (Wright and Taylor 2005). Because of the cost and complexity, PCR cannot replace clinical diagnosis. Recently, a simple and cheap enzyme-based dipstick test has been shown to be correlated with the clinical stages in early trials (Michel et al. 2006).

Epidemiology

Trachoma is the leading cause of preventable blindness. In October 2010, Trachoma was reported as one of the seven most neglected tropical Diseases (NTDs) preventable via mass drug administration (MDA).

Its infliction is primarily aimed at those living in destitute and deprived of treated water and proper sanitation. The number of those involved, either with blindness or severe visual impairment, is estimated eight million worldwide.

When the global effort to eliminate blinding trachoma began, estimated 540 million people were believed to be at risk of trachoma, and about 84 million were thought to have active disease in 57 countries (Resnikoff et al. 2004). In 2009, the number of persons with active disease declined to 41 million and 120 million remain at risk of the disease (Mariotti et al. 2009). At least one person is prone to severe visual loss every 4 min whereas approximately one case of blindness occurs every quarter of an hour owing to this morbid malady (Mariotti et al. 2009) (Fig. 8).

There are groups of individuals implicitly exposed due to poor sanitation or low education status. Children are often prone to infection as they have to provide care to their senior family members who are most likely infected. To make matters worse, patients are often from lower social as well as economic backgrounds. Therefore a serious morbidity such as blindness can seriously jeopardize their

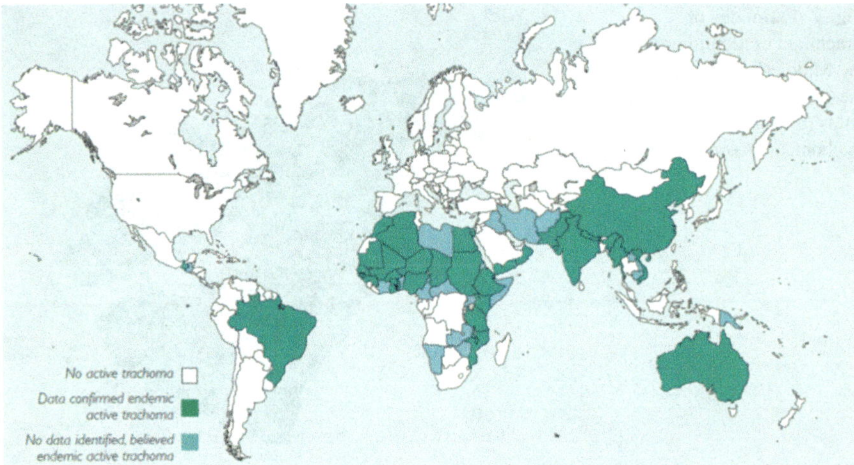

Fig. 8 Estimates of epidemiology based on the Courtesy of Trachoma Atlas (http://www.trachomaatlas.org)

living, which is often accompanied by loss of status, social stigmatization, and alienation (WHO 2003).

Women are more likely to go blind from trachoma than their male counterparts, with a ratio of two to one (Cromwell et al. 2009). A plausible explanation for this propensity is that women are in closer contact with children compared to men, making them highly susceptible to trachoma reinfection (Courtright and West 2004).

Trachoma has been on top of the vision 2020 agenda "the right to sight" as a global initiative to prevent blindness implemented by the WHO and the International Agency for the prevention of blindness in 1999. It has sustained its status as a prime target in the campaign to eliminate blindness by the year 2020.

With disease burden partially mapped, WHO suggests documentation at districts levels, as well as planning for SAFE interventions (a strategic plan designed by WHO for prophylaxis and treatment of Trachoma). This had led to the mapping of 1,115 districts by 2010, of which 559 are tagged endemic at present time.

The heaviest infection load was reported in Africa, Sudan, and Ethiopia in particular, but there are important gaps in data for three large countries: China, India, and Nigeria. Of the 57 countries where trachoma was endemic, eight countries have admittedly met their predetermined elimination targets (Algeria, Ghana, Iran, Libya, Mexico, Morocco, Oman, and Vietnam) (Resnikoff et al. 2004).

Several countries such as Botswana, Djibouti, Laos, Sierra Leone, Rwanda, and the pacific Island nations of Fiji, Kiribati, Nauru, and Vanuatu used to be endemic. There are also nations with missing data gaps (Afghanistan, Cambodia, Cameron, Chad, China, Cote d'Ivoire, Democratic Republic of Congo, Egypt, India, Iraq, Malawi, Mozambique, Pakistan, Somalia, Yemen, Zambia, and Zimbabwe) (Resnikoff et al. 2004). Based on WHO statistics, about half of the remaining

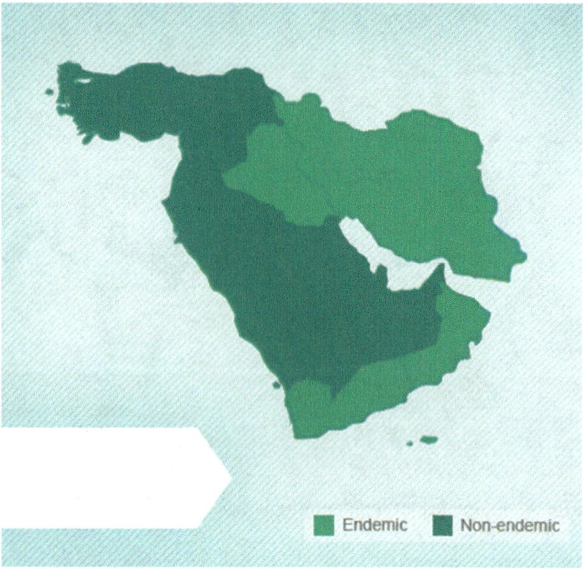

Fig. 9 Estimates of Trachoma epidemiology in the Middle East based on the courtesy of Trachoma Atlas (http://www.trachomaatlas.org)

global burden of active trachoma is probably concentrated in five countries: Ethiopia, Guinea, India, Nigeria, and Sudan (WHO 2010a).

Epidemiologic Features in Middle East and North Africa (MENA)

Formerly endemic nations of Iran, Oman, Saudi Arabia, and United Arab Emirates either have already reached or are nearing their elimination targets, making the geographical distribution of trachoma less consistent in the Middle East. Data is still inadequate in certain regions, namely, Iraq and Yemen (Resnikoff et al. 2004) (Figs. 9, 10, and 11).

Epidemiologic Features in Africa

Sub-Saharan Africa (SSA), especially the Sahel belt countries and East Africa nations, is reported to have had the greatest burden of trachoma. Despite the surveys conducted at national and regional scales over the past decades, there still exist gaps in data pertaining to the current status of the disease, making reliable estimation infeasible (Resnikoff et al. 2004) (Figs. 12 and 13).

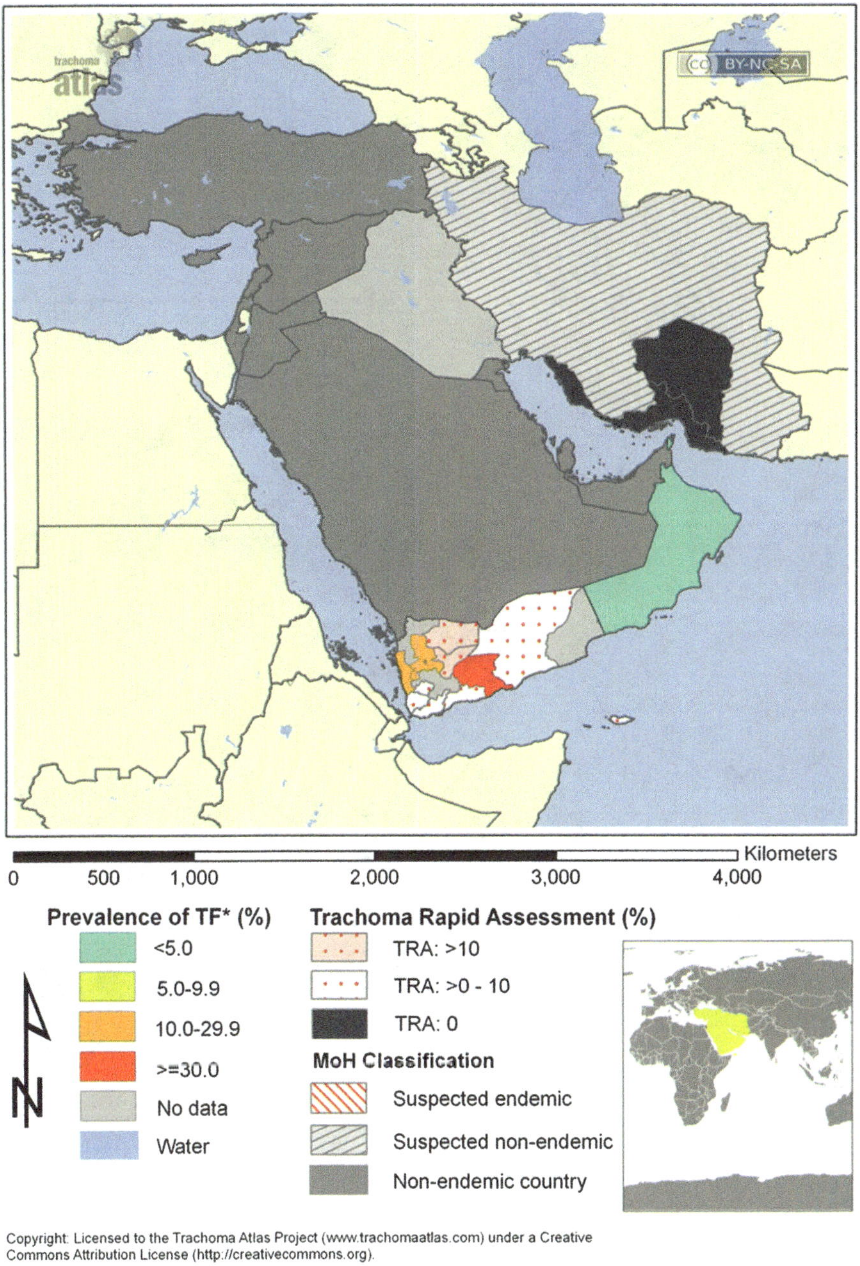

Fig. 10 Estimates of the prevalence of active Trachoma in the Middle East based on the courtesy of Trachoma. Atlas (http://www. trachomaatlas.org)

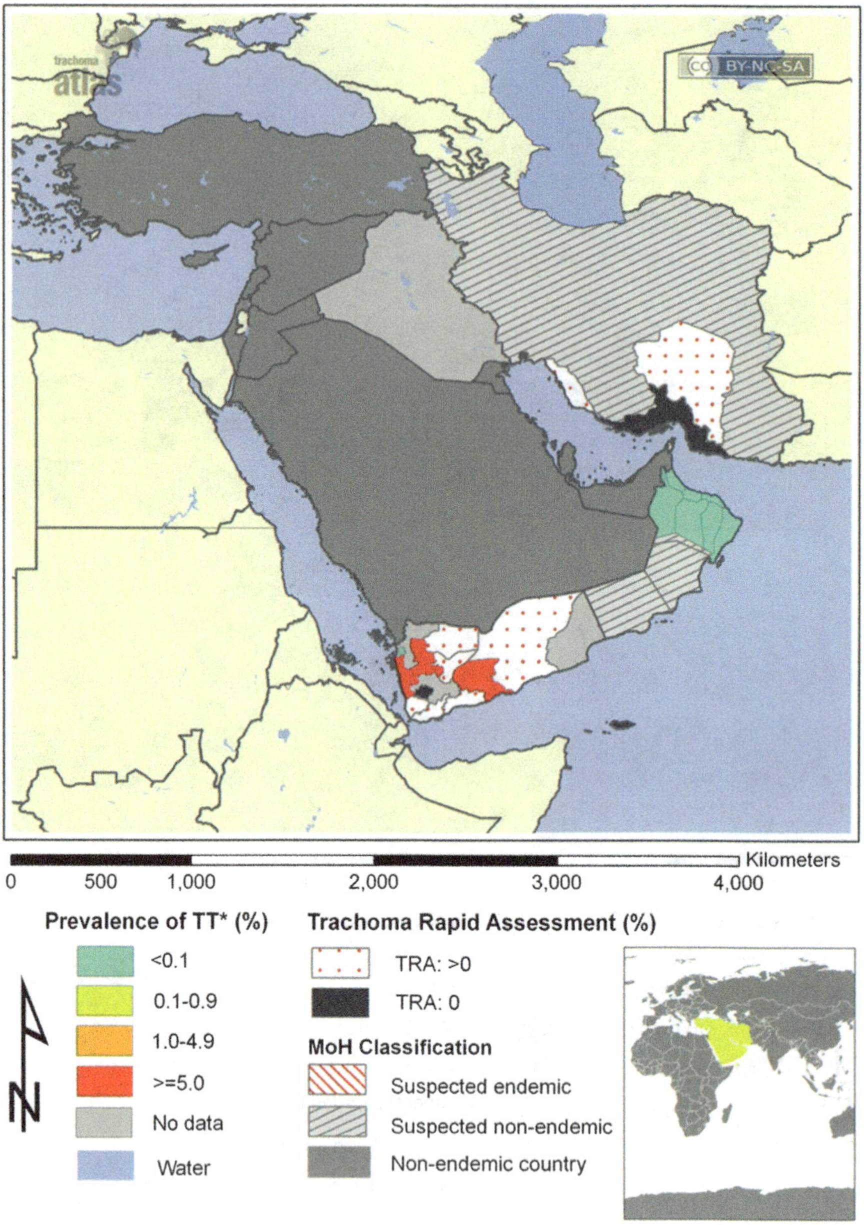

Fig. 11 Estimates of the prevalence of Trichiasis in the Middle East based on the courtesy of Trachoma. Atlas (http://www.trachomaatlas.org)

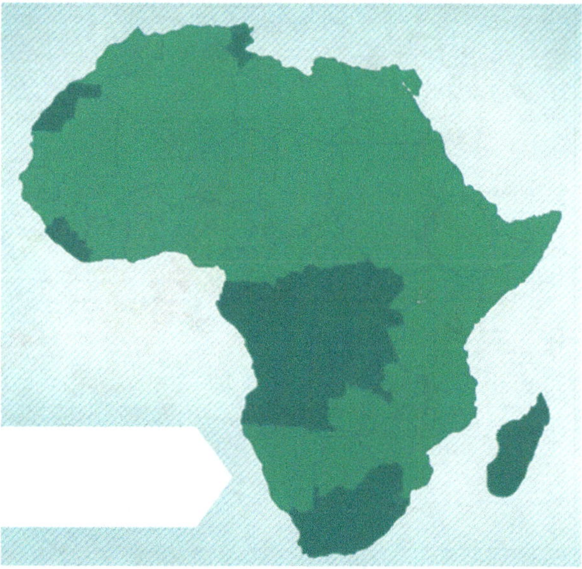

Fig. 12 Estimates of Trachoma epidemiology in Africa are based on the courtesy of Trachoma Atlas (http://www.trachomaatlas.org)

Economic Burden

From an economic perspective, trachoma is a heavy burden to the society. The economic burden of trachoma on the lives of individuals, families, and communities is enormous. Even conservative estimates suggest an annual loss of productivity due to trachoma is between \$3 and \$6 billion (Frick et al. 2003).

From an individualized perspective, vision loss and blindness as results of trachoma can cost the sufferers their social status, leading to their stigmatization. Thus, the eradication of the disease will benefit individuals, family, society, and the nation substantially.

It is estimated that the total intervention will cost a minimal of \$430 million US dollars. This does not include the cost of medication, which will be largely donated by Pfizer and merely includes sanitation measures, such as building toilets and digging boreholes as part of macro-scale development agenda (Frick et al. 2003).

Treatment

Medical Treatment

Antibiotic therapy for trachoma has been used since 1950 in a variety of forms and regiments. Antibiotics can be administered topically to the conjunctiva in the form of ointment. Topical treatment is of lower efficacy because it does not eliminate the organisms in the nasopharynx which serves as a reservoir for the organism. The side

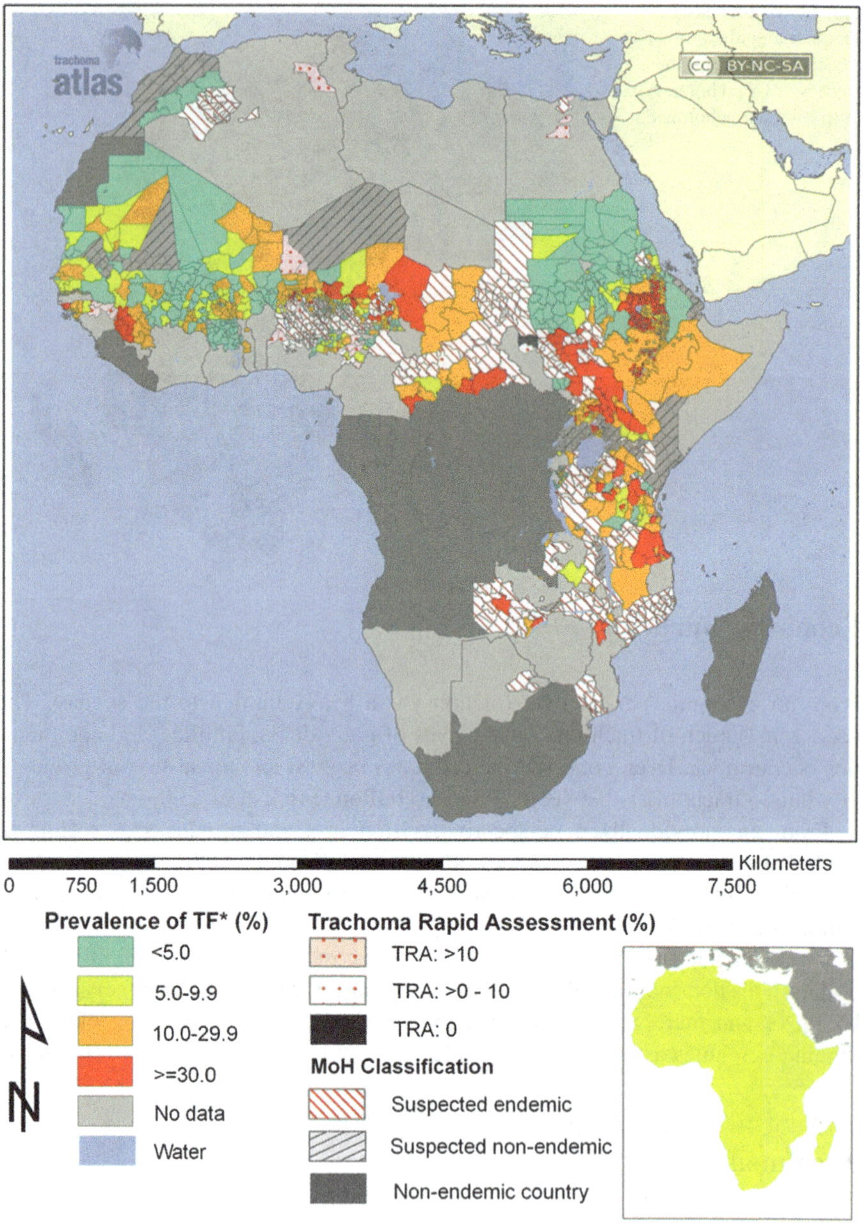

Fig. 13 Estimates of the prevalence of active Trachoma in Africa based on the courtesy of Trachoma. Atlas (http://www.trachomaatlas.org)

effects include burning and blurred vision. Ointment is difficult to be used in children. According to WHO guidelines, 1 % tetracycline ointment must be used twice daily for 6 weeks. Alternatively, it can be applied on 5 consecutive days each month for 6 months. Poor compliance was reported in the latter application owing to length of treatment.

Oral antibiotics can reach a higher systemic therapeutic level. Oral antibiotics are reported with a higher compliance than topical treatment under direct observation. Azithromycin, a single dose of one gram in adults and 20 mg/kg in children, is the antibiotic of choice by WHO. It is confirmed as an effective drug against genital chlamydial infection owing to its long half-life, high intracellular concentration, and low blood level. In pregnant women, erythromycin 250 mg four times a day is the drug of choice. There is evidence that antibiotic use can lower the prevalence of trachoma. There are two types of studies on the efficacy of trachoma treatment. The first compared antibiotics and placebo, which demonstrated a modest lowering of active trachoma prevalence at 3 months but not at 1-year follow-up. These investigations focused on individuals with signs of active infection (Daghfous et al. 1974). It is not feasible to assess the efficacy of any given drug on trachoma in a placebo-controlled trial because reinfection occurs when there is contact with untreated individuals.

The second group of studies, comparing topical and oral antibiotics, reported no difference between the two in terms of efficacy in patients with active trachoma (Bowman et al. 2000; Darougar et al. 1980, 1981; Dawson et al. 1982, 1997; Fraser-Hurt et al. 2001). The Cochrane Review supports the view that oral azithromycin is more effective at clearing infection than topical treatment (Yorston et al. 2006b).

The common belief is that mass community treatment is more effective than targeting individuals, yet this has never been formally established by a randomized placebo-controlled trial. Mass community treatment with azithromycin uses three doses in contrast with a single annual does recommended by WHO (Bailey et al. 1999).

Mass distribution programs are now in progress in many endemic regions. The assumption is that disease control cannot be achieved by only applying antibiotics to subgroups in endemic districts (Bailey et al. 1993; Kamiya 1956; Sutter and Ballard 1983). Thus, every person in the district must receive treatment.

Nevertheless, it is highly unlikely to conduct any research regarding effectiveness of an antibiotic versus a placebo in mass distribution program, mainly on ethical grounds. There are still questions as to the target, duration, and frequency of use, which must be determined via comparative trials. The clinical importance of such studies is still open to debate as the resolution of clinical signs of active disease occurs after the resolution of active infection (Bailey et al. 1999). The ultimate goal is to reduce and prevent blindness, which can only be achieved in 20–30 years after intervention commences.

Surgical Treatment

Lid Surgery

According to the Cochrane Review, there is no clinical trial on the interventions for trichiasis to prevent blindness. Certain interventions have been shown to be more effective at eliminating trichiasis.

The most favorable outcome in surgical procedures for trachoma is tarsal plate full thickness incision and rotation of the lash-bearing lid margin through 180°. There are two main styles in use: bi-lamellar and uni-lamellar tarsal rotation. There is no evidence in terms of difference in outcome between these two procedures, although the latter is said to be varied among experts. Tarsal rotation surgery can be carried out by sufficiently trained ophthalmic nurses or ophthalmologist to achieve the same efficacy. As the number of the ophthalmologists is limited in endemic region, tarsal rotation surgery can be safely delegated to appropriately trained health staff to reduce the surgical backlog. There was also no disparity between surgical outcomes whether procedures are performed in rural areas versus healthcare centers (Yorston et al. 2006a).

There has been no consensus regarding the optimal management for minor cases of trichiasis, in which one to five eyelashes touch the eye. The current recommendation includes both early surgery and epilation.

Nonsurgical techniques such as cryotherapy and electrolysis were shown inferior to tarsal rotation surgery in outcome. Nonsurgical techniques, though, are not clearly assessed in terms of outcome or efficacy. Hence, a randomized controlled trial of these nonsurgical options is needed.

Recurrence still remains a challenge and limiting the effectiveness of surgery in preventing blindness. In addition, no conclusive results were obtained from three studies probing the positive effect of azithromycin on enhancing surgical outcome if prescribed postoperatively (West et al. 2006; Zhang et al. 2006).

Corneal Surgery

Trachomatous keratopathy remains a significant cause of visual morbidity (Wright et al. 2009). Patients with corneal scarring caused by trachoma are high-risk cases for penetrating keratoplasty (PKP) because of poor ocular surface, corneal vascularization, and lid abnormalities. It is important to recognize, however, that the spectrum of post-trachoma sequelae ranges from mild corneal scarring without severe eyelid and ocular surface disease to end-stage corneal scarring and vascularization associated with ankyloblepharon and advanced symblepharon. The prognosis of PKP should also base on the degree of risk factors for graft failures.

Judicious selection of milder cases, combined with strict attention to correction of eyelid abnormalities, such as trichiasis and entropion, and aggressive management of ocular surface disease, such as dry eye syndrome and meibomitis, could improve prognosis for graft survival and achieve good visual outcome for many patients with corneal blindness attributed to chronic trachoma (Monga et al. 2008).

Previous studies have evaluated the role of penetrating keratoplasty as a treatment modality for trachomatous keratopathy (Matta and Zakharia 1973; Al-Fawaz and Wagoner 2008). Matta and Zakharia (1973) analyzed 16 eyes with trachomatous keratopathy that underwent penetrating keratoplasty and reported no deleterious effect of trachoma per se on the graft survival or clarity and final visual outcome. Kocak-Midillioglu et al. (1999) reviewed 16 cases of trachomatous corneal scarring that underwent penetrating keratoplasty. Fourteen of the 16 eyes retained clear grafts (87.5 %), and 13 eyes (81.3 %) achieved 0.1 or better visual acuity after 1 year. Thirteen of the 16 cases (81.3 %) achieved 0.1 or better visual acuity after more than 1 year of follow-up. Al-Fawaz and Wagoner (2008) reviewed the results of penetrating keratoplasty in 127 cases of trachomatous keratopathy. Final visual acuity of 20/160 or better was reported in 67 of 127 cases (53.5 %) of PKP.

Lamellar keratoplasty (LK) is a useful therapeutic modality to treat anterior and mid-stromal corneal opacities in trachoma. It has a lower risk of graft rejection and failure. However, it is technically more demanding, and the visual outcome may be compromised by the presence of interface haze. Use of a microkeratome for LK (automated lamellar therapeutic keratoplasty (ALTK)) enables a smoother and easier dissection of the host and that of the donor lenticule. This could reduce interface-related problems.

Current Strategies

In 1998, WHO Alliance for the Global Elimination of Blinding Trachoma by 2020 (GET 2020) was created to eliminate trachoma worldwide and the Fifty-first World Health Assembly (WHA) called upon its member states to collaborate in the WHO Alliance to eliminate the public health impact of trachoma.

At that time, Pfizer Inc. also committed to donate Zithromax® (Azithromycin) for the preventative antibiotic program that can help stem transmission of the disease. These efforts have encouraged many other organizations to participate in the challenge and seeded a broad community of partners tackling trachoma today.

The SAFE Strategy

The SAFE Strategy is an innovative public health approach designed to treat and prevent trachoma. The value of the WHO-endorsed SAFE strategy has been firmly established and continues to be improved. The components of SAFE are:

- Surgery
- Antibiotics
- Facial cleanliness
- Environmental improvements

The International Trachoma Initiative (ITI) was founded in 1998 in response to WHO's call to eliminate blinding trachoma by 2020. To implement the SAFE strategy for trachoma control, ITI collaborates with governmental and non-governmental agencies at the local, national, and international levels.

Surgery

As the first and foremost measure to prevent blindness, surgery is reserved for reversing the in-turned eyelashes of the patients with trichiasis or entropion. It is a quite simple procedure and can be availably carried out in the community or at healthcare centers. The ubiquitous fear toward surgery has made community-wide surgery more favorable in terms of compliance. Lid surgery takes away the pain of lashes on the eyes but does not remove the scarring or restore sight. It must be done properly as recurrence rates are high. If TF reaches above 10 % in the community, surgery is recommended when district TT prevalence exceeds 0.1 % (Emerson et al. 2000).

Antibiotic Therapy

Antibiotics are applied with the aim of either reducing the infection burden in an affected community or treating an active disease. The administration of topical tetracycline ophthalmic ointment daily for a period of at least 6 weeks, or as an alternative, annual azithromycin tablets or liquid for infants can treat active infection. What defines distribution strategy regards the prevalence of trachoma, the access to drugs, and staff dispensing it.

The WHO commends that all individuals in communities where the prevalence of active trachoma exceeds 10 % of children aged 1–9 be mass-treated with antibiotic therapy. In communities where the prevalence of active disease is between 5 and 10 %, health officials may choose to either mass treat or treat only people with active disease and their families (Chidambaram et al. 2005).

Fortunately, there are a number of antibiotics to which *C. trachomatis* is sensitive (Please see medical treatment). Azithromycin is currently the drug of choice for trachoma. To date, Pfizer's donation of Azithromycin (Zithromax®) through the ITI has reached millions of people in 19 countries. As of 2012, Pfizer has donated 280 million Zithromax® treatments.

Facial Cleanliness

There is a strong correlation between trachoma and facial cleanliness as children with dirty faces can both transmit the disease if infected and catch it if not. Ophthalmic and nasal discharges attract infective flies while rubbing dirty eyes with cloth, sheets, and mother's clothing, namely, shawl, help transmit trachoma at a grand level. Thus, it is strongly recommended that children should have clean faces as part of trachoma control program.

Environmental Improvement

Trachoma cannot be eradicated unless sanitation, waste disposal, and water quality are maintained at an acceptable level. These serve as the pillars to any control program, which seem daunting at times though, and can only be fulfilled via the collaboration with other parts, such as education, clean water, and good sanitation. Therefore, the SAFE strategy addresses poverty and development issues, resulting in improved quality of life for millions of people in the world's poorest countries. The F and E components should be put in place whenever TF prevalence exceeds the 5 % mark in children aged 1–9 years and district-wide distribution of antibiotics should be added in any district with prevalence of TF higher than 10 %.

All four above-mentioned components of the SAFE strategy are absolutely essential in any successful control program. Antibiotics and surgery minus hygiene and sanitation can merely remove symptoms and not the causes of the disease. For sustainable control of trachoma, the F and E components of the SAFE strategy must be present in addition to the S and A components.

A time period of between 4 and a maximum 6 years is needed for SAFE measures to be fully applied. Even so, S, F, and E elements have to be sustained to ensure the prevention of resurgence. This time line will pose a great challenge to the global community owing to the disparity of districts in terms of their conditions, objectives, and resources prior to 2020.

Trachoma Action Plan

At the time of the 14th GET 2020 Alliance meeting in 2010, the ITI was asked by the global alliance to "develop a trachoma action plan (TAP) template for distribution by the alliance" (WHO 2010a).

The Trachoma Action Plan (TAP) was developed to promote specific actions as well as policies with emphasis on the eradication of the disease at national levels by the year 2020. This development accelerates and facilitates the formation of a global strategy, 2020 insight, defining ways to eliminate morbid trachoma by 2020 while establishing coordination and focus in this respect (WHO 2010b).

Although considerable progress has been made, but great action is needed now. A number of countries recently reported having achieved WHO elimination targets. These successes illustrate that trachoma elimination is possible and there is a strong encouragement to continue fighting.

Ultimate Intervention Goals for the Elimination of Blinding Trachoma

The WHO has set ultimate intervention goals (UIGs) for trachoma that indicate the final targets that must be achieved to eliminate trachoma globally.

Trachomatous Trichiasis UIG

Complete eradication is achieved provided that every country manages to reduce the number of infected people to below 1,000 in a district (Emerson and Frost 2006). This cannot be fulfilled unless surgery is available to patients with trichiasis.

Active Trachoma UIG

For blinding trachoma to eventually be eliminated as a public health problem, each country must reduce the number of cases of active trachoma (TF) in children between the ages of 1 and 9 to less than 5 % of the population of children in any district. If TF prevalence is greater than 10 % in children aged 1–9, districts should conduct mass distribution campaigns of topical or oral antibiotic. In any district where TF is between 5 and 10 % in children aged 1–9, targeted treatments may be used instead of mass treatments (Chidambaram et al. 2005).

Facial Cleanliness and Environmental Improvement UIG

The goal is achieved, at any given time, if 80 % of the total populations of children have clean face.

Future Milestones

Post-elimination period should be kept in mind when efforts slow down if not ceased. District performance has to be sustained for at least 3 years post-elimination so that WHO certification can be granted. It is obvious that surveillance, surgical capacity, and preventive measures must be maintained beyond certification on the part of local healthcare systems.

Post-elimination Surveillance

Following achievement of intervention goals, attempts must be made to ensure constant surveillance in order for detecting cases of resurgence as early as possible. It is also regarded as mandatory for WHO to include elimination surveillance in their criteria certification.

Maintained Surgery Capacity for New Cases

As inhabitants of endemic districts will continue to be effected by eyelid scarring as a complication even following elimination, surgery capacity has to be maintained to address these cases effectively (Munoz et al. 1999).

Trachoma Capabilities in the Health System

The local health system shall be empowered to shoulder the responsibility of trachoma detection, treatment, and management so that elimination can be effectively maintained.

Future Researches

Research must focus on the safest strategy leading to cease mass drug use without jeopardizing elimination as well as the best way to identify infection. Several proposed items could improve identification of trachoma including:

- Effective training
- Quality photography with smart phone app such as "Google goggles"
- Pooled PCR tests which reduce the costs
- RNA-based PCR which would be more sensitive but also more expensive
- Rapid diagnostic tests with immediate results and no need for additional lab studies

Animal model will help with a better understanding of the disease dynamics and transmission (Solomon et al. 2003).

References

Albert DM, Edwards DD (eds) (1996) The history of ophthalmology. Blackwell Science, Cambridge, MA

Al-Fawaz A, Wagoner MD (2008) Penetrating keratoplasty for trachomatous corneal scarring. Cornea 27:129–132

Allen SK, Semba RD (2002) The Trachoma 'Menace' in the United States, 1897-1960. Surv Ophthalmol 47(5):500–508

Asbury C, Cline BL, Gammino VM (2000) The Edna McConnell Clark Foundation's Tropical Disease Research Program: a 25-year retrospective review 1974-1999, 15 Aug 2000

Bailey RL, Arullendran P, Whittle HC, Mabey DC (1993) Randomised controlled trial of single-dose azithromycin in treatment of trachoma. Lancet 342:453–456

Bailey R, Duong T, Carpenter R, Whittle H, Mabey D (1999) The duration of human ocular Chlamydia trachomatis is age dependant. Epidemiol Infect 123:479–486

Baneke A (2012) Review: targeting trachoma: strategies to reduce the leading infectious cause of blindness. Travel Med Infect Dis 10(2):92–96

Boldt J (1904) Trachoma. Hodder and Stoughton, London

Bowman RJC, Sillah A, van Debn C, Goode V, Muquit M, Johnson GJ et al (2000) Operational comparison of single-dose azithromycin and topical tetracycline for trachoma. Invest Ophthalmol Vis Sci 41(13):4074–9

Burton MJ, Holland MJ, Faal N et al (2003) Which members of a community need antibiotics to control trachoma? Conjunctival Chlamydia trachomatis infection load in Gambian villages. Invest Ophthalmol Vis Sci 44:4215–4222

Chidambaram JD, Lee DC, Porco TC, Lietman TM (2005) Mass antibiotics for trachoma and the Allee effect. Lancet Infect Dis 5:194–196

Cohen CR, Koochesfahani KM, Meier AS et al (2005) Immunoepidemiologic profile of Chlamydia trachomatis infection: importance of heat-shock protein 60 and interferon-gamma. J Infect Dis 192:591–599

Courtright P, West SK (2004) Contribution of sex-linked biology and gender roles to disparities with trachoma. Emerg Infect Dis 10:2012–2016

Cromwell E, Emerson P, Courtright P (2009) Women and trachoma: achieving gender equity in the implementation of SAFE. The Carter Center, Atlanta

Daghfous T, Dawson CR, Messadi M, Hoshiwara I, Vastine D (1974) Comparative study of the effect of rifampicin and tetracycline versus placebo in the treatment of trachoma [Etude comparative de l'action de la rifampicineet de tetracycline et d'un placebo dans le traitement du trachome]. Int Rev Trach 51:71–80

Darougar S, Jones BR, Viswalingam N, Poirier RH, Allami J, Houshmand A et al (1980) Family-based suppressive intermittent therapy of hyperendemic trachoma with topical oxytetracycline or oral doxycycline. Br J Ophthalmol 64(4):291–295

Darougar S, Viswalingam N, El Sheikh H, Hunter PA, Yearsley P (1981) A double-blind comparison of topical therapy of chlamydial ocular infection (TRIC infection) with rifampicin or chlortetracycline. Br J Ophthalmol 65(8):549–552

Dawson CR, Jones BR, Tarizzo ML (1981) Guide to Trachoma control. World Health Organization, Geneva

Dawson CR, Daghfous T, Hoshiwara I, Ramdhane K, Kamoun M, Yoneda C et al (1982) Trachoma therapy with topical tetracycline and oral erythromycin: a comparative trial. Bull World Health Organ 60(3):347–355

Dawson CR, Schachter J, Sallam S, Sheta A, Rubinstein RA, Washton H (1997) A comparison of oral azithromycin with topical oxytetracycline/polymyxin for the treatment of trachoma in children. Clin Infect Dis 24(3):363–368

Emerson PM, Cairncross S, Bailey RL, Mabey DC (2000) Review of the evidence base for the 'F' and 'E' components of the SAFE strategy for trachoma control. Trop Med Int Health 5: 515–527

Emerson P, Frost L, with Bailey R, Mabey D (2006) Implementing the SAFE Strategy for Trachoma Control, ITI guideline, January 2006

Fraser-Hurt N, Bailey RL, Cousens S, Mabey D, Faal H, Mabey DC (2001) Efficacy of oral azithromycin versus topical tetracycline in mass treatment of endemic trachoma. Bull World Health Organ 79(7):632–640

Frick KD, Basilion EV, Hanson CL, Colchero MA (2003) Estimating the burden and economic impact of trachomatous visual loss. Ophthalmic Epidemiol 10:121–132

Kamiya S (1956) Mass treatment of trachoma. Am J Ophthalmol 42:269–283

Karasch MC (1993) Ophthalmia. In: Kiple K (ed) The Cambridge world history of human disease. Cambridge University Press, New York, pp 897–905

Kocak-Midillioglu I, Akova YA, Kocxak-Altintas AG et al (1999) Penetrating keratoplasty in patients with corneal scarring due to trachoma. Ophthalmic Surg Lasers 30:734–741

Larner A (2004) Ophthalmological observations made during the mid-19th century European encounter with Africa. Arch Ophthalmol 122(2):267–272

Lietman TM, Dawson CR, Osaki SY, Zegans ME (2000) Clinically active trachoma versus actual Chlamydial infection. Med J Aust 172:93–94

Mabey D, Bailey R (1999) Eradication of trachoma worldwide. Br J Ophthalmol 83(11): 1261–1263

MacCallan AF (1936) Trachoma. Butterworth, London

Mariotti SP et al (2009) Trachoma: global magnitude of a preventable cause of blindness. Br J Ophthalmol 93(5):563–8

Markel H (2004) When Germs Travel: six major epidemics that have invaded America since 1900 and the fears they have unleashed. Pantheon, New York

Matta CS, Zakharia HS (1973) Penetrating keratoplasty in trachomatous patients. Br J Ophthalmol 57:115–119

Merriam Webster Dictionary site http://www.merriam-webster.com/dictionary/trachoma

Michel C-EC, Solomon AW, Magbanua JPV et al (2006) Field evaluation of a rapid point-of-care assay for targeting antibiotic treatment for trachoma control: a comparative study. Lancet 367:1585–1590

Millar MI, Lane SD (1988) Ethno-ophthalmology in the Egyptian delta: an historical systems approach to ethnomedicine in the Middle East. Soc Sci Med 26(6):651–657

Monga P, Gupta VP, Dhaliwal U (2008) Clinical evaluation of changes in cornea and tear film after surgery for trachomatous upper lid entropion. Eye (Lond) 22:912–917

Munoz B, West S (1997) Trachoma: the forgotten cause of blindness. Epidemiol Rev 19(2): 205–217

Munoz B et al (1999) Incidence of trichiasis in a cohort of women with and without scarring. Int J Epidemiol 28(6):1167–71

Resnikoff S, Pascolini D, Etya'ale D, Kocur I, Pararajasegaram R, Pokharel GP, Mariotti SP (2004) Global data on visual impairment in the year 2002. Bull World Health Organ 82 (11):844–851

Schachter J, Moncada J, Dawson CR et al (1988) Nonculture methods for diagnosing chlamydial infection in patients with trachoma: a clue to the pathogenesis of the disease? J Infect Dis 158:1347–1352

Solomon AW, Holland MJ, Burton MJ et al (2003) Strategies for control of trachoma: observational study with quantitative PCR. Lancet 362:198–204

Solomon AW, Peeling RW, Foster A et al (2004) Diagnosis and assessment of trachoma. Clin Microbiol Rev 17:982–1011

Stephens RS, Tam MR, Kuo CC, Nowinski RC (1982) Monoclonal antibodies to Chlamydia trachomatis: antibody specifi cs and antigen characterization. J Immunol 128:1083–1089

Sutter EE, Ballard RC (1983) Community participation in the control of trachoma in Gazankulu. Soc Sci Med 17:1813–1817

T'ang FF, Huang YT, Chang HI, Wong KC (1957) Isolation of trachoma virus in chick embryo. J Hyg Epidemiol Microbiol Immunol 1:109–120

Taborisky J (1952) Historic and ethnologic factors in the distribution of trachoma. Am J Ophthalmol 35:1305–1311

Taylor HR (1987) Trachoma grading: a new grading scheme. Rev Int Trach Pathol Ocul Trop Subtrop Sante Publique 64:175–181

Taylor HR (2001) A trachoma perspective. Ophthalmic Epidemiol 8(2):69–72

Taylor HR, Johnson SL, Prendergast RA, Schachter J, Dawson CR, Silverstein AM (1982) An animal model of trachoma II: the importance of repeated reinfection. Invest Ophthalmol Vis Sci 23:507–515

Taylor HR, Agarwala N, Johnson SL (1984) Detection of experimental Chlamydia trachomatis eye infection in conjunctival smears and in tissue culture by use of fl uorescein-conjugated monoclonal antibody. J Clin Microbiol 2:391–395

Thygeson P (1962) Trachoma virus: historical background and review of isolates. Ann N Y Acad Sci 98:6–13

Tower P (1963) The history of trachoma. Arch Ophthalmol 69:157–164

Watts S (2003) The history of trachoma in Egypt. Report for the International Trachoma Initiative (Unpublished), March 2003

Webb S (1990) Prehistoric eye disease (trachoma in Australian Aborigines. Am J Phys Anthropol 81(1):91–100

West SK, West ES, Alemayehu W, Melese M, Munoz B, Imeru A et al (2006) Single-dose azithromycin prevents trichiasis recurrence following surgery. Arch Ophthalmol 124:309–14

Whittum-Hudson JA, Taylor HR, Farazdaghi M, Prendergast RA (1986) Immunohistochemical study of the local inflammatory response to chlamydial ocular infection. Invest Ophthalmol Vis Sci 27(1):64–69

WHO (2003) Alliance for the Global Elimination of Blinding Trachoma by 2020. Report of the 2nd scientific meeting on trachoma, Geneva, 25–27 August, 2003

WHO (2010a) First WHO report on neglected tropical diseases: working to overcome the global impact of neglected tropical disease, Department of Control of Neglected Tropical Diseases, WHO, Geneva

WHO (2010b) Progress report 2000-2009 and strategic plan 2010-2020 of the global programme to eliminate lymphatic filariasis, Department of Control of Neglected Tropical Diseases, WHO, Geneva

World Health Organization (WHO) (1949) World distribution and prevalence of trachoma in recent years. Epidemiological and vital statistics report, Nov–Dec, Geneva

Wright HR, Taylor HR (2005) Clinical examination and laboratory tests for estimation of trachoma prevalence in a remote setting: what are they really telling us? Lancet Infect Dis 5: 313–320

Wright HR, Turner A, Taylor HR (2008) Trachoma. Lancet 371(9628):1945–1954

Wright HR, Keffe JE, Taylor HR (2009) Trachoma, cataracts and uncorrected refractive error are still important contributors to visual morbidity in two remote indigenous communities of the Northern Territory, Australia. Clin Experiment Ophthalmol 37:550–557

Yorston D, Mabey D, Hatt S, Burton M (2006) Interventions for trachoma trichiasis. Cochrane Database Syst Rev (3):CD004008

Yorston D, Mabey D, Hatt S, Burton M (2006) Interventions for trachoma trichiasis. Cochrane Database Syst Rev (3):CD004008.

Zhang H, Kandel RP, Atakari HK, Dean D (2006) Impact of oral azithromycin on recurrence of trachomatoustrichiasis in Nepal over 1 year. Br J Ophthalmol 90:943–948

Cutaneous Leishmaniasis in Middle East and North Africa

Sima Rafati and Farrokh Modabber

Contents

Abstract Cutaneous Leishmaniasis (CL), also known as button d'orient, chiclero's ulcer, Aleppo sore, Delhi's boils, etc., is truly a neglected disease. CL is not life threatening, and this is one of the reasons the disease is truly neglected by most funding agencies and major pharmaceutical companies. Support for research on CL

S. Rafati (✉)
Molecular Immunology and Vaccine Research Lab, Pasteur Institute of Iran, Tehran, Iran
e-mail: Sima-rafatisy@pasteur.ac.ir; s_rafati@yahoo.com

F. Modabber
DNDi, Geneva, Switzerland

Center for Research and Training in Skin Diseases and Leprosy, Tehran University of Medical Sciences, Tehran, Iran
e-mail: farmodabber@dndi.org

M.A. McDowell and S. Rafati (eds.), *Neglected Tropical Diseases - Middle East and North Africa*, Neglected Tropical Diseases, DOI 10.1007/978-3-7091-1613-5_5,

has been more generous basically because its animal models provide good possibilities for studying immunology/immunopathology and host–parasite interactions. Hence, although much information has been generated from these studies, little has been developed for prevention or treatment of CL in decades. However, CL is a disfiguring disease that results in stigma, economic loss, and affects mainly unprivileged populations with limited resources. Scars from CL lesions, particularly on the face, mark the whole life of patients especially girls and women. Yet, currently there is little research and development (R&D) focused on the goal of alleviating the suffering of millions of CL cases, mostly children.

In this chapter, different aspects of cutaneous leishmaniasis including epidemiology and incidence in MENA region, recent knowledge in human and mice immune response, latest achievements in vaccine development, and their obstacle are discussed.

Keywords Cutaneous leishmaniasis • Human vaccine • Leishmanization • Vaccine in animal models • *L. major* • *L. tropica* • Treatment

Cutaneous Leishmaniasis (CL), also known as button d'orient, chiclero's ulcer, Aleppo sore, Delhi's boils, etc., is truly a neglected disease. CL is not life threatening, and this is one of the reasons the disease is truly neglected by most funding agencies and major pharmaceutical companies. Support for research on CL has been more generous basically because its animal models provide good possibilities for studying immunology/immunopathology and host–parasite interactions. Hence, although much information has been generated from these studies, little has been developed for prevention or treatment of CL in decades. However, CL is a disfiguring disease that results in stigma and economic loss and affects mainly unprivileged populations with limited resources. Scars from CL lesions, particularly on the face, mark the whole life of patients especially girls and women. Yet, currently there is little research and development (R&D) focused on the goal of alleviating the suffering of millions of CL cases, mostly children.

History of Cutaneous Leishmaniasis

Cutaneous leishmaniasis is known since ancient times. Bray referred to a tablet in the library of king Ashurbanipal of Assyria in Neinava (Mosul, Iraq), with the description of a painless ulcer most likely CL, translated from a tablet of old Akkadian period in the second or third millennium BC (Bray et al. 1967). There is no record of CL until the Muslim dominance. It is likely that the gap stems from the destruction by fundamentalist Arab Muslims, of most scholarly work, sculptures, and publications such as burning of the largest library of the time at Baghdad built by Persians, than to the late appearance of CL in the area. There are different records on who was the first to describe CL in the tenth–eleventh century. Elgood indicates Abu Mansur (died 991 AD) as the first "Arab" physician to describe oriental sore (Elgood 1934). (It must be noted that the Islamic regime conquered and dominated several nations, including the Persian Empire which extended at times from India to Spain and scholars from the conquered nations often wrote in Arabic as it was the acceptable and widely read scientific language.) Abu Mansur was from Khorasan (North East of Iran) and described a disease compatible with CL, but Bray and Modabber and similarly Killick-Kendrick reported that the earliest accurate description of CL is attributed to the Persian scholar, *Pur-e Sina* (son of Sina), commonly known as IbnSina or by his Latinized name Avicenna (980–1037 AD) in his book "The Canon of Medicine," which was a standard medical text at many medieval universities, including Montpellier and Leuven as late as 1650 (Bray and Modabber 2000; Killick-Kendrick 2010). In a recent book, on Ismail Gorgani (also known as Jorjani), Tajbakhsh indicates that Gorgani (1042–1136) was the first to describe in detail what was prevalent in Balkh and was called "Balkh sore", and around Gorgan, called "Pashegazidegui" (sore of mosquito bite, in Persian language); or "Salak" (meaning one-year sore), as it is now called in Iran (Tadjbakhsh 2005). It is clear that the disease has been known for a millennium and it is remarkable that even in tenth Century AD, CL was thought to be a vector-borne disease calling it Pashegazidegui. Saf'janova described the history of CL in the central Asian regions of USSR and Azerbaijan (Saf'janova 1985). Button d'orient as was called by the French scholars was described in the oasis of Biskra, Algeria, also called button de Biskra (Weber 1876). Hence, CL was obviously known in the region from India to the Eastern Mediterranean and North African countries with different names such as "Delhi Boil," "Baghdad boil," "Aleppo sore," "Balks sore," "Jericho Button," "clou de Biskra," "button d'orient," etc.

In this part of the world, CL is presented in two forms: Zoonotic (caused by *L. major*, ZCL) and anthroponotic (caused by *L. tropica*, ACL). The two forms were first distinguished by Alexander Russel (1715–1768), a Scottish physician who lived in Allepo and described female and male forms of CL. Based on the prevalence in the city vs. adjacent areas, the characteristics of the lesions (later called dry and wet), and the natural history of the two forms (ACL usually lasting more than a year), it is highly probable that they correspond to ACL and ZCL, respectively (Killick-Kendrick 2010). The cause of CL was unknown although parasites were described from sections of a Delhi boil in 1885 by DD Cunningham and 1891 by R. H. Furth. In 1903, J. H. Wright, an American physician published three papers on the discovery of *L. tropica* from the lesion of an Armenian child which strongly indicated the parasitic nature of CL. However, P. F. Borovsky was the first to realize that the parasites were protozoa, but as he had published in a local Russian journal in 1898, the information remained unknown outside Russia until much later till 1938, when it was translated (Killick-Kendrick 2010). *Leishmania* was first grown in Tunisia by Nicolle, who used a modified Novy–MacNeal medium (known for growing *Trypanosomes*) and isolated the parasites from sand flies (Dutta 2008). This medium, now called Novy–MacNeal–Nicolle (NNN) medium, is still used for isolation and growth of *Leishmania*.

Although the notion of vector-borne disease was expressed in the tenth century, and the experiment of Sergent who injected ground-up material of naturally caught *Phlebotomus papatasi* to four volunteers, of whom one developed a self-healing cutaneous lesion much like CL, this was not accepted by many scientists as proof for the vector or the causative agent of CL and it was not until 20 years later, when Adler and Ber managed to infect five volunteers with 27 lesions from bites of 26 sand flies bred and infected experimentally in the laboratory (Adler and Ber 1941). (It should be noted that until identification of *Leishmania* by isoenzyme characterization which was established in 1980s, *Leishmania* isolated from cutaneous lesions in this part of the world were called *L. tropica*, and most publications until then mention *L. tropica*, whereas it was *L. major*.)

Nadim and colleagues described the natural history and life cycles of the two forms in Isfahan and vicinity, which distinguished between dry (ACL) and wet (ZCL) lesions (see Epidemiology, below).

Epidemiology

Cutaneous leishmaniasis has a wide distribution, occurring in the Indian subcontinent, Central and South Western Asia, the Mediterranean region, Africa, and Central and South America. An estimated 1.5 million individuals suffer from different forms of CL globally (Alvar et al. 2012), but only a small percentage receives treatment, due to cost, toxicity, and long duration of therapy. In its different forms, CL in this area (Old World CL, in contrast to Latin America, New World CL) is generally caused by four different species of the protozoan parasite *Leishmania* (*L. major, L. tropica, L. infantum*, and *L. aethiopica*). The reservoir host may be either infected humans (anthroponotic CL, ACL) or mammals (zoonotic CL, ZCL). In the region that is covered by this chapter (Middle East and North Africa or Asia and North Africa), ZCL caused by *L. major* is most frequent followed by ACL caused by *L. tropica*. *L. infantum* (the major cause of visceral leishmaniasis outside Indian subcontinent and East Africa) produces CL to a limited extent. *Leishmania* identification is based on isoenzyme electrophoretic patterns and/or specific DNA probes, none of which are associated with identified biological activity. Although these markers have been very useful for establishing evolutionary tree and interspecies relations, they should not be taken as identifier for the type of disease or speciation. As *Leishmania* are genetically hypervariable and recombination has been shown in many species, variations in the markers mentioned above have been taken as new species (i.e., *Leishmania killiki*), which is now disputed to be a separate species and is believed to be a variant of *L. tropica*. For simplicity, only the major species causing CL are mentioned here.

CL can present in different forms ranging from uncomplicated self-healing skin lesions (most *L. major*) to debilitating large chronic or recurring lesions as in some *L. tropica* lesions (Fig. 1).

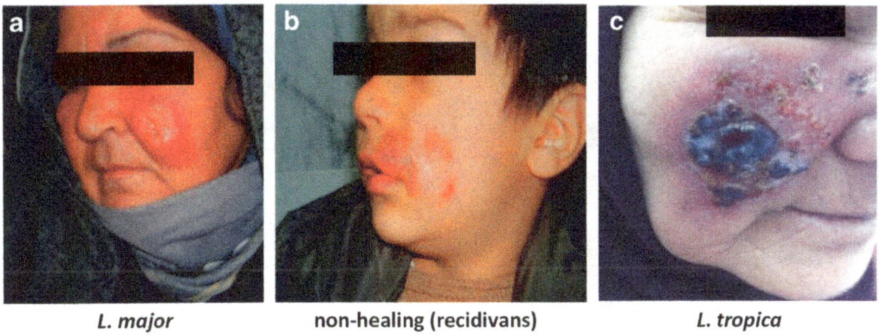

Fig. 1 Pictures of three infected patients in different regions of Iran. (**a**) *L. major* (**b**) non-healing (recidivans), and (**c**) *L. tropic*

Table 1 Latest case numbers of cutaneous leishmaniasis in MENA

Country	Case number
Afghanistan	113,100–226,200
Iran	69,000–113,300
Syria	64,100–105,300
Tunisia	21,400–35,100

Incidence of CL in the Region

CL occurs in countries stretching from India, central, south, and western Asia to countries surrounding the Mediterranean Sea. Recently NTD group of WHO (also see Alvar et al. 2012) reported estimates of global cases of visceral leishmaniasis and CL based on population surveys (in a few countries), publications, and reports of governments to WHO and then multiplied by a factor reflecting estimated underreporting. Admittedly there are gaps and inaccuracies, but this is an important attempt to correct the previous published estimate of incidence of 12 million cases as well as a plea for updates and correction. The estimated CL distribution of few key countries is listed below. Reportedly, in these countries CL is a major public health problem or the most frequent skin disease (Table 1).

These represent CL caused by *L. major* as well as *L. tropica*. The above numbers are reports by respective governments to WHO in 2007. Incidence may change from year to year, but most are stable foci for centuries. Due to recent political unrest, CL mostly caused by *L. tropica* has extended from Syria to Jordan, Turkey, and Iraq by refugees causing an alarming situation in these countries.

Afghanistan (Fig. 2)

The vector of *L. tropica* in Afghanistan is primarily *Phlebotomus sergenti*; human is definitive host although wild animals (stray dogs and foxes) may become infected as secondary hosts.

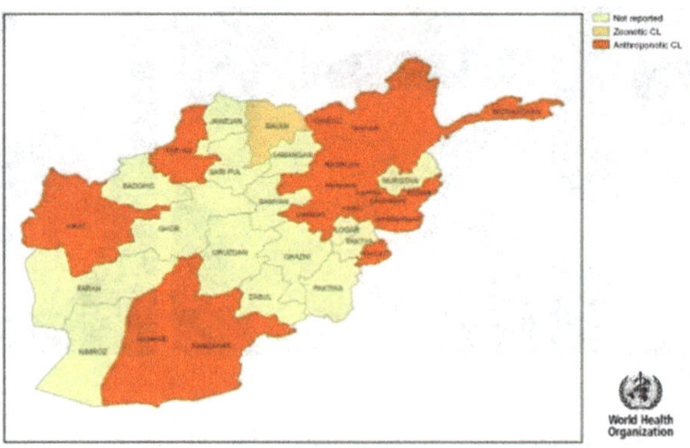

Fig. 2 Geographical distribution of cutaneous leishmaniasis in Afghanistan. With permission from WHO. Copied from Consultative Meeting on Cutaneous Leishmaniasis, Geneva, WHO Headquarters, 30 April to May 2007, Neglected Tropical Diseases Innovative and intensified disease Management, Leishmaniasis Control. WHO/HTM/NTD/IDM/2008.7, 2008. http://www.who.int/leishmaniasis/resources/Cutaneous_leish_cm_2008.pdf

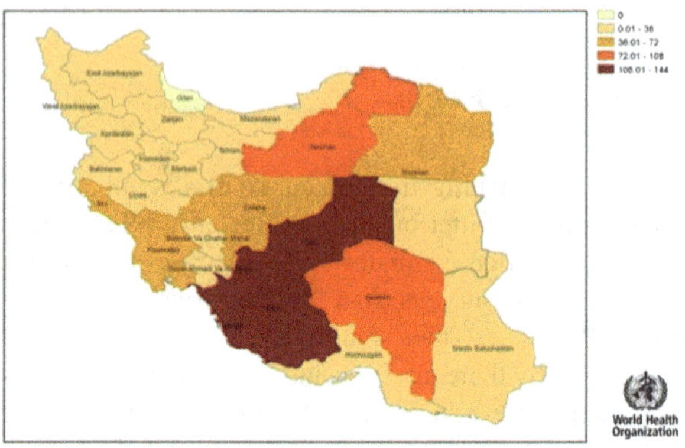

Fig. 3 Incidence of cutaneous leishmaniasis in Iran (Islamic Republic, 2007). With permission from WHO. Copied from Consultative Meeting on Cutaneous Leishmaniasis, Geneva, WHO Headquarters, 30 April to May 2007, Neglected Tropical Diseases Innovative and intensified disease Management, Leishmaniasis Control. WHO/HTM/NTD/IDM/2008.7, 2008. http://www.who.int/leishmaniasis/resources/Cutaneous_leish_cm_2008.pdf

Iran (Fig. 3)

The reservoir of *L. tropica* is infected human cases and the vector is *Ph. sergenti*, but other species, i.e., *Ph. Salehi* have also been reported. The reservoir of *L. major* in central Iran is *Psammomys obesus* while in North East and South West is *Meriones. Tateraindica* has also been reported as possible reservoir around Shiraz. The vector for *L. major* infection is *Ph. papatasi*.

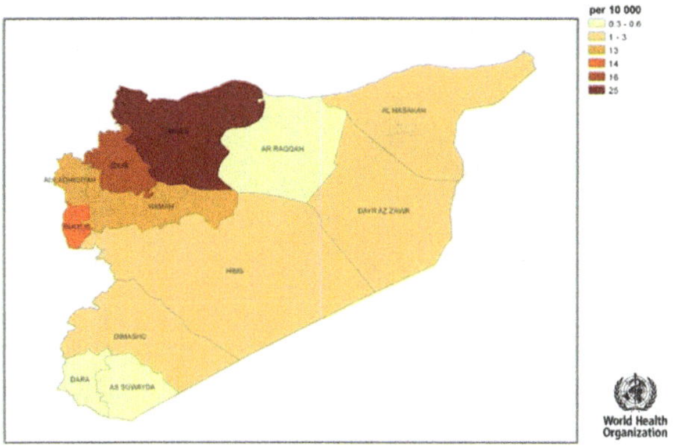

Fig. 4 Incidence of cutaneous leishmaniasis in Syrian Arab Republic, 2007. With permission from WHO. Copied from Consultative Meeting on Cutaneous Leishmaniasis, Geneva, WHO Headquarters, 30 April to May 2007, Neglected Tropical Diseases Innovative and intensified disease Management, Leishmaniasis Control. WHO/HTM/NTD/IDM/2008.7, 2008. http://www.who.int/leishmaniasis/resources/Cutaneous_leish_cm_2008.pdf

Syria (Fig. 4)

CL caused by *L. tropica* is about 90 % of the cases in Syria. Aleppo is known to be an endemic focus of CL for centuries. Humans are the reservoir and no animal reservoir has been identified. The vector is *Ph. sergenti*.

Tunisia (Fig. 5)

ZCL due to *L. major* is the main public health problem with *Psammomys obesus* as reservoir host and *Ph. papatasi* as vector. Sporadic CL due to *L. killicki* occurs further South, sometimes in limited outbreaks and *Ph. sergenti* is the suspected vector. Sporadic CL due to *L. infantum* occurs in towns and villages in the north of the country. The reservoir host is the domestic dog and the vector is *Ph. langeroni*.

Natural History of Cutaneous Leishmaniasis

Cutaneous leishmaniasis starts as a papule at the site of sand fly bite and usually is associated with cell infiltration (induration), followed by ulceration and eventually re-epithelialization and scar. Ulcer may not be observed as it can be covered by a scab. For decades it was believed that the type of lesion can be used to distinguish between "Dry" or urban form caused by *L. tropica* and "Wet" or rural form caused

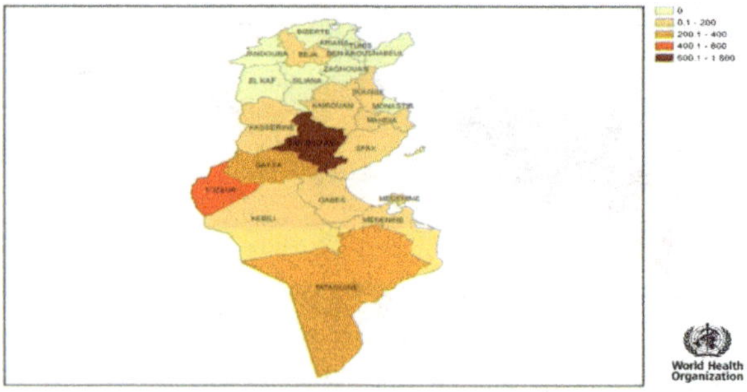

Fig. 5 Incidence of cutaneous leishmaniasis in Tunisia 2007. With permission from WHO. Copied from Consultative Meeting on Cutaneous Leishmaniasis, Geneva, WHO Headquarters, 30 April to May 2007, Neglected Tropical Diseases Innovative and intensified disease Management, Leishmaniasis Control. WHO/HTM/NTD/IDM/2008.7, 2008. http://www.who.int/leishmaniasis/resources/Cutaneous_leish_cm_2008.pdf

by *L. major*. Indeed, in general there is a trend for this correlation. However, with the advent of isoenzme patterns and genetic markers, this simple form of classification has been abandoned. The lesions take 3–18 months to heal in more than 90 % of cases, longer for *L. tropica* lesions. The exact markers/correlates of protection are not known, yet the following observations are pertinent:

1. Asymptomatic Infection (Clinical Resistance) measured by leishmanin skin test: Depending upon the level of exposure (transmission rate) in endemic foci, many individuals become leishmanin skin test (LST) positive without active CL, any scar attributable to leishmaniasis, or known history of CL. There was a general belief that these individuals are asymptomatically infected and are immune to reinfection. There is a controversy as to whether LST-positive response in these individuals is indicative of immunity. Ben Salah and colleagues performed LST before the transmission season followed the subject to evaluate the incidence of CL in an endemic focus of *L. major* in Tunisia. Their conclusion was that there was a predictive value of LST (Ben Salah et al. 2005).

   In contrast, during phase-3 evaluation of first-generation vaccine (Sharifi et al. 1998), children who were LST positive and therefore were excluded from vaccination were not protected (personal communication). Recently, Momeni et al., with a similar objective and design but a different leishmanin, showed that LST had no predictive value (Momeni Boroujeni et al. 2013). It is difficult to know why these results are different in Iran and Tunisia. One reason could be the different leishmanin used. The Tunisia study used one from Italy (made with *L. infantum*) while Iranians have used a standard leishmanin that has been produced at Pasteur Institute of Iran and has been evaluated by TDR/WHO (Alimohammadian et al. 1993).

   It must be pointed out that LST conversion following vaccination with the first-generation vaccines is different from naturally occurring LST positivity.

Since those who convert to positive LST reaction after vaccination have a lower incidence of disease (Antunes et al. 1986; Momeni et al. 1999; Khalil et al. 2000).

2. Spontaneous healing: Those who recover from CL are usually skin test positive and have high resistance to reinfection.
3. Clinical cure following treatment: Complete re-epithelization and resolution of inflammation following successful treatment by antimonials (standard treatment) if without relapse within 6 months posttreatment, is usually associated with immunity.

   These observations are general although much variation exists between different foci and causative organism. Nevertheless, similar phenomena exist in experimental murine models. Resistant mice which recover from CL are highly resistant to live challenge.
4. The pathogenic or non-healing responses are seen in those individuals who have long-lasting lesions for more than 2 years. The edges of lesions are always inflamed. They do not respond to available standard treatment well. Immune response modifiers may be a solution.

Different Aspects of Human Immune Response to *L. major* and *L. tropica*

Development of clinical disease depends on parasite, host, and sand fly factors, dose or route of inoculation, and the maintenance of macrophage in an inert, deactivated state. The complex interactions between many factors triggered by the host's innate and acquired immune response are responsible for pathogenesis of disease. Different cells such as macrophages, neutrophils, natural killer cells, and dendritic cells are responsible for inflammatory responses which cause disease expression and may end up to either symptomless or subclinical infection, self-healing, or chronic leishmaniasis (Kaye and Scott 2011).

Following sand fly transmission or needle inoculation with *L. major* to mice, invading neutrophils rapidly and efficiently capture the parasites. Infiltrating neutrophils cannot destroy the parasites; instead they facilitate infection as depletion of neutrophils prior to infection reduces the parasite load and delays onset of disease (Moradin and Descoteaux 2012; Peters et al. 2008; Ribeiro-Gomes and Sacks 2012; Ribeiro-Gomes et al. 2012). Infection of neutrophils is transient and within a week postinfection macrophages/monocytes take over as the primary host cells. Neutrophils can kill *Leishmania* if activated appropriately. Clearly for a productive infection, *Leishmania* need to establish in macrophages (Afonso et al. 2008). These cells can efficiently take up and incorporate parasites in vacuoles and act as principal antigen-presenting cells in leishmaniasis. Indeed when they enter their hosts, *Leishmania* invade mononuclear monocytes and tissue macrophages (Moradin and Descoteaux 2012). Within the parasitophorous vacuole, the parasites differentiate into amastigotes and replicate only if mononuclear phagocytes are not activated or are deactivated, a process, which is known to be under the control of cytokines (IL-10 and TGF-β). In contrast, if parasitized mononuclear cells are activated by IFN-γ and TNF-α, they are no longer permissive to amastigote replication and survival (Liu and Uzonna 2012). Macrophages recognize pathogens in part by sensing so-called pathogen-associated molecular pattern (PAMP) via their pattern recognition receptors (PPRs). The most analyzed in innate immune receptors are Toll-Like Receptors (TLRs) but another subset of PPRs, the Nucleotide binding Oligomerization Domain (NOD) proteins, have been also implicated in intracellular recognition of pathogens (Singh et al. 2012). Recent results underlined the importance of TLRs and particularly TLR-2 in the recognition of *Leishmania* through their LPG (Becker et al. 2003; de Veer et al. 2003; Flandin et al. 2006) In addition, together with phagocytes, NK cells represent the first line of defense against *Leishmania* by cytolytic destruction of infected cells and secretion of pro-inflammatory cytokines such as IFN-γ and TNF-α. In patients, NK cell number and activity have been mainly associated with protection against or healing of disease. Patients with active cutaneous leishmaniasis have a reduction in the frequency of peripheral NK cells and an increased frequency following immunotherapy (Bogdan 2012).

In humans, the role of Th1 and Th2 $CD4^+$ T cells and the cytokines they produce is not yet well understood. However, since the role of Th1 and Th2 cell responses has been clearly associated with resistance and susceptibility in the murine model of infection with *L. major*, respectively, the comparison of immunological responses in localized cutaneous leishmaniasis (LCL), mucocutaneous leishmaniasis (MCL), and diffuse cutaneous leishmaniasis (DCL) has provided an opportunity to identify such differences in human leishmaniasis (Tacchini-Cottier et al. 2012). Overall, a predominance of Th1 cytokine or high levels of Th2 cytokines were detected in lesions of LCL and DCL patients, respectively, while MCL patients exhibited a Th1/Th2 mixture of cytokine patterns (Bourreau et al. 2003).

It is generally considered that LCL patients present a heterogeneous cellular immune response with a predominant Th1-type immune response but the type of T cells (CD4 or CD8) and their function within the lesions are still debated. Indeed, *Leishmania*-specific $CD4^+$ and $CD8^+$ T cells have been demonstrated both in skin lesions and in PBMC during the acute phase of infection due to New and Old World dermotrophic *Leishmania* such as *L. braziliensis* (Da-Cruz et al. 1994), *L. major* (Gaafar et al. 1999), and *L. aethiopica* (Maasho et al. 1998). Furthermore, an expansion of both $CD4^+$ and $CD8^+$ T cells has been detected in PBMC from healthy subjects stimulated with *Leishmania* such as *L. major* (Rogers and Titus 2004), *L. aethiopica* (Maasho et al. 2000), *L. guyanensis* (Bourreau et al. 2002), and *L. amazonensis* (Russo and Higgins 1999).

A summary of the current understanding of the Th1/Th2 cell response in different forms of cutaneous leishmaniasis is presented in Table 2.

$CD8^+$ T cells also actively participate in the immune response to cutaneous infections in humans (Stager and Rafati 2012). As observed in the low-dose model in mice, *L. major* also induces Th1 and $CD8^+$ T cells in human patients and both

Table 2 Summary of Th-cell differentiation in different clinical forms of cutaneous leishmaniasis

LCL		MCL	DCL	PKDL
Healing	Non-healing			Non-healing
Th1 > Th2	Th2 > Th1	Th1/Th2	Th2	Th1/Th2 or Treg

responses are associated with disease resolution (Sacks and Noben-Trauth 2002). $CD8^+$ T cells were not only observed in large numbers in the lesions of *L. major* patients during the acute phase, but also during the healing process (Nateghi Rostami et al. 2010). The exact role of $CD8^+$ T cells in *L. major* infections in humans is not yet known. A major correlate of protection appears to be the high amounts of IFN-γ produced by $CD8^+$ T cells after restimulation. In vitro studies have also demonstrated that *Leishmania*-specific CTLs are generated upon coculturing human naïve T cells with antigens from *L. amazonensis* promastigotes and IL-12 or with *L. major* parasites (Faria et al. 2009). Moreover, increased granzyme B activity was also found in patients with an active infection and was associated with a good prognosis. In this study, in vitro cytotoxicity by peripheral blood lymphocytes on *L. major*-infected macrophages appeared to be mediated by granzyme B, suggesting that CTL activity may be involved in controlling the parasite growth (Bousoffara et al. 2004). It is possible, though, that the cytotoxic activity not only contributes to disease clearance, but also to the development of skin ulceration, as observed in *L. major*-infected mice.

Current Treatments

There are no satisfactory treatments for any form of CL. Due to lack of a satisfactory treatment for CL caused by *L. major*, WHO recommends no treatment for uncomplicated cases with less than five lesions if lesions are not close to vital organs. This is because 50–60 % of cases self-heal within 6 months and over 90 % in less than a year and patients who recover from *L. major* infection generally have a strong specific immune response and are protected. Available treatments include pentavalent antimonials: sodium stibogluconate (Pentostam, Stibanate) and meglumine antimoniate (Glucantime) which have been used for over half a century (Tiuman et al. 2011). Antimonials are toxic and their efficacy in many regions has diminished. Therapy with antimonials requires daily painful intramuscular injections for 3–4 weeks or intralesional 6–8 injections for 4–6 weeks. Alternative treatments include cryo-therapy and thermo-therapy alone or in combinations with antimonials. All are traumatic, antimonials are toxic and treatment is expensive. Different formulations of topical Paromomycin have been developed which have shown efficacy against *L. major*, but not *L. tropica* lesions. The recent formulation containing paromomycin and gentamycin showed 81 % cure rate (vs. 58 % for vehicle control), giving an efficacy of about 22 %) against *L. major* lesions tested in Tunisia (Ben Salah et al. 2013). This is not much different from the formulation of paromomycin with urea applied for 4 weeks compared with 2 weeks vs. placebo vehicle in trials in Iran (Asilian et al. 2003). We believe that these formulations should become available and used instead of no treatment recommended by WHO until better noninvasive, efficacious safe, and affordable treatments are developed. The "no treatment" recommended by WHO has proven to be difficult to implement, because patients demand treatment. If physicians don't give a treatment, then patients would apply unproven and sometimes harsh traditional treatments (acid, burning, etc.). The only formulation that has been on the market for more than a decade and used for *L. major* lesions is Leishcutan ® by Teva Pharmaceuticals, which contains methylbenzethonium chloride and causes irritation occasionally.

At present there is no satisfactory drug for CL caused by *L. tropica* in Iran and only 15–30 % of patients respond to antimonials. *L. tropica* isolates resistant to antimonial have been isolated from patients' refractory to treatment (Hadighi et al. 2007). Both in Afghanistan and Syria where CL is a major public health problem, *L. tropica* is the predominant species. Being an anthroponotic disease, lack of treatment will perpetuate transmission as it has been seen due to political situations in these two countries and increase in cases.

Several treatment modalities have been tested on CL (Khatami et al. 2007), but unfortunately a major difficulty in evaluating the results from different studies on treatment of CL has been due to inadequacies and inconsistencies in the designs and conduct of clinical trials. A collective attempt (WHO/TDR, WRAIR, DNDi) has been made to standardize clinical trial protocols for treatment of CL (Olliaro et al. 2013).

Human Vaccine

Iran is the most suitable country for clinical development of a vaccine against CL. This is because many sites of stable high incident areas exist which have been extensively studied; several studies and trials from phase-1 to phase-3 have been conducted using killed *L. major* plus BCG (Alimohammadian et al. 2002; Bahar et al. 1996; Mahmoodi et al. 2003; Modabber et al. 2010; Momeni et al. 1999; Noazin et al. 2008; Sharifi et al. 1998).

Many GCP trained scientists are involved in leishmaniasis work, and most importantly and unique to Iran, a live challenge system "Leishmanization" has been standardized by using frozen stabilates and tested (see below).

Leishmanization

Inoculation with the pus of an active CL lesion on the buttocks of young children to prevent lesions on the face was a known practice during the Ottoman Empire. The Russian scientists evaluated this approach in a high-risk population of Uzbekistan (Marzinowsky EI 1924; Moshkovsky 1942; Senekji and Beattie 1941) . Later, when it was possible to grow *L. major* in cell-free culture, it became a practice up until late 1990s. Israeli scientists initiated a series of clinical trials using promastigotes grown in culture and the practice became known as "leishmanization" (Koufman et al. 1978). They determined that in order for leishmanization to be effective, the inoculum must produce an active lesion (called take). After several passages in vitro, the take rate dropped which made the vaccine inefficacious. In addition, due to side effects, the practice was abandoned.

Nadim and colleagues used leishmanization in a high endemic region of Iran and confirmed that in "takes" the incidence of disease was significantly lower than in non-takes and the efficacy was about 80 % (Nadim et al. 1983). Leishmanization was then used during the Iran–Iraq war where reportedly two million people were involved. The follow-up was not possible during the war, but the efficacy was estimated to be very high with only mild lesions developed from natural infection in those who received leishmanization and had a scar (takes). However, later there were at least 50 known cases that developed a large chronic lesion at the site of leishmanization that was very difficult to cure with standard antimonial treatment.

The inoculum for leishmanization has been standardized, and two trials have been done with consistently high take rate (Khamesipour et al. 2005).

There are several advantages in using leishmanization as challenge for evaluation of a new vaccine. Only a few volunteers are needed in an endemic focus compared to phase-2–3 trials, since the take rates are 90–100 % as compared with 2 to maximum 10 % against natural infection per year; it is far more economical. It is possible to fully study the immune responses before challenge unlike field efficacy studies, therefore having a better chance of identifying surrogate markers of immunity. The duration of trial is shorter since all volunteers will be challenged at the same predetermined time which can also be varied to yield information on the duration of immunity. The challenge study however cannot replace field efficacy trials which are required by most regulatory authorities, however, because few volunteers are needed, there are no validated animal models, many vaccine candidates can be tested, and then the best can be selected for field efficacy trials. The most important advantage which will benefit volunteers who participate in a vaccine study is the fact that if the new vaccine candidate would not protect the volunteers, the leishmanization challenge would, unlike for example malaria challenge studies. There are two important points that must be considered: First, leishmanization is only possible now using *L. major*. Due to numerous common antigens between *Leishmania* species, and because usually vaccines are constructed with conserved antigens to be effective against many species, leishmanization may be a good tool to select vaccines against other species of *Leishmania*. Second, at least in mouse model, needle challenge (i.e., leishmanization) is not equivalent to sand fly challenge. Vaccinated mice were protected against needle challenge but not sand fly challenge (Peters et al. 2009). This phenomenon may not be true for humans, but needs to be evaluated.

With all the information generated on biology, genetics of *Leishmania*, host–parasite relations in leishmaniasis, immunology and mechanism of protection, existence of animal models, and human challenge system, why don't we have an effective vaccine for any form of leishmaniasis other than leishmanization which cannot be implemented in large scale? Many of the reasons have been reviewed recently (Modabber 2010). Lack of attention by Pharma and local governments, very limited resources, and lack of commitment for developing a vaccine for bridging research to development are some of the factors. Hotez, P. in his Presidential inaugural speech to ASTMH in 2012 drew attention to leishmaniasis vaccine and suggested to engage in collaboration with Iranian scientists, which would also stimulate better relations between Western countries and Iran, while facilitating vaccine development.

Animal Studies

Short Historical Background in Murine Model

One of the most widely used experimental hosts of *Leishmania* has been the mouse. Nowadays it is clear that the outcome of *Leishmania* infection in mice varies based on parasite strain, genetic background of mice, and the site of inoculation. The genetic background of mice which control the outcome of infection with *L. major*, were shown independently by Behin (Behin et al. 1979) and Nasseri and Modabber (1979). The striking observation was that in six strains (CBA, AKR/J, AKR/cu, C57BL/6, A/J, and C3H) *L. major* inoculation intradermally at the base of the tail produced a localized cutaneous lesion which in most strains healed within 3–4 months, but in contrast, in BALB/c it produced a progressive disease which became generalized and killed 100 % of inoculated animals (Nasseri and Modabber 1979). They showed also that BALB/c failed to produce any delayed hypersensitivity to leishmanin (crude *Leishmania* antigens), tested by footpad reaction (leishmanin skin test, LST), whereas A/J mice in which the lesion healed a strong LST response was demonstrated. Behin et al., also showed that there is different activation threshold for killing of the parasite in vitro and it is higher for cells of non-healer origin (BALB/c mice) (Behin et al. 1979).

Later it was shown that indeed infection in all strains tested (including the resistant strains C57Bl/6) was generalized and live parasites could be isolated from spleen and lymph nodes long after the cutaneous lesion had healed (Leclerc et al. 1981). The mechanism of survival of the parasites in resistant strains still remains elusive; however based on a large induction of monocyte/macrophage precursors during infection which could be infected, the "Safe Target" hypothesis was proposed since these undifferentiated macrophages or other cells infected by *Leishmania* could not be activated to kill the parasite (Mirkovich et al. 1986; Modabber 1987).

The importance of the persistence of live parasites after recovery became apparent in asymptomatic individuals in leishmaniasis endemic foci, who following immune-suppression by HIV or anticancer cytotoxic drugs developed a generalized leishmaniasis. Hence leishmaniasis became known as an opportunistic infection in

HIV-infected patients who had been exposed to *Leishnmania* often without symptoms. Another interesting finding in the *L. major*-BALB/c model was the similarity of signs and symptoms of infected mice to those of human VL patients caused by *L. donovani*. It was therefore of interest to study the apparent immune-suppression believed to be associated with VL. Interestingly, there was no generalized immunosuppression—particularly to thymus-independent antigens if in vivo responses of animals were analyzed (Colle et al. 1983). The apparent suppression to various antigens seen in in vitro responses using spleen cells was simply due to the dilution of lymphocytes and immune competent cells in the spleen by undifferentiated monocytic cells (Leclerc et al. 1982). Still there are some publications that show suppressive response to PHA in spleen cells harvested from infected animals without analyzing the composition of cells in this organ. By removing the irrelevant cells and adjusting the cell concentration on the basis of lymphocytes or looking at in vivo response, immune-suppression will not be seen. This explains why in spite of the dogma that VL is associated with immune-suppression; VL patients usually do not suffer from various infections except at very advanced stages and then it is usually superinfection with *Mycobacteria*, perhaps due to exhaustion of T-cell repertoire or strong Treg expansion.

Vaccine Studies in Animal Models

Different types of live vaccines such as attenuated vaccine (Daneshvar et al. 2010) and recombinant nonpathogenic parasite (Saljoughian et al. 2013) have been studied in dog and murine models, respectively. A large effort has been focused on gp63, which is a major surface protein on *Leishmania* promastigote, because of its protease activity and role in virulence (Badiee et al. 2009a, b, 2012, 2013). BALB/c mice vaccinated with recombinant gp63 encapsulated in cationic liposome with CpG ODN as an adjuvant induced a significant reduction in parasite burden, lowered IL-4 production, and increased Th1 response (Jaafari et al. 2007). Other *Leishmania* proteins that have been targeted for vaccination in the form of protein and DNA vaccination are Cathepsin L line cysteine proteinases (CPA, CPB, and CPC) (Khoshgoo et al. 2008; Rafati et al. 2000, 2001, 2002, 2006a, b). In general, naked DNA has a short half-life within cell as they are degraded by nucleases. Some alternative methods such as coating DNA with nano- or microparticles or gold particles protect plasmids from degradation and increase phagocytic uptake by professional APCs. In this regard, cysteine proteinases have been coated with nano-cationic solid lipid particles (cSLN) and shown to enhance protection in the BALB/c mice model. The used nanoparticle is applicable for both recombinant protein vaccination and DNA vaccination (Doroud et al. 2011a, b, 2011c). Besides all other vaccination strategies, today protective and therapeutic peptide-based vaccine concept has drawn attraction in the field of intracellular infections where multi-CD8 cytotoxic T-cell responses are crucial mediators of immunity. Recently by in silico manipulation and bioinformatics tools, six known proteins from *Leishmania*

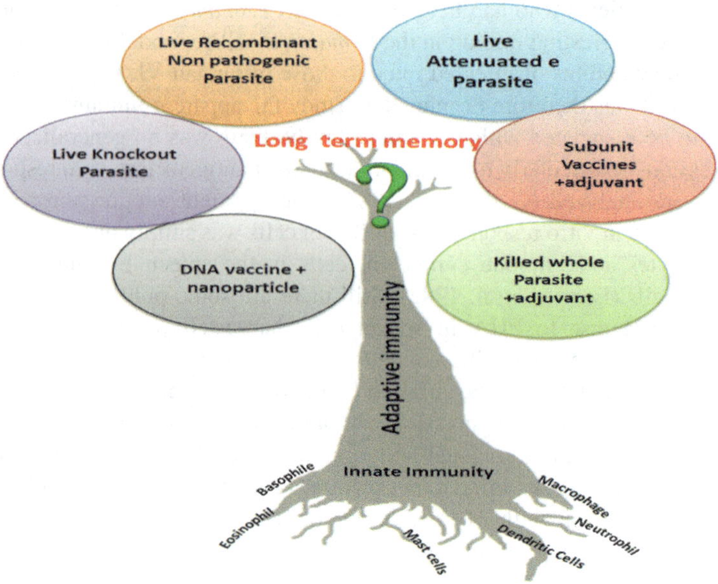

Fig. 6 Current candidate vaccine can be divided into live *Leishmania* parasite (Live knockout parasite, live recombinant nonpathogenic parasite, live-attenuated parasite), killed *Leishmania* preparation, and defined vaccines, including recombinant and DNA vaccine. The essential issue for all types of vaccine is creation of long-term immunity against parasite

(*L.*) *major* were screened for best HLA-A2 binding 9-mer peptides (Seyed et al. 2011). A few peptides from *L. major* Stress-Inducible Protein-1 (LmSTI-1) and Lipophosphoglycan Biosynthetic Protein-3 (LPG-3) were immunogenic after in vitro assays with PBMC from cutaneous leishmaniasis recovered individuals. In HLA-A2 transgenic mice, almost all peptides were able to induce specific CD8 T-cell responses in vivo, indicating the adequate mouse T-cell repertoire before challenge (unpublished data by Seyed et al.). This type of vaccination strategy is another modern view for vaccine development which needs further investigation.

In murine studies, several experimental vaccines are effective, but most of them wane with time. These studies suggest that we need to know more about the requirements for maintenance of anti *Leishmania* immunity in order to better define the correlates of protection. During the past two decades different approaches have been employed such as leishmanization, killed whole parasite, live-attenuated parasite, subunit and DNA vaccines. An important factor for maintenance of immunity is generally believed to be the presence of small number of live parasite in the host. However, live replicating parasites or just persistent antigens are believed to be important for the maintenance of effector memory like T cells but not for central memory T cells. An ideal anti *Leishmania* vaccine must maintain constant turnover of *Leishmania*-specific memory cells in vaccinated host; otherwise repeated booster injections would be required (Okwor et al. 2012; Fig. 6).

Summary

Both anthroponotic (ACL) and zoonotic CL (ZCL) are an important public health problem in this region. The knowledge of the disease goes back several centuries, yet there are no effective and sustainable control measures. Vector and reservoir controls have been tried with short-term moderate success in reducing incidence in Syria, Saudi Arabia, Uzbekistan, and Iran, but they are not sustainable since they require staff, continuous application, and monitoring which are costly. There are no efficacious treatments for either forms and the no treatment recommendation of simple uncomplicated ZCL by WHO cannot be implemented due to patients' demand for treatment. Other than leishmanization which was practiced until recently (Uzbekistan) and in 1980s in Iran, there is no vaccine available. First-generation vaccine (killed *L. major* mixed with BCG as adjuvant) was tested in Iran against ZCL and ACL and in Sudan against VL with minor if any prophylactic efficacy. Addition of alum to this vaccine gave very encouraging results in preliminary proof-of-concept immuno-chemotherapy trial in persistent post kala azar dermal leishmaniasis. This work supported by TDR/WHO was not pursued due to the enthusiasm regarding well-defined second-generation vaccines which were appearing in the horizon, i.e., Leish-111f+MPL-SE. Unfortunately so far there has not been a breakthrough. The most economically reasonable option for control of CL in this region besides safe, effective, and affordable drugs for early treatment of ACL is an effective vaccine. It is up to "Leishmaniacs" in this region to convince vaccine pharmas that an affordable vaccine would find a considerable market in this region.

Furthermore, due to knowledge expansion, dissemination of information is essential in leishmaniasis. Successful vaccine and new drug development will only advance if critical information is shared among researchers and policy makers. During the past two decades, WHO-TDR played important roles in strengthening partnership and involving hundreds of scientists in MENA region. In addition, there have been some attempts for networking, collaboration, and testing new ideas in research on *Leishmania* and leishmaniasis, i.e., Pasteur Institutes and NIH–Canada joint Collaborations; however, more would be needed, particularly now since much burden is imposed due to political situations in this region. In addition to research, collaboration is needed with major pharmas for development of vaccine and drug. The region provides golden opportunities for efficacy studies against CL because of extensive experience in trials of ICH-GCP standards: i.e., in Tunisia, topical paromomycin ointments, and in Iran, several vaccine and drug trials.

The way forward:

1. Involve vaccine industries in those countries where leishmaniasis is a major public health problem.
2. Form not-for-profit Private–Public Partnership and seek assistance (technical and financial) from International industries, donor countries, and philanthropists for leishmaniasis vaccine similar to those working on malaria, TB, etc.
3. Leishmaniasis vaccine requires Advocacy, as has been effective for HIV/AIDS, malaria, and tuberculosis. Donor agencies and national granting authorities must be approached to fund developmental work.

4. Start by testing for therapeutic potentials of candidate vaccines. It takes less funds and much shorter time to evaluate the therapeutic potential of a vaccine than test its prophylactic efficacy. It must be noted, however, that vaccines that can be effective in immune-chemotherapy may not be good prophylactic vaccines, i.e., killed Alum-*L. major* plus BCG (Musa et al. 2008) or killed *L. amazonensis* (Machado-Pinto et al. 2002).
5. Seek support from governments of endemic countries. If endemic country governments become involved and provide financial and infrastructure support, then the cost of vaccine development would be reduced considerably and they will become owners which would facilitate implementation of a vaccine/drug when developed.
6. We need socioeconomic analyses of CL to document impacts of CL in the region.

Acknowledgment We thank WHO for permission to reproduce Figs. 2–5 as indicated. This work was funded by the National Science Foundation of Iran (grant number 91053539) and Pasteur Institute of Iran to SR.

References

Adler S, Ber M (1941) The transmission of L. tropica by the bite of Phlebotomus papatasi. Indian J Med Res 29:803–809

Afonso L, Borges VM, Cruz H, Ribeiro-Gomes FL, DosReis GA, Dutra AN, Clarêncio J, de Oliveira CI, Barral A, Barral-Netto M, Cl B (2008) Interactions with apoptotic but not with necrotic neutrophils increase parasite burden in human macrophages infected with *Leishmania* amazonensis. J Leukoc Biol 84(2):389–396

Alimohammadian MH, Kivanjah M, Pak F, Gaznavia A, Kharazmi A (1993) Evaluation of the efficacy of Iran leishmanin and comparison with leishmanins from Wellcome (UK) and Roma (Italy) in cured cutaneous leishmaniasis patients. Trans R Soc Trop Med Hyg 87(5):550–551

Alimohammadian MH, Khamesipour A, Darabi H, Firooz A, Malekzadeh S, Bahonar A, Dowlati Y, Modabber F (2002) The role of BCG in human immune responses induced by multiple injections of autoclaved *Leishmania* major as a candidate vaccine against leishmaniasis. Vaccine 21(3–4):174–180

Alvar J, Velez ID, Bern C, Herrero M, Desjeux P, Cano J, Jannin J, den Boer M (2012) Leishmaniasis worldwide and global estimates of its incidence. PLoS One 7(5):e35671

Antunes CM, Mayrink W, Magalhaes PA, Costa CA, Melo MN, Dias M, Michalick MS, Williams P, Lima AO, Vieira JB et al (1986) Controlled field trials of a vaccine against New World cutaneous leishmaniasis. Int J Epidemiol 15(4):572–580

Asilian A, Jalayer T, Nilforooshzadeh M, Ghassemi RL, Peto R, Wayling S, Olliaro P, Modabber F (2003) Treatment of cutaneous leishmaniasis with aminosidine (paromomycin) ointment: double-blind, randomized trial in the Islamic Republic of Iran. Bull World Health Organ 81 (5):353–359, Epub 2003

Badiee A, Jaafari MR, Khamesipour A, Samiei A, Soroush D, Kheiri MT, Barkhordari F, McMaster WR, Mahboudi F (2009a) Enhancement of immune response and protection in BALB/c mice immunized with liposomal recombinant major surface glycoprotein of *Leishmania* (rgp63): the role of bilayer composition. Colloids Surf B Biointerfaces 74(1):37–44

Badiee A, Jaafari MR, Khamesipour A, Samiei A, Soroush D, Kheiri MT, Barkhordari F, McMaster WR, Mahboudi F (2009b) The role of liposome charge on immune response generated in BALB/c mice immunized with recombinant major surface glycoprotein of *Leishmania* (rgp63). Exp Parasitol 121(4):362–369

Badiee A, Khamesipour A, Samiei A, Soroush D, Shargh VH, Kheiri MT, Barkhordari F, Robert Mc Master W, Mahboudi F, Jaafari MR (2012) The role of liposome size on the type of immune response induced in BALB/c mice against leishmaniasis: rgp63 as a model antigen. Exp Parasitol 132(4):403–409

Badiee A, Heravi Shargh V, Khamesipour A, Jaafari MR (2013) Micro/nanoparticle adjuvants for antileishmanial vaccines: present and future trends. Vaccine 31(5):735–749

Bahar K, Dowlati Y, Shidani B, Alimohammadian MH, Khamesipour A, Ehsasi S, Hashemi-Fesharki R, Ale-Agha S, Modabber F (1996) Comparative safety and immunogenicity trial of two killed *Leishmania major* vaccines with or without BCG in human volunteers. Clin Dermatol 14(5):489–495

Becker I, Salaiza N, Aguirre M, Delgado J, Carrillo-Carrasco N, Kobeh LG, Ruiz A, Cervantes R, Torres AP, Cabrera N, González A, Maldonado C, Isibasi A (2003) *Leishmania* lipophosphoglycan (LPG) activates NK cells through toll-like receptor-2. Mol Biochem Parasitol 31(130):65–74

Behin R, Mauel J, Sordat B (1979) *Leishmania tropica*: pathogenicity and in vitro macrophage function in strains of inbred mice. Exp Parasitol 48(1):81–91

Ben Salah A, Louzir H, Chlif S, Mokni M, Zaatour A, Raouene M, Ismail RB, Dellagi K (2005) The predictive validity of naturally acquired delayed-type hypersensitivity to leishmanin in resistance to *Leishmania major*-associated cutaneous leishmaniasis. J Infect Dis 192 (11):1981–1987

Ben Salah A, Ben Messaoud N, Guedri E, Zaatour A, Ben Alaya N, Bettaieb J, Gharbi A, BelhadjHamida N, Boukthir A, Chlif S, Abdelhamid K, El Ahmadi Z, Louzir H, Mokni M, Morizot G, Buffet P, Smith PL, Kopydlowski KM, Kreishman-Deitrick M, Smith KS, Nielsen CJ, Ullman DR, Norwood JA, Thorne GD, McCarthy WF, Adams RC, Rice RM, Tang D, Berman J, Ransom J, Magill AJ, Grogl M (2013) Topical paromomycin with or without gentamicin for cutaneous leishmaniasis. N Engl J Med 368(6):524–532

Bogdan C (2012) Natural killer cells in experimental and human leishmaniasis. Front Cell Infect Microbiol 2:69

Bourreau E, Prévot G, Gardon J, Pradinaud R, Hasagewa H, Milon G, Launois P (2002) LACK-specific CD4(+) T cells that induce gamma interferon production in patients with localized cutaneous leishmaniasis during an early stage of infection. Infect Immun 70(6):3122–3129

Bourreau E, Gardon J, Pradinaud R, Pascalis H, Prévot-Linguet G, Kariminia A, Pascal L (2003) Th2 responses predominate during the early phases of infection in patients with localized cutaneous leishmaniasis and precede the development of Th1 responses. Infect Immun 71 (4):2244–2246

Bousoffara T, Louzir H, Ben Salah A, Dellagi K (2004) Analysis of granzyme B activity as a surrogate marker of *Leishmania*-specific cell-mediated cytotoxicity in zoonotic cutaneous leishmaniasis. J Infect Dis 189(7):1265–1273

Bray RS, Modabber F (2000) The history of leishmaniasis. In: Gilles HM (ed) Protozoal diseases. Hodder Arnold Publisher, New York, pp 414–419

Bray RS, Abdul-Rahim GF, Tajelin S (1967) The present state of leishmaniasis in Iraq. Proc Natl Acad Sci U S A 2:171–186

Colle JH, Truffa-Bachi P, Chedid L, Modabber F (1983) Lack of general immunosuppression during visceral *Leishmania tropica* infection in BALB/c mice: augmented antibody response to thymus-independent antigens and polyclonal activation. J Immunol 131(3):1492–1495

Da-Cruz AM, Conceição-Silva F, Bertho AL, Coutinho SG (1994) *Leishmania*-reactive CD4+ and CD8+ T cells associated with cure of human cutaneous leishmaniasis. Infect Immun 62 (6):2614–2618

Daneshvar H, Molaei MM, Kamiabi H, Burchmore R, Hagan P, Stephen Phillips R (2010) Gentamicin-attenuated *Leishmania* infantum: cellular immunity production and protection of dogs against experimental canine leishmaniasis. Parasite Immunol 32(11–12):722–730 de Veer MJ, Curtis JM, Baldwin TM, DiDonato JA, Sexton A, McConville MJ, Handman E, Schofield L (2003) MyD88 is essential for clearance of *Leishmania major*: possible role for lipophosphoglycan and Toll-like receptor 2 signaling. Eur J Immunol 33(10):2822–2831

Doroud D, Zahedifard F, Vatanara A, Najafabadi AR, Rafati S (2011a) Cysteine proteinase type I, encapsulated in solid lipid nanoparticles induces substantial protection against *Leishmania major* infection in C57BL/6 mice. Parasite Immunol 33(6):335–348

Doroud D, Zahedifard F, Vatanara A, Najafabadi AR, Taslimi Y, Vahabpour R, Torkashvand F, Vaziri B, Rafati S (2011b) Delivery of a cocktail DNA vaccine encoding cysteine proteinases type I, II and III with solid lipid nanoparticles potentiate protective immunity against *Leishmania major* infection. J Control Release 153(2):154–162

Doroud D, Zahedifard F, Vatanara A, Taslimi Y, Vahabpour R, Torkashvand F, Vaziri B, Rouholamini Najafabadi A, Rafati S (2011c) C-terminal domain deletion enhances the protective activity of cpa/cpb loaded solid lipid nanoparticles against *Leishmania major* in BALB/c mice. PLoS Negl Trop Dis 5(7):e1236

Dutta A (2008) Pursuit of medical knowledge: Charles Donovan (1863-1951) on kala-azar in India. J Med Biogr 16(2):72–76

Elgood C (1934) The early history of Baghdad boil. J R Asiat Soc 3:519–533

Faria DR, Souza PE, Durães FV, Carvalho EM, Gollob KJ, Machado PR, Dutra WO (2009) Recruitment of CD8+ T cells expressing granzymeA is associated with lesion progression in human cutaneous leishmaniasis. Parasite Immunol 31(8):432–439

Flandin JF, Chano F, Descoteaux A (2006) RNA interference reveals a role for TLR2 and TLR3 in the recognition of *Leishmania* donovani promastigotes by interferon-gamma-primed macrophages. Eur J Immunol 36(2):411–420

Gaafar A, Veress B, Permin H, Kharazmi A, Theander TG, el Hassan AM (1999) Characterization of the local and systemic immune responses in patients with cutaneous leishmaniasis due to *Leishmania major*. Clin Immunol 91(3):314–320

Hadighi R, Boucher P, Khamesipour A, Meamar AR, Roy G, Ouellette M, Mohebali M (2007) Glucantime-resistantLeishmania tropica isolated from Iranian patients with cutaneous leishmaniasis are sensitive to alternative antileishmania drugs. Parasitol Res 101(5):1319–1322

Jaafari MR, Badiee A, Khamesipour A, Samiei A, Soroush D, Kheiri MT, Barkhordari F, McMaster WR, Mahboudi F (2007) The role of CpG ODN in enhancement of immune response and protection in BALB/c mice immunized with recombinant major surface glycoprotein of *Leishmania* (rgp63) encapsulated in cationic liposome. Vaccine 25(32):6107–6117

Kaye P, Scott P (2011) Leishmaniasis: complexity at the host–pathogen interface. Nat Rev Microbiol 9:604–615

Khalil EA, El Hassan AM, Zijlstra EE, Mukhtar MM, Ghalib HW, Musa B, Ibrahim ME, Kamil AA, Elsheikh M, Babiker A, Modabber F (2000) Autoclaved *Leishmania major* vaccine for prevention of visceral leishmaniasis: a randomised, double-blind, BCG-controlled trial in Sudan. Lancet 356(9241):1565–1569

Khamesipour A, Dowlati Y, Asilian A, Hashemi-Fesharki R, Javadi A, Noazin S, Modabber F (2005) Leishmanization: use of an old method for evaluation of candidate vaccines against leishmaniasis. Vaccine 23(28):3642–3648

Khatami A, Firooz A, Gorouhi F, Dowlati Y (2007) Treatment of acute Old World cutaneous leishmaniasis: a systematic review of the randomized controlled trials. J Am Acad Dermatol 57 (2):335.e1–335.e29, Epub 2007

Khoshgoo N, Zahedifard F, Azizi H, Taslimi Y, Alonso MJ, Rafati S (2008) Cysteine proteinase type III is protective against *Leishmania* infantum infection in BALB/c mice and highly antigenic in visceral leishmaniasis individuals. Vaccine 26(46):5822–5829

Killick-Kendrick B (2010) Oriental sore: an ancient tropical disease and hazard for European travelers. Feature article, Welcome History 43:4–5

Koufman Z, Egoz N, Greenblatt CL, Handman E, Montilio B, Even-Paz Z (1978) Observations on immunization against cutaneous leishmaniasis in Israel. Isr J Med Sci 14(2):218–222

Leclerc C, Modabber F, Deriaud E, Cheddid L (1981) Systemic infection of *Leishmania tropica* (major) in various strains of mice. Trans R Soc Trop Med Hyg 75(6):851–854

Leclerc C, Modabber F, Deriaud E, Djoko-Tamnou J, Chedid L (1982) Visceral *Leishmania tropica* infection of BALB/c mice: cellular analysis of in vitro unresponsiveness to sheep erythrocytes. Infect Immun 37(3):895–902

Liu D, Uzonna JE (2012) The early interaction of *Leishmania* with macrophages and dendritic cells and its influence on the host immune response. Front Cell Infect Microbiol 2:83

Maasho K, Sanchez F, Schurr E, Hailu A, Akuffo H (1998) Indications of the protective role of natural killer cells in human cutaneous leishmaniasis in an area of endemicity. Infect Immun 66 (6):2698–2704

Maasho K, Satti I, Nylén S, Guzman G, Koning F, Akuffo H (2000) A *Leishmania* homologue of receptors for activated C-kinase (LACK) induces both interferon-gamma and interleukin-10 in natural killer cells of healthy blood donors. J Infect Dis 182(2):570–578

Machado-Pinto J, Pinto J, da Costa CA, Genaro O, Marques MJ, Modabber F, Mayrink W (2002) Immunochemotherapy for cutaneous leishmaniasis: a controlled trial using killed *Leishmania* (Leishmania) amazonensis vaccine plus antimonial. Int J Dermatol 41(2):73–78

Mahmoodi M, Khamesipour A, Dowlati Y, Rafati S, Momeni AZ, Emamjomeh M, Hejazi H, Modabber F (2003) Immune response measured in human volunteers vaccinated with autoclaved *Leishmania* major vaccine mixed with low dose of BCG. Clin Exp Immunol 134 (2):303–308

Marzinowsky EI SA (1924) Oriental sore and immunity against it. Trans R Soc Trop Med Hyg 18:67–69

Mirkovich AM, Galelli A, Allison AC, Modabber FZ (1986) Increased myelopoiesis during *Leishmania major* infection in mice: generation of 'safe targets', a possible way to evade the effector immune mechanism. Clin Exp Immunol 64(1):1–7

Modabber F (1987) A model for the mechanism of sensitivity of BALB/c mice to L. major and premunition in leishmaniasis. Ann Inst Pasteur Immunol 138(5):781–786

Modabber F (2010) Leishmaniasis vaccines: past, present and future. Int J Antimicrob Agents 36 (Suppl 1):58–61

Modabber F, Coler R, Reed S (2010) Vaccine against *Leishmania*. In: Levine MM, Dougan M, Nabel G, Good MF, Nataro J, Liu MA et al (eds) New generation vaccines, 4th edn. Informa Healthcare, New York, pp 790–807

Momeni Boroujeni A, Aminjavaheri M, Moshtaghian B, Momeni A, Momeni AZ (2013) Reevaluating leishmanin skin test as a marker for immunity against cutaneous leishmaniasis. Int J Dermatol 52(7):827–830

Momeni AZ, Jalayer T, Emamjomeh M, Khamesipour A, Zicker F, Ghassemi RL, Dowlati Y, Sharifi I, Aminjavaheri M, Shafiei A, Alimohammadian MH, Hashemi-Fesharki R, Nasseri K, Godal T, Smith PG, Modabber F (1999) A randomised, double-blind, controlled trial of a killed L. major vaccine plus BCG against zoonotic cutaneous leishmaniasis in Iran. Vaccine 17 (5):466–472

Moradin N, Descoteaux A (2012) *Leishmania* promastigotes: building a safe niche within macrophages. Front Cell Infect Microbiol 2012:2(121)

Moshkovsky CD (1942) The principle of re-inoculation in oriental sore in relation to prophylactic vaccination. Med Parasitol Parasit Dis 11:66–75

Musa AM, Khalil EA, Mahgoub FA, Elgawi SH, Modabber F, Elkadaru AE, Aboud MH, Noazin S, Ghalib HW, El-Hassan AM, Leishmaniasis Research Group/Sudan (2008) Immunochemotherapy of persistent post-kala-azar dermal leishmaniasis: a novel approach to treatment. Trans R Soc Trop Med Hyg 102(1):58–63

Nadim A, Javadian E, Tahvildar-Bidruni G, Ghorbani M (1983) Effectiveness of leishmanization in the control of cutaneous leishmaniasis. Bull Soc Pathol Exot Filiales 76(4):377–383

Nasseri M, Modabber F (1979) Generalized infection and lack of delayed hypersensitivity in BALB/c mice infected with *Leishmania tropica* major. Infect Immun 26(2):611–614

Nateghi Rostami M, Keshavarz H, Edalat R, Sarrafnejad A, Shahrestani T, Mahboudi F, Khamesipour A (2010) CD8+ T cells as a source of IFN-γ production in human cutaneous leishmaniasis. PLoS Negl Trop Dis 4(10):e845

Noazin S, Modabber F, Khamesipour A, Smith PG, Moulton LH, Nasseri K, Sharifi I, Khalil EA, Bernal ID, Antunes CM, Kieny MP, Tanner M (2008) First generation leishmaniasis vaccines: a review of field efficacy trials. Vaccine 26(52):6759–6767

Okwor I, Mou Z, Liu D, Uzonna J (2012) Protective immunity and vaccination against cutaneous leishmaniasis. Front Immunol 3:128

Olliaro P, Vaillant M, Arana B, Grogl M, Modabber F, Magill A, Lapujade O, Buffet P, Alvar J (2013) Methodology of clinical trials aimed at assessing interventions for cutaneous leishmaniasis. PLoS Negl Trop Dis 7(3):e2130

Peters NC, Egen JG, Secundino N, Debrabant A, Kimblin N, Kamhawi S, Lawyer P, Fay MP, Germain RN, Sacks D (2008) In vivo imaging reveals an essential role for neutrophils in leishmaniasis transmitted by sand flies. Science 15(321):970–974

Peters NC, Kimblin N, Secundino N, Kamhawi S, Lawyer P, Sacks DL (2009) Vector transmission of *leishmania* abrogates vaccine-induced protective immunity. PLoS Pathog 5(6):e1000484

Rafati S, Baba AA, Bakhshayesh M, Vafa M (2000) Vaccination of BALB/c mice with *Leishmania major* amastigote-specific cysteine proteinase. Clin Exp Immunol 120(1):134–138

Rafati S, Salmanian AH, Hashemi K, Schaff C, Belli S, Fasel N (2001) Identification of *Leishmania major* cysteine proteinases as targets of the immune response in humans. Mol Biochem Parasitol 113(1):35–43

Rafati S, Kariminia A, Seyde-Eslami S, Narimani M, Taheri T, Lebbatard M (2002) Recombinant cysteine proteinases-based vaccines against *Leishmania major* in BALB/c mice: the partial protection relies on interferon gamma producing CD8(+) T lymphocyte activation. Vaccine 20 (19–20):2439–2447

Rafati S, Zahedifard F, Nazgouee F (2006a) Prime-boost vaccination using cysteine proteinases type I and II of *Leishmania infantum* confers protective immunity in murine visceral leishmaniasis. Vaccine 24(12):2169–2175

Rafati S, Ghaemimanesh F, Zahedifard F (2006b) Comparison of potential protection induced by three vaccination strategies (DNA/DNA, Protein/Protein and DNA/Protein) against *Leishmania major* infection using Signal Peptidase type I in BALB/c mice. Vaccine 24 (16):3290–3297

Ribeiro-Gomes FL, Sacks D (2012) The influence of early neutrophil-*Leishmania* interactions on the host immune response to infection. Front Cell Infect Microbiol 2012:2(59)

Ribeiro-Gomes FL, Peters NC, Debrabant A, Sacks DL (2012) Efficient capture of infected neutrophils by dendritic cells in the skin inhibits the early anti-*Leishmania* response. PLoS Pathog 8(2):e1002536

Rogers KA, Titus RG (2004) Characterization of the early cellular immune response to *Leishmania major* using peripheral blood mononuclear cells from *Leishmania*-naive humans. Am J Trop Med Hyg 71(5):568–576

Russo DM, Chakrabarti P, Higgins AY (1999) *Leishmania*: naive human T cells sensitized with promastigote antigen and IL-12 develop into potent Th1 and CD8(+) cytotoxic effectors. Exp Parasitol 93(3):161–170

Sacks D, Noben-Trauth N (2002) The immunology of susceptibility and resistance to *Leishmania* major in mice. Nat Rev Immunol 2(11):845–858

Saf'janova VM (1985) Leishmaniasis in the USSR. In: Change K-P, Bray RS (eds) Leishmaniasis. Elsevier, Amsterdam

Saljoughian N, Taheri T, Zahedifard F, Taslimi Y, Doustdari F, Bolhassani A, Doroud D, Azizi H, Heidari K, Vasei M, Namvar Asl N, Papadopoulou B, Rafati S (2013) Development of novel prime-boost strategies based on a tri-gene fusion recombinant L. tarentolae vaccine against experimental murine visceral leishmaniasis. PLoS Negl Trop Dis 7(4):e2174

Senekji HA, Beattie CP (1941) Artificial infection and immunization of man with cultures of L. tropica. Trans R Soc Trop Med Hyg 34:415–419

Seyed N, Zahedifard F, Safaiyan S, Gholami E, Doustdari F, Azadmanesh K, Mirzaei M, Saeedi Eslami N, Khadem Sadegh A, Eslami Far A, Sharifi I, Rafati S (2011) In silico analysis of six known *Leishmania major* antigens and in vitro evaluation of specific epitopes eliciting HLA-A2 restricted CD8 T cell response. PLoS Negl Trop Dis 5(9):e1295

Sharifi I, FeKri AR, Aflatonian MR, Khamesipour A, Nadim A, Mousavi MR, Momeni AZ, Dowlati Y, Godal T, Zicker F, Smith PG, Modabber F (1998) Randomised vaccine trial of single dose of killed *Leishmania major* plus BCG against anthroponotic cutaneous leishmaniasis in Bam, Iran. Lancet 351(9115):1540–1543

Singh RK, Srivastava A, Singh N (2012) Toll-like receptor signaling: a perspective to develop vaccine against leishmaniasis. Microbiol Res 167(8):445–451

Stager S, Rafati S (2012) CD8(+) T cells in *Leishmania* infections: friends or foes? Front Immunol 3:5

Tacchini-Cottier F, Weinkopff T, Launois P (2012) Does T helper differentiation correlate with resistance or susceptibility to infection with L. major? Some insights from the murine model. Front Immunol 3:32

Tadjbakhsh H (2005) Al-Aghraz al-Tabi'a va al-mabahith al-Aliyya by Ismail Jurjani. University of Tehran, Tehran

Tiuman TS, Santos AO, Ueda-Nakamura T, Filho BP, Nakamura CV (2011) Recent advances in leishmaniasis treatment. Int J Infect Dis 15(8):e525–e532. doi:10.1016/j.ijid.2011.03.021

Weber E (1876) Etudes sur le clou de Bistra. Rec Mem Med Milit (III ser) 32:44–59

Visceral Leishmaniasis: Immune Mechanisms and New Insights in Vaccine Development and Control

Sarfaraz Ahmad Ejazi and Nahid Ali

Contents

Abstract Visceral leishmaniasis (VL), an emerging and sustainable fatal disease, claims significant proportion of lives predominantly in the marginalised areas of developing countries. The impact of existing interventions to control the disease is insufficient. Epidemics and resurgence of the disease can be correlated with expansion of the vector habitat, emergence of co-infections, poor socio-economic condition, mass migration due to natural calamities or civil war, and laxity in policymaking. Since vaccine is unavailable, early diagnosis and successful treatment

S.A. Ejazi • N. Ali (✉)
Infectious Diseases and Immunology Division, Indian Institute of Chemical Biology, 4 Raja S. C. Mullick Road, Kolkata 700032, India
e-mail: sarfaraz.ejazi@hotmail.com; nali@iicb.res.in

M.A. McDowell and S. Rafati (eds.), *Neglected Tropical Diseases - Middle East and North Africa*, Neglected Tropical Diseases, DOI 10.1007/978-3-7091-1613-5_6,

are vital for VL management. Complete treatment of post kala-azar dermal leishmaniasis (PKDL) patients and effective surveillance of asymptomatic individuals need implementation in VL control programme. In zoonotic VL, screening of animal reservoirs is of utmost important. Revolutionising research perspective is highly recommended to build upon our knowledge on parasite, its mode of action, and the immune status of the host during infection, to support the VL control programme. A multidisciplinary and comprehensive effort would be imperative among scientists, medical professionals, and policymakers in order to control and eliminate the disease.

Keywords Leishmaniasis • Visceral leishmaniasis • Clinical manifestations • Immunology • Vaccine • Control

Introduction

Leishmaniasis continues to be one of the major neglected tropical diseases, endemic in tropical, subtropical, and Mediterranean regions. The disease constitutes varied and complex clinical manifestations when an obligate protozoan parasite *Leishmania* infects human. According to the clinical heterogeneity leishmaniasis broadly confers as cutaneous leishmaniasis (CL), mucocutaneous leishmaniasis (MCL), and visceral leishmaniasis (VL). CL manifests as numerous skin lesions mainly on the uncovered body parts, whereas MCL shows disfiguring and devastated facial parts like nose and mouth (Chappuis et al. 2007). VL or in general kala-azar is the severe form of leishmaniasis, certainly fatal if not treated properly and promptly. This visceral form of disease is caused by *Leishmania donovani* in the Indian subcontinent and East Africa, where infection is anthroponotic (man to man) with the sole reservoir, humans (Giorgobiani et al. 2011). *Leishmania infantum* (syn. *Leishmania chagasi*) is predominantly the causative parasite for VL in South America, Central and West Asia, Europe, and the Mediterranean basin including Middle East and North Africa (MENA) (Sundar and Chakravarty 2012). In these regions the disease is zoonotic (animal to man) where reservoirs other than humans are present such as canines. The vectors for natural transmission of VL are served by the female sand fly of genus *Phlebotomus* for *L. donovani* and *Lutzomyia* for *L. infantum* (Giorgobiani et al. 2011). Sand fly harbours an elongated, flagellar, metacyclic promastigote form of the parasite in their midgut and subsequently invades the parasite into the human bloodstream while feeding. Parasite resides initially in the peripheral macrophages and while in the cellular host transforms into oval, non-flagellated and non-motile amastigote form. Sand fly when sucks the blood meal of infected humans or canines gets infected by these amastigotes. Inside the gut, amastigote replicates into promastigote and comes to the saliva of sand fly and hence continues the transmission cycle.

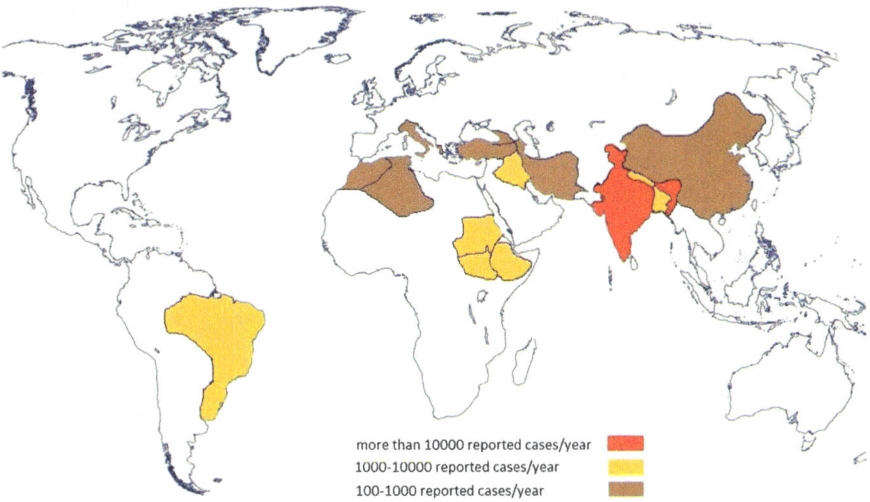

Fig. 1 Country wise reported cases of visceral leishmaniasis per year (Alvar et al. 2012). Copyright: © 2012 Alvar et al. This is an open-access article, free of all copyright, and may be freely reproduced, distributed, transmitted, modified, built upon, or otherwise used by anyone for any lawful purpose. The work is made available under the Creative Commons CC0 public domain dedication

Epidemiology of VL Infection

VL is endemic in over five continents of the world having maximum disease burden in South Asia followed by MENA region (including Sudan), East Africa, and South Americas (Fig. 1). After 20 years of leishmaniasis update, World Health organization (WHO) convened the meetings and collected the disease information from country representatives and literature search in between 2007 and 2010. Around 0.2 billion people live with a risk of VL infection globally with estimated 0.2–0.4 million cases occurring each year (Alvar et al. 2013). Deaths due to leishmaniasis have been tentatively estimated to be 20,000–40,000 per year (Alvar et al. 2012). In the last 5 years more than 58,000 VL cases per year have been officially counted (Ejazi and Ali 2013). Endemicity of VL is found in more or less 80 countries of the world which are climatically favourable for vector habitat and socially marginalised areas within the developing nations (Jha 2006). More than 90 % of the global VL is shared by the eight countries: India, Bangladesh, Sudan, Brazil, Ethiopia, South Sudan, Iraq, and Nepal. India with the highest reported cases each year (about 35,000/year) bears approximately 60 % of world's burden of VL (Croft et al. 2006). Morocco, Iran, Georgia, and Algeria are the second most affected countries in the MENA region with 100–1,000 reported VL cases annually (Alvar et al. 2012). Over the last two decades, geographical spread of the disease has been shown in newer areas of the world such as Southern Europe, South West Asia, and Central Africa (Palatnik-de-Sousa and Day 2011). HIV/VL co-infections contributed largely to increase in the incidence of VL as probability of the disease got higher (100–2,320 times) in the immune suppressed HIV (Alvar et al. 2008). WHO has reported 1.1 million new cases of HIV/VL co-infection which is approximately 2–12 % of all VL cases (Ejazi and Ali 2013). Most of the reported HIV/VL co-infections are being observed from South Western Europe (Sundar and Chakravarty 2012). Incidence of co-infections is also being reported progressively in the Indian subcontinent, Brazil, and sub-Saharan Africa which were already endemic for VL with increasing emergence of HIV (Sundar and Chakravarty 2012). HIV/VL co-infections are considerably increasing due to population interference between VL endemic regions and the HIV pandemic areas.

Clinical Manifestations in VL

In humans, following an incubation period (about 2–6 months), symptoms of the disease appear when parasite is established in the visceral organs such as spleen, lymph nodes, bone marrow, and the liver. The clinical symptoms of VL are characterised by persistent systemic infection including prolonged fever, weight loss, fatigue, loss of appetite, cough, abdominal pain, oedema, and diarrhoea as well as enlargement of spleen, lymph nodes and liver, pancytopenia, anaemia, and hypergammaglobulinaemia (Chappuis et al. 2007). Symptoms of the disease may vary in different endemic areas depending upon infective species, population race, and also the immune profile of infected host. For example, enlargement of the lymph nodes is a common feature in Sudan but is not commonly found in the Indian subcontinent. Irregular and long-lasting fever is the usual scenario in kala-azar patients often presented with rigour and chills. During the advance stage of the disease enlarged spleen and liver cause concomitant abdominal pain. Haematopoiesis in the early infection is normal but later on affected as life span of leucocytes and erythrocytes gets reduced (WHO 1984). Persistent enlarged spleen is also responsible for destruction of red blood cells (RBCs) leading to anaemia-associated fatigue and weakness. Production of prothrombin is decreased in the liver which may cause mucosal haemorrhage when associated with thrombocytopenia. Oedema is caused due to hypoalbuminaemia. Diarrhoeal symptoms may occur due to parasitisation of the intestine leading to ulceration and secondary enteritis. In the advanced stage of the disease secondary bacterial infections like tuberculosis, pneumonia, and dysentery and massive bleeding could be the reason of death (Malla and Mahajan 2006).

Infection with *L. donovani* or *L. infantum* does not always result in clinical sign and symptoms. Instead a population retains the parasite without disease manifestation. Although the reason for this natural protection is not well understood it is suggested that immunological status of the host may play an important role. However, at any stage of the infection if immunity weakens this asymptomatic or subclinical VL can emerge as a full-blown disease. Same is the case with canine VL that also shows a great number of asymptomatic dogs within the population. Both human and canine having asymptomatic infection without showing pathology serve as reservoirs for transmission of the disease by sand fly.

Despite the successful course of treatment, VL (mainly due to *L. donovani* infection) may also result in a skin-related disease called post kala-azar dermal leishmaniasis (PKDL). Maculopapular or nodular rashes and ulceration on the skin are the common symptoms in PKDL, which is often confused with other skin diseases like leprosy and vitiligo. About 5–20 % of treated VL patients in the Indian subcontinent develop PKDL usually after 2–7 years period whereas 50–60 % in Sudan show PKDL simultaneously or within 6 months of VL (Salotra and Singh 2006). Moreover, PKDL was also found in individuals who never had prior history of VL, thus suggesting their subclinical infection. Though PKDL is not a fatal disease but as the skin lesions of the infected host harbour invasive parasites, they can serve as a reservoir for VL transmission.

Diagnosis of VL

As the parasite invades the reticulo-endothelial system, detection of the parasite microscopically from tissues of spleen, bone marrow, or lymph nodes is generally used as the gold standard test. Parasitological examination of splenic aspirates is more common in the Indian subcontinent and East Africa, whereas bone marrow aspiration is a usual practice in the MENA regions (Guerin et al. 2002). Direct agglutination test (DAT), enzyme-linked immunosorbent assay (ELISA), and strip tests are some of the principal antibody detection tests presently in use for the diagnosis of VL in different endemic regions. DAT has been routinely used in many regions of VL endemicity. However, differences in antigen quality, long incubation time, need for serial serum dilution, and variations in the cut-off titre across the centres limited its applicability. Detection of serum antibody through ELISA is a frequently used technique for VL diagnosis, but the sensitivity and specificity of ELISA significantly depend on the coated antigen. Crude antigens showed cross reactivity so several recombinant antigens have been cloned for VL diagnosis. rK39 which is a 39 amino acid *L. chagasi* peptide was found very appealing for VL diagnosis. rK39-ELISA has been further evolved into a rapid immunochromatographic strip test in which recombinant antigen is impregnated onto nitrocellulose membrane, such as rK39 strips which are commercially available and extensively used in the Indian subcontinent (Cunningham et al. 2012). The strips showed moderately to lower sensitivities and specificities in Sudan, Ethiopia, and Brazil (Ejazi and Ali 2013). However, rK39 strips are being extensively tested in the region other than the Indian subcontinent. Our lab has already reported an *L. donovani* promastigote membrane antigen (LAg)-based ELISA and dipstick for detection of serum antibodies of Indian and Brazilian VL and PKDL subjects (Saha et al. 2011). PCR gives very good specificity and sensitivity, but the test is subjective and requires a laboratory facility which is not available in field conditions (Srivastava et al. 2011). KAtex is the only urine-based diagnostic tool commercially available for VL though poor sensitivity and need of urine to be boiled are their limitations (Hatam et al. 2009). Thus there is a need for a better, easily affordable, simple, and rapid test for early diagnosis of the disease as a control measure.

Treatment Options for VL

Pentavalent antimonials, sodium stibogluconate (SSG), and meglumine antimoniate are being used as a first line of treatment for almost 70 years in VL therapeutics. They are extensively used for VL as well as PKDL treatment all over the world, except the Indian subcontinent where the parasite has become unresponsive to the drug. However, the adverse effects such as pancreatitis, cardiac, renal, and hepatic toxicity and arthralgia are associated with the drug (Sundar and Chakravarty 2010). Moreover, intravenous and intramuscular route of drug administration and long treatment regimen support need of a safe and oral drug, but the requirement is not yet fulfilled. Amphotericin B is a highly effective but toxic second line of treatment, given intravenously. Several liposomal formulations of this drug are also available commercially such as AmBisome that has already been registered in Brazil, India, and Egypt (Sundar et al. 2011). AmBisome is a comparatively safe drug, but their high cost is a big limitation for general treatment. However, WHO has provided the drug at lower price especially for developing countries, though it is still expensive and beyond the reach of poor people (Mondal et al. 2010). Miltefosine which is the first orally effective drug for VL is considered as a first line of treatment in antimony-resistant region such as India and Bangladesh (Sundar and Chakravarty 2012). However, it is a teratogenic drug and cannot be used on females of childbearing age (Dorlo et al. 2012). Several other anti-leishmanial drugs such as sitamaquine, pentamidine, and paromomycin are in different phases of clinical trials (Mondal et al. 2010). Combination therapy with already existing drugs of VL has shown great safety and efficacy. Combination of SSG and paromomycin has shown a great relief during VL epidemics of South Sudan in the early 1990s (Melaku et al. 2007). Drugs with several combinations are currently under trial mainly in the Indian subcontinent, Sudan, Ethiopia, and Brazil (http://clinicaltrials.gov/).

Immunology of VL

In order to develop vaccines for visceral leishmaniasis, comprehensive knowledge of the immune response and pathogenesis is particularly important. Resistance and susceptibility to VL infection have been determined by the genetic and immunological status of infected host and the species- and strain-specific variations in the infecting parasites. Synchronised interplay between innate and adaptive host immune components is crucial for defence against the parasite. Unlike CL, protective immune response for VL is not associated with a skewed Th1 polarisation, but rather characterised by a mixed Th1/Th2 immune response. Immunosuppressive conditions during active VL are mainly due to elevated levels of IL-4 and IL-10 and lower expression of IL-12 which downregulates IFNγ production and causes loss of lymphoproliferation (Bhattacharya and Ali 2013). The key role of IL-10 in VL progression has been established through experimental evidences though source of the cytokine is still not confirmed.

Humoral Response

Presence of elevated levels of *Leishmania* antigen-specific immunoglobulin isotypes IgG, IgM, IgE, and IgG subclasses is the characteristics of active human VL (Anam et al. 1999; Ghosh et al. 1995). IFNγ-mediated upregulation of IgG1 and IgG3 and stimulation of IgG4 secretion through IL-4 have been postulated in VL, which suggests Th1/Th2-type response (Saha et al. 2006). Prevalence of antigen-specific IgG subclasses was shown to be varied in different VL endemic regions, maybe due to differences in parasite genotypes and ethnic variations among the population. Study in India showed predominance of IgG1 and revealed that IgG3 is specifically associated with the disease (Anam et al. 1999). Moreover, it has been observed that IL-10 helps in the survival of B cells and proliferation of plasma cells to switch towards IgG1 and IgG3 (Caldas et al. 2005). Significant decrease in subclasses IgG2 and IgG3 and enhancement of IgG1 were revealed against leishmanial antigen in antimony-resistant patients from India (Anam et al. 1999). However, role of B cells and elevated antibody levels with either disease protection or pathogenesis is still debatable. Consensually, it has been observed that *Leishmania*-specific antibodies participate in progression of the disease rather than protection and mice lacking B cells were shown to be less susceptible to *L. donovani* infection (Galvao-Castro et al. 1984; Smelt et al. 2000). Again, prevalence of these antibodies in asymptomatic healthy individuals in VL endemic areas and persistence of antibodies years after cure suggest their probable role in protection (Kumar and Nylen 2012).

Apart from immunology, antibodies are extremely valuable in the diagnosis of VL. Major diagnostic tools such as DAT, ELISA, and rK39 strip tests are based on polyclonal antigen-specific antibody detection in the patient.

Innate Immunity

The bite of a sand fly propagates *Leishmania* in the peripheral blood along with the saliva which contains a variety of active immunomodulatory components performing vasodilation and inhibition of coagulation, thus altering the environment of the feeding site (Sacks and Kamhawi 2001). Several salivary molecules have also been observed to attract and recruit neutrophils and macrophages at the site of infection (Zer et al. 2001). Neutrophils are the earliest cells that reach the site of infection where they phagocytose the parasites. *Leishmania* succeeds to extend the life span of neutrophils and can survive for hours to days inside these cells. Infected neutrophils secrete chemokines such as CXCL8 or IL-8 which attracts further neutrophils to the infected site (Laufs et al. 2002). Anti-leishmanial activities of neutrophils were implicated by the mechanism of oxidative burst and through microbicidal proteins. On the other hand, *Leishmania* also induce the secretion of macrophage inflammatory protein-1β (MIP-1β) along with the saliva products of sand fly to recruit monocytes at the site of infection (van Zandbergen et al. 2004). Infected neutrophils undergo apoptosis and are phagocytosed by the macrophages through receptor-mediated endocytosis. This leads to the suppression of normal microbicidal activity of macrophages through transforming growth factor (TGF) β (Ribeiro-Gomes et al. 2004). Promastigotes outside the neutrophils exploit the complement-mediated opsonisation pathway for uptake through dermal macrophage receptors such as CR3. CR3-dependent mechanism facilitates phagocytosis, prevents oxidative burst, and inhibits IL-12 essential for cell-mediated immunity (Marth and Kelsall 1997). *Leishmania* parasites evade humoral immune response by populating in the phagolysosomes of macrophages where they modulate several signalling processes such as JAK2/STAT1 cascade, Protein Kinase C, and MAP Kinase pathways that favour their growth and survival in host cells (Shadab and Ali 2011). After phagocytosis, promastigotes retain themselves in the parasitophorous vacuoles of the macrophage. Transformation of promastigotes into amastigotes takes place in this acidic environment rich in hydrolases. Inside macrophages, *Leishmania* has evolved several adaptive mechanisms to survive such as inhibition of hydrolytic enzymes, calcium chelation, phagosome-endosome fusion, host signalling pathway, pro-inflammatory cytokine production, NO production, and oxidative burst (Cunningham 2002). *Leishmania* also secrete chemotactic factors that induce macrophages to produce several chemo-attractants for monocytes and macrophages such as CXCL2, CCL-2, and MCP-1 (van Zandbergen et al. 2002; Ritter and Moll 2000; Muzio et al. 2000). Macrophage also exerts leishmanicidal activity through iNOS-dependent manner in the spleen and liver; however, *L. donovani* induce arginase 1 to overcome their iNOS-dependent killing in an IL-10-mediated fashion (Bhattacharya and Ali 2013). Hence macrophages serve dual function during infection: act as an effector cell, preventing the early *Leishmania* growth, as well as provide residence for parasite replication too.

Previously, NK cells were considered superfluous for the control of VL. Recent studies suggested their role in both immunosuppression and immunoprotection. NK cells are an essential source of early IFNγ that can mount protective response through activating macrophages. In human VL, suppression in numbers and activity of NK cells has been demonstrated during chronic infections and recovery of the cells and its activity following treatment (Cenini et al. 1993). NK cells have been shown to reach the infected site as early as 12 h in *L. infantum*-injected C57BL/6 mice. TLR-dependent activation of DCs and IL-12 was found obligatory for NK cell activation (Liese et al. 2008). The immigration of NK cells at the site of infection was found to be linked with chemokines IP-10 or CXCL9 and CXCL10 and suggested the immunopathogenic role of these chemokines during active disease (Vester et al. 1999; Hailu et al. 2004). Through cell depletion and adaptive transfer experiments, it was shown that NK cells which express IL-10 antagonise the control of *L. donovani* infection in liver and spleen (Maroof et al. 2008). More studies are needed to address the role of NK cells in VL infection either beneficial or dispensable.

DCs also play an imperative role in phagocytosis and antigen processing and presentations during early *Leishmania* infection. *Leishmania*-specific IgG opsonised the infected parasites and, through Fcγ receptors of DCs, promote their engulfment (Peters et al. 1995). Antigen presentation by DCs and thus through MHC I and MHC II pathways leads to stimulation of $CD8^+$ and $CD4^+$ T cell responses, essential for resistance against *Leishmania* (Vanloubbeeck and Jones 2004). Large repertoires of pattern-recognition receptors (PRRs), such as toll-like receptor (TLRs), are expressed by DCs on their cell surface which recognises specific and conserved pathogen structures, PAMPs.

Pathogen-associated molecular patterns (PAMPs) are the conserved motifs present in a class of microbes, which are recognised by TLRs.

Activation of DCs through PRR induces the maturation of DCs into MHC class II antigen-presenting cells, leading to $CD4^+$ T cell activation. DCs can also endocytose the transfected dead cells and express the parasite antigen through MHC class I pathway by cross presentation, leading to the activation of $CD8^+$ T cells (Vanloubbeeck and Jones 2004). Myeloid DCs phagocytose *L. infantum* and exert cytotoxic effects by inducing NK cells to release IFNγ. Plasmacytoid DCs can also induce NK cells in later stage of the disease but are unable to phagocytose the parasites (Schleicher et al. 2007). However, it has also been observed that DCs secreted IL-6, regulating the expansion of IL-10 producing T cells during *L. donovani* infection, suggesting its immune suppressive role (Stager et al. 2006).

Adaptive Immunity

Th1/Th2 Paradigm

Suppressed cell-mediated immunity is the characteristic feature of VL, as patients of active VL show unresponsiveness to the Leishmanin skin test (LST) or Montenegro test. The protective or pathogenic condition of the disease is mediated by distinguished subpopulations of helper T cells and secreted proinflammatory and immunosuppressive cytokines from specific cluster of differentiations, mainly $CD4^+$ T cells. The immunosuppression in the active stage of the disease can be reversed by effective chemotherapy. The T cell subpopulations derived from PBMCs of active VL patients showed lesser $CD4^+$ T cells than $CD8^+$ T cells and vice versa in cured individuals (Saha et al. 2006). Protective immunity of murine *L. donovani* infection has been observed as Th1 driven with upregulation of IFNγ and IL-12. However, pathogenesis in human VL is not implicated as a clearly polarised Th2 response. Rather it is a mixed Th1/Th2 response. Both pro- and anti-inflammatory cytokines such as IFNγ, IL-12, TNFα, IL-6, IL-1, IP-10, IL-4, IL-10, IL-15, and TGFβ were found upregulated in the serum of active VL patients. Therefore, it has been suggested that establishment of the disease in human VL is not because of deprived secretion of pro-inflammatory cytokines rather their unresponsiveness (Hailu et al. 2004). IFNγ, a signature cytokine of Th1 cells, is potent macrophage stimulating factor leading to leshmanicidal activity. Enhanced IFNγ mRNA is found in the lymphoid organs such as spleen and bone marrow because antigen-specific lymphocytes are trapped in these organs (Banerjee et al. 2008; Nylen and Sacks 2007). Reports have also suggested the dual nature of this cytokine as IFNγ when coupled with TNFα promotes amastigote progression (Belkaid et al. 2001). In experimental VL, TNFα seems to be decisive both for *L. donovani* resistance and resolution as the cytokine exerts cytotoxic effect on the parasite and its receptors (TNFR) were related to VL pathogenesis (Medeiros et al. 2000). Clinical manifestations such as weakness, weight loss, and anaemia during active VL are suggested to mediate by upregulation of TNFα. IL-12 which is secreted by innate immune cells is the key player for generating Th1-stimulated IFNγ production. Absence of IL-12 showed downregulation of IFNγ, TNFα, and iNOS production ultimately leading to increase in parasite load (Adhikari et al. 2012). IL-4 which is secreted by a distinct Th2 cell population was once considered as marker for active VL. IL-4 acts against Th1 development and secretion of IFNγ so drives the early infection environment towards Th2 polarisation. In contrast, IL-4 also has its effect on differentiation and stimulation of other $CD4^+$ T cells, B cells, and macrophages. IL-10 a pleiotropic cytokine was initially regarded as Th2 cytokine. Later reports however suggested its source to be cells other than T cells, like macrophages. IL-10 is a foremost regulatory cytokine found upregulated in the serum of active VL patients, and elevated levels of its mRNA in different tissue aspirates such as bone marrow, lymph nodes, spleens, and PBMCs (Bhattacharya and Ali 2013). After successful course of treatment, the levels of IL-10 and IL-10 mRNA decrease (Ghalib et al. 1993; Kenney et al. 1998). IL-10 as an anti-inflammatory cytokine suppresses the pro-inflammatory immune condition by inhibiting macrophage and DC, thus promoting disease progression. It hinders macrophage responsiveness to the activation signal as well as downregulates CCR7 expression, thereby impairing influx of DCs to draining lymph nodes. In mice model rapid parasite killing and upregulation of IL-12, iNOS, and NO were observed after inhibition of IL-10 (Caldas et al. 2005; Murray et al. 2002, 2005). Therefore, IL-10 is clearly an immunosuppressive cytokine inhibition of which promotes effective Th1 response. Apart from IL-10, the role of cytokine TGFβ was also investigated in disease progression. It has been found that levels of IL-10 and TGFβ go down in response to chemotherapy during systemic VL, but retention of these cytokines was detected in some SAG-treated patients (Saha et al. 2007). It is suggested that in host, parasite promotes the conversion of latent TGFβ into active TGFβ as its surviving strategy (Saha et al. 2006). Consequently, human VL is characteristic of elevated Th2 response along with the insufficient and unresponsive Th1 response.

Regulatory T Cells (Tregs)

Tregs are a dedicated population of T cells that maintain homeostasis and self-tolerance by suppressing a range of immune cells such as Th1, Th2, and Th17 cells and APCs. Though the concept of suppressor cells is almost half a century old, Sakaguchi and his group in the mid-1990s reported $CD4^+$ T cells with IL-2 receptor α-chain ($CD25^+$) as potent regulatory cells (Sakaguchi et al. 1995). Further, discovery of forkhead box protein3 (Foxp3), transcription factor for Tregs, paved the way to identify this regulatory cell population and its role in disease conditions. However, CD25 and FoxP3 are not the definitive markers for these cells; rather many cell surface and nuclear markers along with the co-stimulatory molecules have been proposed as Treg marker like glucocorticoid-induced TNF receptor family related protein (GITR), cytotoxic T-lymphocyte antigen 4 (CTLA4), and low CD127 (Workman et al. 2009). The role of Tregs in immune regulation has been found beneficial in various autoimmune diseases, whereas disease progression has been reported by these cells in infectious diseases including *Leishmania*. Based on the site of generation, Tregs have been divided into thymus-derived natural Treg cells (nTregs) and antigen-specific induced or adaptive Treg cells (iTregs) derived in the periphery. $CD4^+CD25^+FoxP3^+$ nTregs when exported to peripheral environment has been proposed to suppress the effector T cell (Teffs) functions. Under antigenic stimulation by APCs, iTregs ($Foxp3^+$ or $Foxp3^-$) can be divided into IL-10 secreting type1 (Tr1) cells and TGFβ secreting T helper3 (Th3) cells (Bilate and Lafaille 2012). However, both subsets are induced by the same cytokines which they are secreting, i.e. IL-10 and TGFβ. Tregs show a variety of regulatory functions to suppress the immune cells. It has been suggested that Tregs suppress Teffs through direct cell-to-cell contact mechanism (McHugh et al. 2002). The anti-inflammatory cytokines such as IL-10, TGFβ, and IL-35 are expressed by the Tregs for their suppressive activity (Shimizu et al. 2002). Tregs also interact with DCs and can abrogate their antigen presentation and stimulation of Teffs. It has also been shown that Tregs can kill Teffs directly through secreting cytolysing granzyme B and perforin.

Earlier, involvement of Tregs was shown in disease progression of CL and frequency of these cells was found to increase with healing. However, role of Tregs in VL is still controversial, with further confusion in the source of IL-10 during infection. In BALB/c mice infected with *L. infantum* $CD4^+CD25^+FoxP3^+$ Tregs were found to be enriched in lymph nodes and spleen with enhanced levels of Treg-secreted IL-10 (Rodrigues et al. 2009). In a study on Indian kala-azar patients it was observed that $CD4^+CD25^+FoxP3^+$ Tregs neither accumulated in the spleen nor were a chief source of IL-10. Rather, $CD4^+CD25^-FoxP3^-$ cells were reported as a source of IL-10 in the pathogenesis of human VL (Nylen et al. 2007). The same group further assessed $CD4^+CD25^+FoxP3^+$ T cells in spleen and PBMCs of VL patients and found there were no changes in the frequencies of these cells during pre- and post-treatment when compared with PBMCs of endemic healthy controls (Maurya et al. 2010). These observations supported their earlier findings that $CD4^+CD25^+FoxP3^+$ Treg cells were not related to human VL. Contradictory to these results, previously Saha et al. had reported elevated levels of suppressive $CD4^+CD25^+$ cells in the PBMCs of pre-treated human VL as compared to the post-treatment (Saha et al. 2007). Recently a study with immunocompromised alymphoblastic mice showed increased $CD4^+Foxp3^+$ T cell persistence in the liver when inoculated with *L. donovani* promastigotes (Tiwananthagorn et al. 2012). Subsequently, reduction in both $CD4^+Foxp3^+$ T cells and overall parasite burden was observed when treated with anti-CD25. Recently, Rai et al. have reported a persistence of $CD4^+CD25^+Foxp3^+$ Tregs in the bone marrow of confirmed human VL patients and these cells were reported to be potent source of IL-10 and which downregulated the Teff cell activation IL-10 (Rai et al. 2012). Regulatory mechanism of Tregs is not well understood in infectious diseases; thus more investigations are needed to say the definitive role of Tregs in immunosuppressive disease conditions like VL.

Th17 Cells

Th17 cells are unique and an independent subset of T cells dedicated to produce IL-17A and IL-17F with co-expression of IL-22 and IL-21. Induction of Th17 cells from naive $CD4^+$ T cells has been demonstrated by a range of cytokines such as IL-6, TGFβ, IL-23, IL-1β, and IL-21, in various combinations in mice and human. Development of a distinct Th17 cell population in mice is induced by the synergistic effect of IL-6 and TGFβ, while IL-1β and IL-23 are responsible for their expansion. However, for development of human Th17 cells the role of TGFβ is controversial, whereas exclusive expression of CD161 was reported by Th17 cells. Distinct populations of Th17 cells express transcription factor retinoic acid-related orphan receptorγt (RORγt) and the chemokine receptor CCR6. Differentiation of Th17 cells were suggested to be driven through signal transducer and activator of transcription3 (STAT3) as a signalling component. IL-17 cytokines encourage the recruitment of APCs at the site of infection through induction of various cell types such as neutrophils to produce chemokines and cytokines, G-CSF and IL-8. IL-22 is IL-10 family cytokine which enhances the innate immunity to protect the tissues from damage, whereas IL-21 performs the autocrine function for Th17 cells.

Both protective and pathogenic roles for IL-17 have been reported from different disease conditions in leishmaniasis. Lack of consensus has also been observed within the species as in *L. braziliensis* infection IL-17 shown to be promoting pathogenesis in human MCL while offering protection from CL in human and mice (Boaventura et al. 2010; Novoa et al. 2011; Vargas-Inchaustegui et al. 2008). Varying reports have also been observed from studies on *L.major*-infected mice. IL-17 was reported to promote disease progression via regulation of neutrophils in the absence of IL-10. Enhancement of protective immunity has also been shown through IL-17 in CL (Lopez Kostka et al. 2009; Gonzalez-Lombana et al. 2013; Wu et al. 2010).

A study with IL-27R-deficient mice when infected with *L. major* showed enhanced IL-17 producing $CD4^+$ T cells associated with more severe lesions (Anderson et al. 2009). Immunosuppressive functions of IL-27 and IL-21 were demonstrated in human VL wherein IL-27 was found elevated in the plasma. IL-27 mRNA, IL-27R mRNA, and IL-21mRNA in the spleen of pre-treated VL patients were demonstrated to be enhanced as compared to post-treated biopsies (Ansari et al. 2011). The first detailed knowledge of Th17 cells in VL was reported by Pitta et al. in *L. donovani*-infected Sudanese patients. Differentiation of Th17 cells was found to be stimulated in *L. donovani* infection and their cytokines IL-17 and IL-22 independently and strongly protected from VL. Thus protective role of Th17 cells was observed in human VL and defects in these cells are supposed to encourage the risk of VL infection (Pitta et al. 2009). Recently in *L. donovani*-infected BABL/c mice, recombinant IL-17 was shown to restrain the parasite burden via generation of IFNγ and NO by the phagocytic cells (Ghosh et al. 2013). Picture of Th17 cells and their cytokine IL-17 in VL infection is however still unclear.

Immunity in PKDL vs. VL

Since PKDL is a sequel of VL, knowledge of PKDL immunology is of utmost importance to unravel the factors responsible for dermal outcome. The question as to why prevalence of PKDL is only in limited endemic areas of VL and in a very few individuals within the prior infected VL population remains unanswered. However, several experiments have been conducted and role of different cell types and cytokines investigated during PKDL. In humoral immunity during PKDL, elevated levels of IgG1, IgG3, and IgG4 were found in the serum of Indian PKDL patients (Ansari et al. 2008a; Ganguly et al. 2008). Initially in the early 1990s, Ghalib et al. reported mRNA of both IL-10 and IFNγ in PKDL lesions, suggesting both Th1 and Th2 type of response (Ghalib et al. 1993). Later on, predominance of IL-10 along with IFNγ was demonstrated in PKDL lesions. PBMCs of all PKDL patients increasingly responded to leishmanial antigen by the production of IFNγ (Ismail et al. 1999; Gasim et al. 2000). A study was conducted with follow up VL patients who later on emerged with or without PKDL. The patients who developed PKDL showed IL-10 in the keratinocytes and/or sweat glands as compared to the group who did not develop PKDL with the absence of IL-10 in the skin cells during VL (Gasim et al. 1998). Moreover, elevated levels of IL-10 and TGFβ were also found in leishmanial antigen-stimulated PBMCs' culture supernatant of PKDL patients as compared to VL, suggesting their probable role in reactivation of the disease in the form of PKDL (Saha et al. 2007). Role of TNFα has also been reported in PKDL pathogenesis wherein elevated levels of the cytokine and its mRNA were depicted in the lesions and serum of the patients as compared to VL (Ansari et al. 2006a). Furthermore, the ratio of intralesional TNFα and IL-10 was shown to be substantially higher when compared with VL (Ansari et al. 2008b). Elevated levels of immunoregulatory cell population as $CD4^+CD25^+$ were first reported by Saha et al. in the PBMCs of PKDL patients. Interestingly percent population of these cells were slightly less than pre-treated VL and significantly higher than the post-treated VL as well as healthy controls suggested its suppressive activity in PKDL (Saha et al. 2007). mRNA analysis of PKDL lesions revealed the upregulation of nTreg markers such as Foxp3, CD25, and CTLA4 along with IL-10 before treatment which receded in post-treatment lesions (Ansari et al. 2006b). Ganguli et al. have reported a different regulatory population during the disease which decreases after treatment. Peripheral $CD8^+CD28^-$ cells were found to be elevated and source of enhanced IL-10 production when stimulated with the antigen (Ganguly et al. 2010). Study of Th17 cells has revealed the concomitant expression of IL-17 and IL-23 during active PKDL. Enhanced expression of IL-17 mRNA and IL-23 mRNA was observed in the lesions with upregulation of these cytokines in antigen-stimulated culture supernatants of PBMCs. Moreover, IL-17 producing cells in the peripheral blood and the cytokine in plasma of pre-treated PKDL patients were reduced in post-treatment (Katara et al. 2012). Although efforts have been put during the last decade to understand the basic immunology of PKDL we are still lacking the key answers. Immunological research based on paired samples and investigations right from VL till the PKDL follow-up is needed to discover the bridge between these two different manifestations of the same infection.

Vaccine Candidates for VL

Vaccine development greatly benefits the control and management of infectious diseases. Some forms of leishmaniasis show natural resistance to reinfection. This provides the justification for the development of anti-leishmanial vaccines which should preferentially be safe and induce long-lasting immunity. The current strategy for vaccination is based on the understanding of parasite infection mechanism and anti-leishmanial immune responses of the host. Studies based on infected human and experimental models suggest a role for both innate and adaptive responses for anti-leishmanial immunity. Induction of macrophages, dendritic cells, $CD4^+$ and $CD8^+$ T lymphocytes are vital in *Leishmania* infection. It has been established that IFNγ producing Th1 immune response by leishmanial antigen-specific $CD4^+$ T cells is pivotal for resistance to leishmaniasis, so target of *Leishmania* vaccine is to promote these cells for protection. Most of the research and trials for *Leishmania* vaccine have been conducted with *Leishmania major* against CL. Live-attenuated parasite, defined parasitic proteins, and their DNA constructs have been evaluated for vaccination in VL, but we are still lacking the availability of an effective and safe vaccine for human VL. However, the severity associated with VL demands priority in developing vaccine against this lethal disease. It is not clear yet whether a single vaccine could be feasible for both forms of the disease, though some experiments suggested cross presentation between the *Leishmania* species. But indeed, any successes in CL vaccination will definitely benefit the VL immunisation research.

Live Parasite as Vaccine

Although there is no leishmaniasis vaccine commercially available today, the history of leishmaniasis vaccine is very old and has been traditionally used in the Middle East from ancient times. Some of the tribal societies there used to expose their children to sand fly bites to protect them from skin lesions in later stage of life (Handman 2001). The developmental background of leishmaniasis vaccine started from live vaccines, subsequently killed vaccines, and subunit vaccines and then reached to DNA vaccines. The concept of vaccination with live and virulent *Leishmania* parasite, termed leishmanisation, has been practised for centuries in the Middle East and central Asia where pus of cutaneous lesion from CL-infected individuals was deliberately injected to uninfected person (Nadim et al. 1983). In the early 1900s, when Nicolle and Manceau established the culture conditions for *Leishmania* promastigote to support the parasite growth in vitro, leishmanisation technique became more précised to deliver controlled infection. During 1970s and 1980s extensive leishmanisation trials were successfully carried out in Iran, Israel, and former Soviet Union where this strategy was found 100 % efficacious for large-scale prevention of CL (Khamesipour et al. 2005, 2006; Kellina 1981). A different approach using live, non-attenuated, non-virulent *Leishmania tarentolae* as vaccine was shown by Breton et al. in BALB/c mice against infectious challenge with *L. donovani* and found promising effectiveness of this candidate (Breton et al. 2005). But inoculation of live parasite is not acceptable as a twenty-first century vaccine because of several adverse effects such as persistence of lesions, lack of vaccine virulence quality control, immunosuppression, emergence of HIV in leishmaniasis endemic area, and of course the ethical issues; therefore leishmanisation has never been trialled for lethal human VL.

Although the two most successful vaccines, small pox and polio, are live pathogens; there is only one prophylactic live vaccine currently in use for leishmaniasis i.e. live virulent *L. major* combined with killed parasite. The vaccine is also registered in Uzbekistan (Mutiso et al. 2013).

First-Generation Vaccines

Safety concerns associated with live parasite vaccines subsequently forced the researchers to shift their focus towards safer alternatives of using killed *Leishmania* parasites called first-generation vaccines. 1940s was the beginning of trials conducted with killed *Leishmania* as vaccines mainly in the South America for CL. Before this, the strategy using killed *Leishmania* parasite had been proposed for therapeutic purposes in CL. Later on, Mayrink et al. in Brazil and Convit et al. in

Venezuela developed killed *L. amazonensis*-based vaccine and autoclaved *L. mexicana*-based vaccine with the adjuvant BCG respectively, and its immunotherapeutic and prophylactic efficacies were evaluated for CL (Alvar et al. 2013). Adjuvants (in Latin = "to help") are a group of carbohydrates, lipids, peptides, or nucleic acid structures that act as a PRR agonist which targets the host DCs to elicit the immune response. Adjuvants are used in many vaccines to deliver candidate antigens of the pathogen to boost the humoral as well as cellular immune response. For example, Bacillus Calmette Guerine (BCG) from *Mycobacterium tuberculosis*, Aluminium salts (alum), IL-12, and CpG deoxynucleotides target various PRRs (particularly TLRs) of DCs and stimulate antigen presentation and T cell activation. After extensive trials in South America it was found that both Mayrink's and Convit's vaccines were effective as an adjunct therapeutic to reduce the load on antimonials, but results were inconclusive for the prophylactic use of these vaccines (Mutiso et al. 2013). Phase 1 and phase 2 trials of safety, efficacy, and potency of the autoclaved, killed, *L. major* (ALM) with or without BCG have been performed in zoonotic as well as anthroponotic foci of Iran against CL (Bahar et al. 1996; Momeni et al. 1999; Sharifi et al. 1998) and in Sudan against VL (Khalil et al. 2000). The vaccine was found to be safe but not very immunogenic with BCG. In order to enhance its immunogenicity, ALM was adsorbed on adjuvant alum (aluminium salt) and then mixed with BCG or IL-12. Phase 2 and phase 3 trials of alum-ALM+BCG vaccine have been completed in Iran with positive and negative LST responded volunteers (http://clinicaltrials.gov/). Vaccination of Indian langur monkeys with ALM+BCG, or alum-ALM+BCG against *L. donovani* infections was reported as a good vaccine candidate (Dube et al. 1998; Misra et al. 2001). In Sudan, alum-ALM+BCG along with SSG showed impressive immunotherapeutics for PKDL treatment as compared to SSG alone (Musa et al. 2008). In a recent approach in Kenya, sonicated *L. donovani* promastigotes were delivered intradermally to vervet monkeys with several adjuvant combinations such as alum+BCG, montainde ISA 720 (MISA), and monophosphoryl lipid A (MPLA). The results suggested that sonicated *L. donovani*+MISA was safe and protective in vervet monkeys (Mutiso et al. 2012). Immunogenicity of killed ALM+BCG was evaluated in canine VL (CVL) and found protective against *L. infantum* infection (Lasri et al. 1999). Recently, a newer strategy was applied by Bruhn et al in which killed *L. infantum chagasi* promastigotes were treated with psoralen compound amotosalen (S-59) and low dose of ultraviolet A radiation and then injected to BALB/c mice. Treated *Leishmania* strain was avirulent but metabolically active and protects mice from *L. infantum* challenge (Bruhn et al. 2012). In one way, all the killed *Leishmania* vaccines showing good safety were rather weak immunogens and performed appreciably good as therapeutic vaccines but poor in prophylaxis. Moreover, in the non-standardised protocol, poorly defined and inconsistent potencies are the major shortcomings of this approach.

Second-Generation Vaccines

Advancement in identification and characterisation of *Leishmania* antigens that induce appropriate immune response led to the development of second-generation vaccines. A variety of different antigens, when delivered alone or in combination with other antigens, with or without adjuvants, have been shown to provide protective immunity in different *Leishmania* experimental models. Over the last two decades several native and recombinant *Leishmania* proteins have been tested with varying degrees of success. Fucose mannose ligand (FML), a fractionated *L. donovani* surface protein, was tested with different formulations of adjuvants such as saponin, alum, BCG, and IL-12 and demonstrated protection against *L. donovani* challenge (Palatnik-de-Sousa et al. 1994; Santos et al. 1999, 2002). Series of experiments were conducted with different experimental models in Brazil where FML+saponin is shown to be more immunogenic with no side effects. In Phase 3 trials FML with QuilA saponin when vaccinated in dogs induced strong and long-lasting immunity against CVL (Borja-Cabrera et al. 2002). Owing to the prophylactic efficacy of FML+saponin, this formulation was licensed in 2003 as "Leishmune®" in Brazil for CVL vaccination (Palatnik-de-Sousa 2012). After the availability of Leishmune® since 2004, incidence of CVL and human VL was observed to decrease in the endemic areas of Brazil where dogs were vaccinated with Leishmune® (Palatnik-de-Sousa et al. 2009). In India, Ali and her group tested *L. donovani* promastigote membrane antigens (LAg), encapsulated in different formulations of liposomes against *L. donovani* challenge in susceptible BALB/c mice and hamsters (Afrin and Ali 1997). Liposomes are lipid-bilayer membranes consisting of natural or synthetic phospholipids that can entrap antigens and act as proficient vaccine delivery vehicles as well as adjuvant. LAg when encapsulated with cationic liposomes was found to promote protective immunity in mice and it was also effective in VL therapeutics (Mazumdar et al. 2004; Dey et al. 2000). Among the three cationic liposomal formulations evaluated with LAg, reverse-phase evaporation vesicles (REV), multilamellar vesicles (MLV), and dehydration–rehydration vesicles (DRV), MLV, and DRV induced complete protection against VL (Bhowmick et al. 2010). Protective efficacy of LAg in cationic liposomes was compared with LAg associated with adjuvants such as BCG and monophosphoryl lipid (MPL) A along with trehalose dicorynomycolate (TDM). The result showed maximum protection of LAg encapsulated in cationic liposomes against murine VL (Ravindran et al. 2010). Soluble leishmanial antigens (SLA) were also evaluated with several formulations of liposomes and adjuvants such as MPL-TDM and non-coding plasmid DNA (pDNA). It was found that these formulations modulate the immune system towards Th1 and therefore induce durable protection against experimental VL (Bhowmick et al. 2007; Ravindran et al. 2012; Mazumder et al. 2007). To investigate the different antigenic components of LAg, peptides of LAg and SLA were electroeluted and entrapped separately with cationic liposomes and evaluated for vaccination against VL in BALB/c mice (Afrin et al. 2002; Bhowmick and Ali 2009). Glycoprotein 63 (GP63), conserved in all *Leishmania* species, was identified as a potent antigen for vaccination when encapsulated with cationic liposomes (Afrin et al. 2002; Bhowmick et al. 2008). Subsequently, recombinant GP63 (rGP63) when entrapped in cationic liposomes and conjugated with MPL-TDM boosted Th1 immune response significantly after *L. donovani* challenge (Mazumder et al. 2011a). *L. donovani* GP63 in conjugation with heat shock protein 70 (Hsp70) when primed in BABL/c mice demonstrated elevated level of protection than GP63 alone (Kaur et al. 2011a). Protective efficacy of *L. donovani* Hsp70 + Hsp83 was evaluated with adjuvants MPLA and in combination with (autoclaved *L. donovani*) ALD. Hsp70+ Hsp83+ MPLA was shown to increase level of protective cytokine IL-2 with IgG2a, and IFNγ (Kaur et al. 2011b).

BALB/c mice when immunised with recombinant *L. donovani* hydrophilic acylated surface protein B1 (HASPB1) without an adjuvant induced IL-12 production from dendritic cells resulting in reduced spleen parasite burden (Stager et al. 2000). Amastigote-specific protein A2 is considered as important virulence factor for *L. donovani* survival in mammalian host. Mice when immunised with this protein conferred protection against VL challenge through increased IFNγ production (Ghosh et al. 2001). Two recombinant proteins of *L. donovani*, recombinant open reading frame (rORFF) and recombinant biopterin transporter 1 (BT1), conferred partial protection either alone or in combination with each other against VL infection (Dole et al. 2000). However, protein rORFF, when immunised with CpG-ODN prior to challenge with *L. donovani*, induced Th1 response and enhanced production of IgG2a and IFNγ in a dose-dependent manner (Tewary et al. 2004). In a study, C57BL/6 mice were immunised with *L. infantum*-isolated recombinant sterol 24-c-methyltransferase (rSMT), a highly conserved protein of *Leishmania*. rSMT along with MPL in stable emulsion illustrated significant reduction in parasite burden in both spleen and liver of mice (Goto et al. 2007). Recently, through immunoproteomic approach, an amastogote-stage-specific hypothetical *L. infantum* protein (LiHyp1), having homology with the superoxygenase gene family, has been identified, cloned, and evaluated for protection against *L. infantum* challenge with saponin in BALB/c mice. Result showed significant parasite reduction in spleen, liver, and bone marrow of the immunised mice and suggested as a potent vaccine candidate for CVL (Martins et al. 2013). In addition, several other proteins such as glycoprotein36 (GP36), *Leishmania* homologue of receptors for activated C kinase (LACK), recombinant protein (rF14), 78KDa *L. donovani* protein, ribosomal proteins, *L. donovani* gamma-glutamyl cysteine synthetase (LdγGCS), *L. donovani* elongation factor-2 (rLe1F-2), *L. donovani* protein disulfide isomerase (LdPDI), *L. infantum* thiol-dependent reductase 1 (TDR1), *L. donovani* triose phosphate isomerase (LdTPI), and secretory serine protease (pSP) have been evaluated for vaccine efficacies against VL infection in experimental models and PBMCs of infected patients (Das and Ali 2012). However, none of the proteins succeeded to reach the clinical trials. Several urinary proteins from infected VL patients were isolated and evaluated for vaccination studies. Protein rLi-ntf2 with adjuvant MPLA-SE showed potent Th1 response in BALB/c mice with decrease of parasite burden in spleen at 40 days post challenge of *L. infantum chagasi* (Kashino et al. 2012).

A breakthrough has been achieved in *Leishmania* vaccine research with recombinant polyprotein Leish111f, comprising tandem fusion of three antigens, *Leishmania* elongation initiation factor (LeIF), thiol-specific antioxidant (TSA), and *Leishmania major* stress-inducible protein 1 (LmSTI1). The polyprotein was evaluated for CL prophylaxis in murine and non-human primate models with adjuvant IL-12 (Coler et al. 2002). Polyprotein Leish111f has been proved to be safe, immunogenic, and effective in preclinical trials. Later trials of Leish111f were conducted with MPL-SE since IL-12 is not recommendable for human use (Skeiky et al. 2002). Therefore Leish111f or Leish-F1 became the first defined vaccine for leishmaniasis to reach human clinical trials. Leish111f was also found to be useful against VL infection. Mice when immunised with Leish111f+MPL-SE showed strong humoral and Th1 immune responses with significant increase in IFNγ, IL-2, and TNF against *L. infantum* challenge (Coler et al. 2007). A Phase1 clinical trial of Leish-F1+MPL-SE has been conducted in India on healthy individuals with and without the history of VL. After three doses of the vaccine on day 0, 28, and 56, subjects were followed up to 168 days. After this period, vaccine was safe and well tolerated in both the subjects with elevated IFNγ response produced by the induced T cells (Chakravarty et al. 2011). Phase 1 and Phase 2 clinical trials have also been successfully completed in Peru, Brazil, Columbia, and India to assess the safety, tolerability, and immunogenicity of the vaccine (http://clinicaltrials.gov/). However a Phase 3 trial of Leish-111f in Italy was found unsuccessful in dogs against *L. infantum* infection under field conditions (Gradoni et al. 2005). Leish110f, a slightly modified version of Leish111f, was evaluated with Glucantime drug as an immunotherapeutic vaccine for CVL and observed higher survival chances, reduction in number of deaths, and enhanced cellular reactivity to the vaccines (Miret et al. 2008). Leish-F1 vaccine has also been evaluated as an adjunctive therapeutics in combination with pentavalent antimonials in human CL and MCL and was shown to shorten the time of treatment with early clinical cure (Nascimento et al. 2010; Llanos-Cuentas et al. 2010). Several clinical trials of Leish-111F as the name Leish-F1, Leish-F2, and Leish-F3 are in process. Phase 1 and Phase 2 trials of Leish-F2 in combination with SSG have been recently completed in Sudan and Peru for adjunctive therapeutic vaccine candidate. A Phase 1 trial of Leish-F3 is being conducted on healthy volunteers in the USA to access the prophylaxis ability of this vaccine (clinical trials.gov).

The host immune system is modulated not only by *Leishmania* molecules but also by the salivary peptides and molecules injected into the feeding pool. These pharmologically active salivary gland antigens have been investigated as potential anti-leishmanial vaccine candidates. However, the function of most sand fly salivary peptides is unknown. Current research shows their role in disease progression as well as protection. Earlier it was shown that prior exposure to sand fly bites provide significant protection in mice against *L. major* which was correlated with enhanced DTH response and IFNγ production (Kamhawi et al. 2000). 15 KDa proteins were extracted from the salivary gland of *Phlebotomus papatasi* and shown to protect vaccinated mice (Valenzuela et al. 2001). Again salivary proteins such as maxadilan (a vasodilatory peptide) and LJM19 (11 KDa protein) were isolated from

Lutzomyia longipalpis and found to confer strong protection in experimental models of CL and VL (Morris et al. 2001; Gomes et al. 2008). Salivary gland homogenates of *L. longipalpis* have been shown to induce IL-6, IL-8, and IL-12p40, and inhibit IL-10 and TNFα in in vitro stimulated monocytes and DCs of healthy human (Costa et al. 2004). Healthy volunteers from endemic areas of Brazil were experimentally exposed to laboratory-reared uninfected *L. longipalpis* displayed increase in anti-salivary gland IgG1, IgG4, IgGE, and several salivary gland proteins in the serum. PBMC from the volunteers, one year after the first exposure, showed recall IFNγ response and suggesting the feasibility of human immunisation against sand fly saliva (Vinhas et al. 2007). Despite these studies on sand fly salivary proteins, more efforts are needed to define the salivary gland proteins as well as the quantification and standardisation of infectious dose (Ready 2013).

Third-Generation Vaccines

In recent years research using DNA vaccines have been emerged with great interest and success. Genes encoding the antigenic protein as vaccine candidates are cloned in mammalian expression vector and injected through skin or muscles. Subsequently, the DNA vaccine is engulfed by the cells and transported to the nucleus where the desired protein is expressed by the host cells. DNA vaccines can be multiplexed and may afford gene sequences of more than one antigenic protein as well as the adjuvants sequence in the plasmid carrier. Immunological findings related to DNA vaccination suggested induction of both $CD4^+$ and $CD8^+$ T cell response through MHC class II and class I presentation, respectively (Vanloubbeeck and Jones 2004). Many recombinant antigens earlier which had tested as subunit protein vaccines were further evaluated as DNA vaccines too. The LACK (*Leishmania* homologue of receptors for activated C kinase)-based DNA vaccine which was evaluated as potentially protective against experimental CL showed no protective effect against systemic *L. donovani* infection (Melby et al. 2001). However, vaccination of mice with DNA construct of kinesin protein from the locus of *L. donovani* microtubule demonstrated Th1 immune response through significant increase in IFNγ and IL-2 levels without increasing IL-4 level (Dey et al. 2008). Gene of *L. donovani* proteophosphoglycans (PPGs), a secretory and surface-bound protein of both promastigotes and amastigotes, has been identified and the N-terminal domain of 1.6 kb was cloned and constructed as a DNA vaccine. This vaccine showed 80 % prophylactic efficacy in golden hamsters which survived for six months after *L. donovani* challenge (Samant et al. 2009). After evaluation of the immunogenicity of two recombinant antigens LdTPI and LdPDI, DNA construct was assessed and found protective from VL infection in experimental models in India (Kushawaha et al. 2012a, b). A DNA construct of kinetoplastid membrane protein-11 (KMP-11) has been shown to protect hamsters from both pentavalent antimony responsive and resistant forms of *L. donovani* challenge through upregulation of IFNγ, TNFα, and IL-12 and downregulation of IL-10 (Basu et al. 2005). Heterologous prime-boost vaccination strategy has been shown to be successful in many studies where experimental models were immunised with DNA and boosted with the recombinant proteins. BALB/c mice when immunised with the plasmid carrying ORFF gene (F/pcDNA3.1) and given a booster dose of recombinant (rORFF) protein showed enhanced production of IFNγ, IgG2, and antigen-specific T cells for protection against VL (Tewary et al. 2005). A heterologous prime-boost vaccination strategy was conducted in India and compared with DNA vaccination strategy (DNA/DNA) and recombinant protein-based vaccination strategy (protein/protein). In DNA/protein heterologous vaccination BALB/C mice were immunised with the DNA construct and subsequently boosted with the recombinant gp63. DNA-prime/Protein-boost vaccination significantly reduced the parasite burden by 10^7- to 10^{10}-fold in liver and spleen, respectively. Moreover *L. donovani* challenge induced polarised Th1 response with upregulation of IFNγ, IL-12, NO, IgG2a/IgG1 ratio and downregulation of IL-4 and IL-10 responses compared to other vaccination strategies. Therefore, it was suggested that gp63-based heterologous DNA/protein vaccination induced immune response and protection against *L.donovani* challenge (Mazumder et al. 2011b). The advantages of DNA vaccines over other vaccination strategies are that DNA is a stable biomolecule, can be produced rapidly, and does not require cold chain storage conditions for its efficacy, but the safety concerns related to DNA vaccination must be addressed properly.

Live-Attenuated Parasite Vaccines

A promising approach to mimic natural infection is to alter virulence genetic factor before vaccination. This kind of targeted attenuation retains the complete spectrum of antigens necessary for immune response by the host cells. However, studies based on live-attenuated parasites are very few in leishmaniasis and mainly in VL. *L. donovani* biopterin transporter1 (BT1) gene has been disrupted and mice were immunised with this live mutant parasite. Splenocytes of LdBTI null mutant immunised mice showed enhanced IFNγ production when stimulated with *L. donovani* promastigotes and, therefore, suggested as a vaccine candidate (Papadopoulou et al. 2002). In a study, single allele knockout of *L. infantum* silent information regulatory 2 (LiSIR2) gene was prepared and BALB/c mice were immunised with it. Vaccinated mice elicited complete protection as demonstrated both Th1- and Th2-type response with polarisation of high IFNγ/low IL-10 ratio (Carrion et al. 2011). Centrin is a calcium-binding protein, involves in centrosome duplication in higher eukaryotes. Mice when immunised with null mutant of centrin (*LdCEN*$^{-/-}$) showed enhancement of cytokines IL-2, IFNγ, and TNFα as well as IgG2a and NO production in macrophages suggesting a protective Th1 immune response (Selvapandiyan et al. 2009). A genetically mutated *L. donovani* protein 27 (Ldp27$^{-/-}$) parasite when vaccinated into mice induced both pro- and anti-inflammatory response against *L. donovani* challenge and also showed cross protection when challenged with *L. major* and *L. braziliensis* (Dey et al. 2013). Studies are required in non-human primates and dogs to correctly assess these live-attenuated vaccines.

Although in the last few years great boost has been observed in Leishmania research we are still far from a safe and effective vaccine. Increase in the knowledge of host–parasite interaction, identification of novel vaccine candidates through proteomics, development of viral delivery systems, and newer adjuvants will definitely channelise the vaccination research for VL.

Control Strategies for Elimination of VL

VL is significantly an important public health issue mainly across the developing world within the population that are socially underprivileged and economically deprived. Spread of the disease in new areas and reemergence in the previously endemic foci have focused the way to develop and implement VL control programme effectively. Multiple factors are responsible for increasing VL incidents such as emergence of HIV/VL co-infection, expansion of the sand fly habitats, migration of people and for zoonotic VL dogs in non-endemic areas, as well as poor economic conditions, famine, civil war, and lack of strategy in policymaking for the disease. The control and elimination programme should ensure several aspects such as active and passive case detection, successful treatment, vector control, and animal reservoir control in zoonotic locus.

Effective disease surveillance system is vital for early diagnosis of the disease. Government settings of primary health care in most of the underdeveloped regions are restricted to poor source and amenities so many cases become underestimated as comes under private or non-government hospitals. Therefore, actual disease incidence is expected to be much more than the official figure. Active case detection strategies such as door to door screening, arrangement of temporary camps, and disease awareness in the endemic area could be beneficial to provide a more realistic data on the VL burden (Singh et al. 2011). Gathering disease information through door to door search is considered as gold standard of screening, but it requires more efforts and manpower. This type of screening for a single disease is not feasible but can be supplemented with other ongoing programmes. Temporary camps dedicated to VL could manage to detect the disease in one shot though this is a tough task to promote the people to come to the camps. An incentive could play a better role for this situation. To educate the local people in the endemic area is equally important to raise the awareness and basic knowledge of the disease. A large gap between the asymptomatic individuals and the case finding strategy should also be addressed properly. Field adaptable rK39 strips were found imperative to enable the screening of large number of asymptomatic VL reservoirs in short course of time (Topno et al. 2010). Diagnosis of PKDL is also important because of its competence to being a carrier for the disease.

Since no prophylactic or therapeutic vaccines are available for VL, treatment is the only way to benefit the disease sufferers. Treatment options for VL are very limited while antimonials are still the first line of drug in most parts of the world. Emergence of antimonial resistance in the Indian subcontinent is alarming for rest of the world. Need of hospitalisation for a long time hampers the complete course of treatment as most of the sufferers are poor daily wage labourers. Oral miltefosine is effective in treating VL without hospitalisation; however, drug must be provided free or at least in subsidised price by the government to avoid the incompliance of treatment by the patients. Complete treatment of PKDL is also important; however, there is no drug specific to PKDL only. The therapeutics of PKDL largely depends on antimonials but with long treatment regimen than VL.

Treatment can only supplement the disease management since it is beneficial for an individual and not for the whole population. Therefore eradication of sand fly vector is an integrated approach for any VL management programme. Geographically different endemic regions comprise different environmental characteristics for sand fly. Knowledge, attitude, and practice (KAP) studies are conducted in many VL endemic areas to assess the livelihood conditions and other factors which could be favourable for sand fly propagation. These findings are very important to understand the disease condition better and accordingly to give appropriate recommendations to the control programme. The sand fly control mainly depends on indoor residual spraying (IRS) and insecticide-treated bed nets (ITB) (Picado et al. 2012). IRS strategy with DDT had significantly reduced the disease burden in India at past. Despite their high cost, distribution of ITB was assessed for protection against the disease in the Indian subcontinents and Iran. More sustainable strategy needs to be developed for vector control programme.

In zoonotic VL, screening of infected and asymptomatic canine population is also important to address the control programme. Insecticide-impregnated dog collars have been used in endemic regions like Brazil as a vector control strategy (Ribeiro et al. 2013). Apart from vaccination, diagnosis and treatment of infected dogs or euthanasia should also be carried out properly, to benefit the human VL ultimately. Clinical and active research could also supplement the VL elimination programme in many ways. Development of a rapid non invasive diagnostic, safe oral drug, and effective vaccine can boost the disease management. Political commitment is also very important to control the disease as to provide better funding opportunities as well as to implement better policies against disease outbreak. Thus, the need of the hour is to seek new prevention strategies, better research priority, and an improved political commitment by the policymakers to completely eradicate the disease from the globe.

Acknowledgements The authors thank Amrita Das, Pradyot Bhattacharya, and Md. Asad for valuable suggestions and proofreading of the manuscript.

Financial and Competing Interests Disclosure: The authors were supported by the Council of Scientific and Industrial Research, Government of India (BSC 0114). The authors have no other relevant affiliations or financial involvement with any organisation or entity with a financial interest in or financial conflict with the subject matter or materials discussed in the manuscript apart from those disclosed.

No writing assistance was utilised in the production of this manuscript.

References

Adhikari A et al (2012) Mycobacterium indicus pranii (Mw) re-establishes host protective immune response in Leishmania donovani infected macrophages: critical role of IL-12. PLoS One 7(7): e40265

Afrin F, Ali N (1997) Adjuvanticity and protective immunity elicited by Leishmania donovani antigens encapsulated in positively charged liposomes. Infect Immun 65(6):2371–2377

Afrin F et al (2002) Characterization of Leishmania donovani antigens encapsulated in liposomes that induce protective immunity in BALB/c mice. Infect Immun 70(12):6697–6706

Alvar J et al (2008) The relationship between leishmaniasis and AIDS: the second 10 years. Clin Microbiol Rev 21(2):334–359, table of contents

Alvar J et al (2012) Leishmaniasis worldwide and global estimates of its incidence. PLoS One 7 (5):e35671

Alvar J et al (2013) Case study for a vaccine against leishmaniasis. Vaccine 31(Suppl 2):B244–B249

Anam K et al (1999) Differential decline in Leishmania membrane antigen-specific immunoglobulin G (IgG), IgM, IgE, and IgG subclass antibodies in Indian kala-azar patients after chemotherapy. Infect Immun 67(12):6663–6669

Anderson CF et al (2009) IL-27 regulates IL-10 and IL-17 from CD4+ cells in nonhealing Leishmania major infection. J Immunol 183(7):4619–4627

Ansari NA, Ramesh V, Salotra P (2006a) Interferon (IFN)-gamma, tumor necrosis factor-alpha, interleukin-6, and IFN-gamma receptor 1 are the major immunological determinants associated with post-kala azar dermal leishmaniasis. J Infect Dis 194(7):958–965

Ansari NA, Saluja S, Salotra P (2006b) Elevated levels of interferon-gamma, interleukin-10, and interleukin-6 during active disease in Indian kala azar. Clin Immunol 119(3):339–345

Ansari NA et al (2008a) Elevated levels of IgG3 and IgG4 subclass in paediatric cases of kala azar. Parasite Immunol 30(8):403–409

Ansari NA et al (2008b) Evidence for involvement of TNFR1 and TIMPs in pathogenesis of post-kala-azar dermal leishmaniasis. Clin Exp Immunol 154(3):391–398

Ansari NA et al (2011) IL-27 and IL-21 are associated with T cell IL-10 responses in human visceral leishmaniasis. J Immunol 186(7):3977–3985

Bahar K et al (1996) Comparative safety and immunogenicity trial of two killed Leishmania major vaccines with or without BCG in human volunteers. Clin Dermatol 14(5):489–495

Banerjee A, De M, Ali N (2008) Complete cure of experimental visceral leishmaniasis with amphotericin B in stearylamine-bearing cationic liposomes involves down-regulation of IL-10 and favorable T cell responses. J Immunol 181(2):1386–1398

Basu R et al (2005) Kinetoplastid membrane protein-11 DNA vaccination induces complete protection against both pentavalent antimonial-sensitive and -resistant strains of Leishmania donovani that correlates with inducible nitric oxide synthase activity and IL-4 generation: evidence for mixed Th1- and Th2-like responses in visceral leishmaniasis. J Immunol 174 (11):7160–7171

Belkaid Y et al (2001) The role of interleukin (IL)-10 in the persistence of Leishmania major in the skin after healing and the therapeutic potential of anti-IL-10 receptor antibody for sterile cure. J Exp Med 194(10):1497–1506
Bhattacharya P, Ali N (2013) Involvement and interactions of different immune cells and their cytokines in human visceral leishmaniasis. Rev Soc Bras Med Trop 46(2):128–134
Bhowmick S, Ali N (2009) Identification of novel Leishmania donovani antigens that help define correlates of vaccine-mediated protection in visceral leishmaniasis. PLoS One 4(6):e5820
Bhowmick S, Ravindran R, Ali N (2007) Leishmanial antigens in liposomes promote protective immunity and provide immunotherapy against visceral leishmaniasis via polarized Th1 response. Vaccine 25(35):6544–6556
Bhowmick S, Ravindran R, Ali N (2008) gp63 in stable cationic liposomes confers sustained vaccine immunity to susceptible BALB/c mice infected with Leishmania donovani. Infect Immun 76(3):1003–1015
Bhowmick S et al (2010) Comparison of liposome based antigen delivery systems for protection against Leishmania donovani. J Control Release 141(2):199–207
Bilate AM, Lafaille JJ (2012) Induced CD4+Foxp3+ regulatory T cells in immune tolerance. Annu Rev Immunol 30:733–758
Boaventura VS et al (2010) Human mucosal leishmaniasis: neutrophils infiltrate areas of tissue damage that express high levels of Th17-related cytokines. Eur J Immunol 40(10):2830–2836
Borja-Cabrera GP et al (2002) Long lasting protection against canine kala-azar using the FML-QuilA saponin vaccine in an endemic area of Brazil (Sao Goncalo do Amarante, RN). Vaccine 20(27–28):3277–3284
Breton M et al (2005) Live nonpathogenic parasitic vector as a candidate vaccine against visceral leishmaniasis. Infect Immun 73(10):6372–6382
Bruhn KW et al (2012) Killed but metabolically active Leishmania infantum as a novel whole-cell vaccine for visceral leishmaniasis. Clin Vaccine Immunol 19(4):490–498
Caldas A et al (2005) Balance of IL-10 and interferon-gamma plasma levels in human visceral leishmaniasis: implications in the pathogenesis. BMC Infect Dis 5:113
Carrion J et al (2011) Leishmania infantum HSP70-II null mutant as candidate vaccine against leishmaniasis: a preliminary evaluation. Parasit Vectors 4:150
Cenini P et al (1993) Mononuclear cell subpopulations and cytokine levels in human visceral leishmaniasis before and after chemotherapy. J Infect Dis 168(4):986–993
Chakravarty J et al (2011) A clinical trial to evaluate the safety and immunogenicity of the LEISH-F1+MPL-SE vaccine for use in the prevention of visceral leishmaniasis. Vaccine 29(19):3531–3537
Chappuis F et al (2007) Visceral leishmaniasis: what are the needs for diagnosis, treatment and control? Nat Rev Microbiol 5(11):873–882
Coler RN et al (2002) Immunization with a polyprotein vaccine consisting of the T-Cell antigens thiol-specific antioxidant, Leishmania major stress-inducible protein 1, and Leishmania elongation initiation factor protects against leishmaniasis. Infect Immun 70(8):4215–4225
Coler RN et al (2007) Leish-111f, a recombinant polyprotein vaccine that protects against visceral Leishmaniasis by elicitation of CD4+ T cells. Infect Immun 75(9):4648–4654
Costa DJ et al (2004) Lutzomyia longipalpis salivary gland homogenate impairs cytokine production and costimulatory molecule expression on human monocytes and dendritic cells. Infect Immun 72(3):1298–1305
Croft SL, Sundar S, Fairlamb AH (2006) Drug resistance in leishmaniasis. Clin Microbiol Rev 19 (1):111–126
Cunningham AC (2002) Parasitic adaptive mechanisms in infection by leishmania. Exp Mol Pathol 72(2):132–141
Cunningham J et al (2012) A global comparative evaluation of commercial immunochromatographic rapid diagnostic tests for visceral leishmaniasis. Clin Infect Dis 55(10):1312–1319
Das A, Ali N (2012) Vaccine development against Leishmania donovani. Front Immunol 3:99

Dey T et al (2000) Antileishmanial activities of stearylamine-bearing liposomes. Antimicrob Agents Chemother 44(6):1739–1742
Dey A et al (2008) Kinesin motor domain of Leishmania donovani as a future vaccine candidate. Clin Vaccine Immunol 15(5):836–842
Dey R et al (2013) Live attenuated Leishmania donovani p27 gene knockout parasites are nonpathogenic and elicit long-term protective immunity in BALB/c mice. J Immunol 190 (5):2138–2149
Dole VS et al (2000) Immunization with recombinant LD1 antigens protects against experimental leishmaniasis. Vaccine 19(4–5):423–430
Dorlo TP et al (2012) Translational pharmacokinetic modelling and simulation for the assessment of duration of contraceptive use after treatment with miltefosine. J Antimicrob Chemother 67 (8):1996–2004
Dube A et al (1998) Vaccination of langur monkeys (Presbytis entellus) against Leishmania donovani with autoclaved L. major plus BCG. Parasitology 116(Pt 3):219–221
Ejazi SA, Ali N (2013) Developments in diagnosis and treatment of visceral leishmaniasis during the last decade and future prospects. Expert Rev Anti Infect Ther 11(1):79–98
Galvao-Castro B et al (1984) Polyclonal B cell activation, circulating immune complexes and autoimmunity in human american visceral leishmaniasis. Clin Exp Immunol 56(1):58–66
Ganguly S et al (2008) Increased levels of interleukin-10 and IgG3 are hallmarks of Indian post-kala-azar dermal leishmaniasis. J Infect Dis 197(12):1762–1771
Ganguly S et al (2010) Enhanced lesional Foxp3 expression and peripheral anergic lymphocytes indicate a role for regulatory T cells in Indian post-kala-azar dermal leishmaniasis. J Invest Dermatol 130(4):1013–1022
Gasim S et al (1998) High levels of plasma IL-10 and expression of IL-10 by keratinocytes during visceral leishmaniasis predict subsequent development of post-kala-azar dermal leishmaniasis. Clin Exp Immunol 111(1):64–69
Gasim S et al (2000) The development of post-kala-azar dermal leishmaniasis (PKDL) is associated with acquisition of Leishmania reactivity by peripheral blood mononuclear cells (PBMC). Clin Exp Immunol 119(3):523–529
Ghalib HW et al (1993) Interleukin 10 production correlates with pathology in human Leishmania donovani infections. J Clin Invest 92(1):324–329
Ghosh AK, Dasgupta S, Ghose AC (1995) Immunoglobulin G subclass-specific antileishmanial antibody responses in Indian kala-azar and post-kala-azar dermal leishmaniasis. Clin Diagn Lab Immunol 2(3):291–296
Ghosh A, Zhang WW, Matlashewski G (2001) Immunization with A2 protein results in a mixed Th1/Th2 and a humoral response which protects mice against Leishmania donovani infections. Vaccine 20(1–2):59–66
Ghosh K et al (2013) Successful therapy of visceral leishmaniasis with curdlan involves T-helper 17 cytokines. J Infect Dis 207(6):1016–1025
Giorgobiani E et al (2011) Epidemiologic aspects of an emerging focus of visceral leishmaniasis in Tbilisi, Georgia. PLoS Negl Trop Dis 5(12):e1415
Gomes R et al (2008) Immunity to a salivary protein of a sand fly vector protects against the fatal outcome of visceral leishmaniasis in a hamster model. Proc Natl Acad Sci USA 105(22):7845–7850
Gonzalez-Lombana C et al (2013) IL-17 mediates immunopathology in the absence of IL-10 following Leishmania major infection. PLoS Pathog 9(3):e1003243
Goto Y et al (2007) Protective immunization against visceral leishmaniasis using Leishmania sterol 24-c-methyltransferase formulated in adjuvant. Vaccine 25(42):7450–7458
Gradoni L et al (2005) Failure of a multi-subunit recombinant leishmanial vaccine (MML) to protect dogs from Leishmania infantum infection and to prevent disease progression in infected animals. Vaccine 23(45):5245–5251
Guerin PJ et al (2002) Visceral leishmaniasis: current status of control, diagnosis, and treatment, and a proposed research and development agenda. Lancet Infect Dis 2(8):494–501

Hailu A et al (2004) Elevated plasma levels of interferon (IFN)-gamma, IFN-gamma inducing cytokines, and IFN-gamma inducible CXC chemokines in visceral leishmaniasis. Am J Trop Med Hyg 71(5):561–567

Handman E (2001) Leishmaniasis: current status of vaccine development. Clin Microbiol Rev 14 (2):229–243

Hatam GR et al (2009) Improvement of the newly developed latex agglutination test (Katex) for diagnosis of visceral leishmaniasis. J Clin Lab Anal 23(4):202–205

Ismail A et al (1999) Immunopathology of post kala-azar dermal leishmaniasis (PKDL): T-cell phenotypes and cytokine profile. J Pathol 189(4):615–622

Jha TK (2006) Drug unresponsiveness & combination therapy for kala-azar. Indian J Med Res 123 (3):389–398

Kamhawi S et al (2000) Protection against cutaneous leishmaniasis resulting from bites of uninfected sand flies. Science 290(5495):1351–1354

Kashino SS et al (2012) Identification of Leishmania infantum chagasi proteins in urine of patients with visceral leishmaniasis: a promising antigen discovery approach of vaccine candidates. Parasite Immunol 34(7):360–371

Katara GK et al (2012) Evidence for involvement of Th17 type responses in post kala azar dermal leishmaniasis (PKDL). PLoS Negl Trop Dis 6(6):e1703

Kaur T, Sobti RC, Kaur S (2011a) Cocktail of gp63 and Hsp70 induces protection against Leishmania donovani in BALB/c mice. Parasite Immunol 33(2):95–103

Kaur J, Kaur T, Kaur S (2011b) Studies on the protective efficacy and immunogenicity of Hsp70 and Hsp83 based vaccine formulations in Leishmania donovani infected BALB/c mice. Acta Trop 119(1):50–56

Kellina OI (1981) Problem and current lines in investigations on the epidemiology of leishmaniasis and its control in the U.S.S.R. Bull Soc Pathol Exot Filiales 74(3):306–318

Kenney RT et al (1998) Splenic cytokine responses in Indian kala-azar before and after treatment. J Infect Dis 177(3):815–818

Khalil EA et al (2000) Autoclaved Leishmania major vaccine for prevention of visceral leishmaniasis: a randomised, double-blind, BCG-controlled trial in Sudan. Lancet 356(9241):1565–1569

Khamesipour A et al (2005) Leishmanization: use of an old method for evaluation of candidate vaccines against leishmaniasis. Vaccine 23(28):3642–3648

Khamesipour A et al (2006) Leishmaniasis vaccine candidates for development: a global overview. Indian J Med Res 123(3):423–438

Kumar R, Nylen S (2012) Immunobiology of visceral leishmaniasis. Front Immunol 3:251

Kushawaha PK et al (2012a) Evaluation of Leishmania donovani protein disulfide isomerase as a potential immunogenic protein/vaccine candidate against visceral Leishmaniasis. PLoS One 7 (4):e35670

Kushawaha PK et al (2012b) Leishmania donovani triose phosphate isomerase: a potential vaccine target against visceral leishmaniasis. PLoS One 7(9):e45766

Lasri S et al (1999) Immune responses in vaccinated dogs with autoclaved Leishmania major promastigotes. Vet Res 30(5):441–449

Laufs H et al (2002) Intracellular survival of Leishmania major in neutrophil granulocytes after uptake in the absence of heat-labile serum factors. Infect Immun 70(2):826–835

Liese J, Schleicher U, Bogdan C (2008) The innate immune response against Leishmania parasites. Immunobiology 213(3–4):377–387

Llanos-Cuentas A et al (2010) A clinical trial to evaluate the safety and immunogenicity of the LEISH-F1+MPL-SE vaccine when used in combination with sodium stibogluconate for the treatment of mucosal leishmaniasis. Vaccine 28(46):7427–7435

Lopez Kostka S et al (2009) IL-17 promotes progression of cutaneous leishmaniasis in susceptible mice. J Immunol 182(5):3039–3046

Malla N, Mahajan RC (2006) Pathophysiology of visceral leishmaniasis—some recent concepts. Indian J Med Res 123(3):267–274

Maroof A et al (2008) Posttranscriptional regulation of Il10 gene expression allows natural killer cells to express immunoregulatory function. Immunity 29(2):295–305
Marth T, Kelsall BL (1997) Regulation of interleukin-12 by complement receptor 3 signaling. J Exp Med 185(11):1987–1995
Martins VT et al. (2013) Correction: antigenicity and protective efficacy of a leishmania amastigote-specific protein, member of the super-oxygenase family, against visceral leishmaniasis. PLoS Negl Trop Dis 7(4)
Maurya R et al (2010) Human visceral leishmaniasis is not associated with expansion or accumulation of Foxp3+ CD4 cells in blood or spleen. Parasite Immunol 32(7):479–483
Mazumdar T, Anam K, Ali N (2004) A mixed Th1/Th2 response elicited by a liposomal formulation of Leishmania vaccine instructs Th1 responses and resistance to Leishmania donovani in susceptible BALB/c mice. Vaccine 22(9–10):1162–1171
Mazumder S et al (2007) Non-coding pDNA bearing immunostimulatory sequences co-entrapped with leishmanial antigens in cationic liposomes elicits almost complete protection against experimental visceral leishmaniasis in BALB/c mice. Vaccine 25(52):8771–8781
Mazumder S, Maji M, Ali N (2011a) Potentiating effects of MPL on DSPC bearing cationic liposomes promote recombinant GP63 vaccine efficacy: high immunogenicity and protection. PLoS Negl Trop Dis 5(12):e1429
Mazumder S et al (2011b) Potency, efficacy and durability of DNA/DNA, DNA/protein and protein/protein based vaccination using gp63 against Leishmania donovani in BALB/c mice. PLoS One 6(2):e14644
McHugh RS et al (2002) CD4(+)CD25(+) immunoregulatory T cells: gene expression analysis reveals a functional role for the glucocorticoid-induced TNF receptor. Immunity 16(2):311–323
Medeiros IM et al (2000) Circulating levels of sTNFR and discrepancy between cytotoxicity and immunoreactivity of TNF-alpha in patients with visceral leishmaniasis. Clin Microbiol Infect 6 (1):34–37
Melaku Y et al (2007) Treatment of kala-azar in southern Sudan using a 17-day regimen of sodium stibogluconate combined with paromomycin: a retrospective comparison with 30-day sodium stibogluconate monotherapy. Am J Trop Med Hyg 77(1):89–94
Melby PC et al (2001) Leishmania donovani p36(LACK) DNA vaccine is highly immunogenic but not protective against experimental visceral leishmaniasis. Infect Immun 69(8):4719–4725
Miret J et al (2008) Evaluation of an immunochemotherapeutic protocol constituted of N-methyl meglumine antimoniate (Glucantime) and the recombinant Leish-110f + MPL-SE vaccine to treat canine visceral leishmaniasis. Vaccine 26(12):1585–1594
Misra A et al (2001) Successful vaccination against Leishmania donovani infection in Indian langur using alum-precipitated autoclaved Leishmania major with BCG. Vaccine 19(25–26):3485–3492
Momeni AZ et al (1999) A randomised, double-blind, controlled trial of a killed L. major vaccine plus BCG against zoonotic cutaneous leishmaniasis in Iran. Vaccine 17(5):466–472
Mondal S, Bhattacharya P, Ali N (2010) Current diagnosis and treatment of visceral leishmaniasis. Expert Rev Anti Infect Ther 8(8):919–944
Morris RV et al (2001) Sandfly maxadilan exacerbates infection with Leishmania major and vaccinating against it protects against L. major infection. J Immunol 167(9):5226–5230
Murray HW et al (2002) Interleukin-10 (IL-10) in experimental visceral leishmaniasis and IL-10 receptor blockade as immunotherapy. Infect Immun 70(11):6284–6293
Murray HW et al (2005) Antagonizing deactivating cytokines to enhance host defense and chemotherapy in experimental visceral leishmaniasis. Infect Immun 73(7):3903–3911
Musa AM et al (2008) Immunochemotherapy of persistent post-kala-azar dermal leishmaniasis: a novel approach to treatment. Trans R Soc Trop Med Hyg 102(1):58–63
Mutiso JM et al (2012) Leishmania donovani whole cell antigen delivered with adjuvants protects against visceral leishmaniasis in vervet monkeys (Chlorocebus aethiops). J Biomed Res 26 (1):8–16

Mutiso JM et al (2013) Development of Leishmania vaccines: predicting the future from past and present experience. J Biomed Res 27(2):85–102
Muzio M et al (2000) Differential expression and regulation of toll-like receptors (TLR) in human leukocytes: selective expression of TLR3 in dendritic cells. J Immunol 164(11):5998–6004
Nadim A et al (1983) Effectiveness of leishmanization in the control of cutaneous leishmaniasis. Bull Soc Pathol Exot Filiales 76(4):377–383
Nascimento E et al (2010) A clinical trial to evaluate the safety and immunogenicity of the LEISH-F1+MPL-SE vaccine when used in combination with meglumine antimoniate for the treatment of cutaneous leishmaniasis. Vaccine 28(40):6581–6587
Novoa R et al (2011) IL-17 and regulatory cytokines (IL-10 and IL-27) in L. braziliensis infection. Parasite Immunol 33(2):132–136
Nylen S, Sacks D (2007) Interleukin-10 and the pathogenesis of human visceral leishmaniasis. Trends Immunol 28(9):378–384
Nylen S et al (2007) Splenic accumulation of IL-10 mRNA in T cells distinct from CD4+CD25+ (Foxp3) regulatory T cells in human visceral leishmaniasis. J Exp Med 204(4):805–817
Palatnik-de-Sousa CB (2012) Vaccines for canine leishmaniasis. Front Immunol 3:69
Palatnik-de-Sousa CB, Day MJ (2011) One Health: the global challenge of epidemic and endemic leishmaniasis. Parasit Vectors 4:197
Palatnik-de-Sousa CB et al (1994) Experimental murine Leishmania donovani infection: immunoprotection by the fucose-mannose ligand (FML). Braz J Med Biol Res 27(2):547–551
Palatnik-de-Sousa CB et al (2009) Decrease of the incidence of human and canine visceral leishmaniasis after dog vaccination with Leishmune in Brazilian endemic areas. Vaccine 27 (27):3505–3512
Papadopoulou B et al (2002) Reduced infectivity of a Leishmania donovani biopterin transporter genetic mutant and its use as an attenuated strain for vaccination. Infect Immun 70(1):62–68
Peters C et al (1995) The role of macrophage receptors in adhesion and uptake of Leishmania mexicana amastigotes. J Cell Sci 108(Pt 12):3715–3724
Picado A et al (2012) Vector control interventions for visceral leishmaniasis elimination initiative in South Asia, 2005-2010. Indian J Med Res 136(1):22–31
Pitta MG et al (2009) IL-17 and IL-22 are associated with protection against human kala azar caused by Leishmania donovani. J Clin Invest 119(8):2379–2387
Rai AK et al (2012) Regulatory T cells suppress T cell activation at the pathologic site of human visceral leishmaniasis. PLoS One 7(2):e31551
Ravindran R et al (2010) Comparison of BCG, MPL and cationic liposome adjuvant systems in leishmanial antigen vaccine formulations against murine visceral leishmaniasis. BMC Microbiol 10:181
Ravindran R, Maji M, Ali N (2012) Vaccination with liposomal leishmanial antigens adjuvanted with monophosphoryl lipid-trehalose dicorynomycolate (MPL-TDM) confers long-term protection against visceral leishmaniasis through a human administrable route. Mol Pharm 9 (1):59–70
Ready PD (2013) Biology of phlebotomine sand flies as vectors of disease agents. Annu Rev Entomol 58:227–250
Ribeiro VM et al (2013) Control of visceral leishmaniasis in Brazil: recommendations from Brasileish. Parasit Vectors 6(1):8
Ribeiro-Gomes FL et al (2004) Macrophage interactions with neutrophils regulate Leishmania major infection. J Immunol 172(7):4454–4462
Ritter U, Moll H (2000) Monocyte chemotactic protein-1 stimulates the killing of leishmania major by human monocytes, acts synergistically with IFN-gamma and is antagonized by IL-4. Eur J Immunol 30(11):3111–3120
Rodrigues OR et al (2009) Identification of regulatory T cells during experimental Leishmania infantum infection. Immunobiology 214(2):101–111
Sacks D, Kamhawi S (2001) Molecular aspects of parasite-vector and vector-host interactions in leishmaniasis. Annu Rev Microbiol 55:453–483

Saha S et al (2006) Immune responses in kala-azar. Indian J Med Res 123(3):245–266
Saha S et al (2007) IL-10- and TGF-beta-mediated susceptibility in kala-azar and post-kala-azar dermal leishmaniasis: the significance of amphotericin B in the control of Leishmania donovani infection in India. J Immunol 179(8):5592–5603
Saha S et al (2011) Easy test for visceral Leishmaniasis and post-kala-azar dermal leishmaniasis. Emerg Infect Dis 17(7):1304–1306
Sakaguchi S et al (1995) Immunologic self-tolerance maintained by activated T cells expressing IL-2 receptor alpha-chains (CD25). Breakdown of a single mechanism of self-tolerance causes various autoimmune diseases. J Immunol 155(3):1151–1164
Salotra P, Singh R (2006) Challenges in the diagnosis of post kala-azar dermal leishmaniasis. Indian J Med Res 123(3):295–310
Samant M et al (2009) Immunization with the DNA-encoding N-terminal domain of proteophosphoglycan of Leishmania donovani generates Th1-type immunoprotective response against experimental visceral leishmaniasis. J Immunol 183(1):470–479
Santos WR et al (1999) Vaccination of Swiss Albino mice against experimental visceral leishmaniasis with the FML antigen of Leishmania donovani. Vaccine 17(20–21):2554–2561
Santos WR et al (2002) Saponins, IL12 and BCG adjuvant in the FML-vaccine formulation against murine visceral leishmaniasis. Vaccine 21(1–2):30–43
Schleicher U et al (2007) NK cell activation in visceral leishmaniasis requires TLR9, myeloid DCs, and IL-12, but is independent of plasmacytoid DCs. J Exp Med 204(4):893–906
Selvapandiyan A et al (2009) Intracellular replication-deficient Leishmania donovani induces long lasting protective immunity against visceral leishmaniasis. J Immunol 183(3):1813–1820
Shadab M, Ali N (2011) Evasion of host defence by Leishmania donovani: subversion of signaling pathways. Mol Biol Int 2011:343961
Sharifi I et al (1998) Randomised vaccine trial of single dose of killed Leishmania major plus BCG against anthroponotic cutaneous leishmaniasis in Bam, Iran. Lancet 351(9115):1540–1543
Shimizu J et al (2002) Stimulation of CD25(+)CD4(+) regulatory T cells through GITR breaks immunological self-tolerance. Nat Immunol 3(2):135–142
Singh SP et al (2011) Options for active case detection of visceral leishmaniasis in endemic districts of India, Nepal and Bangladesh, comparing yield, feasibility and costs. PLoS Negl Trop Dis 5(2):e960
Skeiky YA et al (2002) Protective efficacy of a tandemly linked, multi-subunit recombinant leishmanial vaccine (Leish-111f) formulated in MPL adjuvant. Vaccine 20(27–28):3292–3303
Smelt SC et al (2000) B cell-deficient mice are highly resistant to Leishmania donovani infection, but develop neutrophil-mediated tissue pathology. J Immunol 164(7):3681–3688
Srivastava P et al (2011) Diagnosis of Indian visceral leishmaniasis by nucleic acid detection using PCR. PLoS One 6(4):e19304
Stager S, Smith DF, Kaye PM (2000) Immunization with a recombinant stage-regulated surface protein from Leishmania donovani induces protection against visceral leishmaniasis. J Immunol 165(12):7064–7071
Stager S et al (2006) Distinct roles for IL-6 and IL-12p40 in mediating protection against Leishmania donovani and the expansion of IL-10+ CD4+ T cells. Eur J Immunol 36 (7):1764–1771
Sundar S, Chakravarty J (2010) Antimony toxicity. Int J Environ Res Public Health 7(12):4267–4277
Sundar S, Chakravarty J (2012) Recent advances in the diagnosis and treatment of kala-azar. Natl Med J India 25(2):85–89
Sundar S et al (2011) Comparison of short-course multidrug treatment with standard therapy for visceral leishmaniasis in India: an open-label, non-inferiority, randomised controlled trial. Lancet 377(9764):477–486
Tewary P et al (2004) Immunostimulatory oligodeoxynucleotides are potent enhancers of protective immunity in mice immunized with recombinant ORFF leishmanial antigen. Vaccine 22 (23–24):3053–3060

Tewary P et al (2005) A heterologous prime-boost vaccination regimen using ORFF DNA and recombinant ORFF protein confers protective immunity against experimental visceral leishmaniasis. J Infect Dis 191(12):2130–2137
Tiwananthagorn S et al (2012) Involvement of CD4(+) Foxp3 (+) regulatory T cells in persistence of Leishmania donovani in the liver of alymphoplastic aly/aly mice. PLoS Negl Trop Dis 6(8): e1798
Topno RK et al (2010) Asymptomatic infection with visceral leishmaniasis in a disease-endemic area in bihar, India. Am J Trop Med Hyg 83(3):502–506
Valenzuela JG et al (2001) Toward a defined anti-Leishmania vaccine targeting vector antigens: characterization of a protective salivary protein. J Exp Med 194(3):331–342
van Zandbergen G et al (2002) Leishmania promastigotes release a granulocyte chemotactic factor and induce interleukin-8 release but inhibit gamma interferon-inducible protein 10 production by neutrophil granulocytes. Infect Immun 70(8):4177–4184
van Zandbergen G et al (2004) Cutting edge: neutrophil granulocyte serves as a vector for Leishmania entry into macrophages. J Immunol 173(11):6521–6525
Vanloubbeeck Y, Jones DE (2004) The immunology of Leishmania infection and the implications for vaccine development. Ann N Y Acad Sci 1026:267–272
Vargas-Inchaustegui DA, Xin L, Soong L (2008) Leishmania braziliensis infection induces dendritic cell activation, ISG15 transcription, and the generation of protective immune responses. J Immunol 180(11):7537–7545
Vester B et al (1999) Early gene expression of NK cell-activating chemokines in mice resistant to Leishmania major. Infect Immun 67(6):3155–3159
Vinhas V et al (2007) Human anti-saliva immune response following experimental exposure to the visceral leishmaniasis vector, Lutzomyia longipalpis. Eur J Immunol 37(11):3111–3121
WHO (1984) The leishmaniases. Report of a WHO Expert Committee. World Health Organ Tech Rep Ser 701:1–140
Workman CJ et al (2009) The development and function of regulatory T cells. Cell Mol Life Sci 66 (16):2603–2622
Wu W, Huang L, Mendez S (2010) A live Leishmania major vaccine containing CpG motifs induces the de novo generation of Th17cells in C57BL/6 mice. Eur J Immunol 40(9):2517–2527
Zer R et al (2001) Effect of sand fly saliva on Leishmania uptake by murine macrophages. Int J Parasitol 31(8):810–814

Leprosy

Alireza Firooz, Yahya Dowlati, and Azin Ayatollahi

Contents

A. Firooz
Center for Research & Training in Skin Diseases & Leprosy, Tehran University of Medical Sciences, Tehran, Iran

415 Taleghani Avenue, Tehran 14166, Iran
e-mail: firozali@sina.tums.ac.ir

Y. Dowlati (✉) • A. Ayatollahi
Center for Research & Training in Skin Diseases & Leprosy, Tehran University of Medical Sciences, Tehran, Iran
e-mail: dowlatiy@yahoo.com; azinay@gmail.com

M.A. McDowell and S. Rafati (eds.), *Neglected Tropical Diseases - Middle East and North Africa*, Neglected Tropical Diseases, DOI 10.1007/978-3-7091-1613-5_7,

Abstract Leprosy is one of the most primitive diseases with which human beings have been confronted for thousands of years. It primarily affects skin and peripheral nervous system. The cause of disease, *Mycobacterium leprae*, was discovered by Armauer Hansen in 1873. After the introduction of Multi Drug Therapy (MDT) by WHO in 1982 as the first effective treatment for leprosy, the prevalence of the disease has decreased significantly in most parts of the world.

Keywords Leprosy • Multidrug therapy • Elimination • Reaction • Cell-mediated immunity • Disability

Definition

Leprosy is a chronic infectious disease caused by *Mycobacterium leprae*. It prominently affects the skin and peripheral nerves of the extremities; however, during the advanced stage of multibacillary leprosy, the *M. leprae* will be found in most of the body tissues and organs with the exception of central nervous system, gastrointestinal tract, and lungs (James et al. 2011).

Historical Overview

One of the most primitive diseases with which human beings have been confronted for their survival is Hansen's disease or leprosy. As a brief allusion to the provenance of this disease in the history of medicine, one can exemplify the epistle that has been found in India, which is also known to be the oldest of its kind.

In addition, it should be noted that the first-proven documented case of this disease has been discovered in mummies of ancient Egypt. Based on the previously mentioned historical references, it seems that the origin of the disease could be traced back to the Far East and India. Later it expanded to the Middle East and Europe through the ancient Macedonian army during fourth century BC (Browne 1985). After its inception throughout the European countries, the prevalence of leprosy reached to its highest level during the medieval era. Although no treatment for leprosy was available until the twentieth century, it has been widely believed that the occurrence of the disease had been reduced markedly because of noticeable improvements in socioeconomic and healthcare conditions as well as a reduction in population density of families. Leprosy was transmitted to Africa and Canada via European immigrants and later on to the USA through African migrants (Monot et al. 2009).

Epidemiology

Before 1981, when there was no definite treatment for leprosy, patients had to use Dapsone throughout the remainder of their lifetime. The prevalence of this disease was estimated to embrace 11 million of the world population (Daumerie 2003). However, in 1982 after the introduction of Multidrug Therapy (MDT) by World Health Organization (WHO), its occurrence was decreased significantly and in the beginning of 2012 reached 181,941 (0.34 cases/10,000 population). The tropical regions of South-East Asia (India, Bangladesh, Indonesia, Nepal, and Myanmar), South America (Brazil), and Africa (Congo, Ethiopia, Madagascar, Tanzania) have the highest prevalence of leprosy due to their poverty and overcrowded populations. The majority of cases (117,147) are from South-East Asia and 7,368 cases are reported from Eastern Mediterranean (WHO 2013). Due to the long latency period, the exact incidence of the disease is not clear.

Subclinical infection is prevalent in endemic regions and there is a possibility of spontaneous cure especially in children with paucibacillary disease. New detected cases of leprosy are registered and reported annually. The annual case detection rate has decreased from 600,000 to 800,000 during pre- MDT to 219,075 (4.06 cases/100,000 population) in 2011. The number of cases reported from Eastern Mediterranean region in this year (January to December, 2011) is 4,346 (0.71/100,000).

In 1991 WHO introduced the goal of elimination (but not eradication) of leprosy in the world. WHO defined this goal as a decrease in leprosy prevalence to less than one case per 10,000 population. This target was globally met in 2000, but due to the latest WHO data, in eight countries, Marshall Islands, Micronesia, Kiribati, Kumori, South Sudan, Burundi, Brazil, and Liberia, the prevalence of leprosy is still more than 1/10,000. Among these countries, Marshall Islands have the highest prevalence of leprosy with the prevalence of 25/10,000 population. In the Eastern Mediterranean region, the highest numbers are reported from South Sudan and Sudan (WHO 2013).

In the years1992 and 1997, Iran reached the goal of leprosy elimination across the whole country and its associated townships, respectively. In 2011, prevalence of leprosy was limited to 78 cases (0.017/10,000) and the new registered cases were 36 (0.04/100,000). Most of the patients were male aged 50–70 years with multibacillary type of leprosy (WHO 2013). Provinces where the most cases of leprosy occurred were the following: Qazvin, Hormozgan, Gilan, Ardebil, West Azerbaijan, Golestan, Sistan and Balouchestan, and Khorasan Razavi (Dowlati et al. 2012).

The leprosy latency period is long and the mean duration of incubation period for the paucibacillary leprosy and the multibacillary leprosy are 4 and 8 years, respectively. In most parts of the world, leprosy is more prevalent in male than female patients and the paucibacillary type is more common. Although leprosy can occur at any age, most patients in endemic areas are younger than 35 years, and the mean age of paucibacillary patients is lower than multibacillary, due to the shorter incubation period.

Etiology

In 1873, a Norwegian scientist, Armauer Hansen, identified *Mycobacterium leprae* as the definitive cause of leprosy. This weakly acid-fast bacillus has the following unique characteristics:

1. Human is its only known reservoir. However, in some recent studies the possibility that armadillo might be the reservoir of *M. leprae* and transmit it to humans is proposed (Truman et al. 2011).
2. This bacillus is considered an obligate intracellular microorganism (mainly within macrophages and Schwann cells) with which the body confronts through cell-mediated immunity (CMI).
3. It has a relatively prolonged period of cell division, approximately 12–14 days that in turn results in a lengthy latency period.
4. An ideal ambient temperature for its growth is 30 °C that allows the bacteria to proliferate in cutaneous and peripheral nerves—zones with relatively lower temperatures compared to other tissues.
5. Despite various efforts that have taken place to discover a synthetic culture medium, no such environment has been found thus far. Furthermore, it is not possible to culture *M. leprae* in vitro. Instead, armadillo or mouse footpad is used for proliferation, which can introduce difficulties in laboratory diagnosis, study of susceptibility and resistance to antibiotics, and discovery of new drugs.
6. Mutation rate of this bacillus, in contrast to *Mycobacterium tuberculosis* bacillus, is relatively low and genetic sequences of bacilli that have been isolated from patients all over the world (including Iran) are quite similar (Monot et al. 2009). Recently some cases of diffuse lepromatous leprosy (LL) have been reported which is caused by a related but genetically different mycobacterium. This new mycobacterium is called *M. lepromatosis* (Han et al. 2008).
7. The cell wall of *M. leprae* consists of several antigenic molecules which are used as substrates for serological tests. The most important antigens include Phenolic Glycolipid 1 (PGL-1) and Lipoarabinomannan (LAM).

Pathogenesis

The main reservoir of the *M. leprae* is the upper respiratory tract of the untreated patient with multibacillary leprosy. The disease is transmitted through the nasal secretion of patients to those with a long close contact with them (Job et al. 2008).

Normal skin is resistant to the bacillus penetration but there are rare cases of leprosy caused by inoculation (vaccination, penetrating trauma, tattooing, surgery) (Ghorpade 2009). There is no case of congenital leprosy since *M. leprae* cannot be transmitted through the placenta.

There are rare cases of leprosy reported in the USA that have no history of close contact with leprosy patients or of travelling to endemic areas. However, these individuals have had exposure to armadillos, despite the fact that the role of armadillo in the transmission of the disease is not yet defined (Truman et al. 2011).

Nearly 90 % of people are genetically resistant to leprosy and their cell-mediated immune system will kill the bacillus after infection. Susceptible individuals will contract the disease after a long household contact with patients having multibacillary leprosy. If the cell-mediated immune function fails to eradicate the microorganism, the bacilliproliferate slowly in the upper respiratory tract and after a relatively lengthy incubation period, the bacilli are disseminated throughout the body and initially inhabit the skin and peripheral nerves, due to their lower temperature.

Genetics

The susceptibility of developing leprosy is under the control of genetic factors. The coincidence of leprosy among monozygotic twins and dizygotic twins are 60–80 % and 15–25 %, respectively, emphasizing the role of genetic factors (Mohammed and Ramanujam 1966). However, a single gene effect has yet to be defined. Different genes have been identified from different parts of the world, suggesting that multiple genes might be involved in susceptibility to leprosy. The type of the disease depends on the CMI response that is controlled by the major histocompatibility complex (MHC) genes. Multiple studies revealed that the HLA-DR2 and HLA-DR3 are more prevalent in tuberculoid leprosy, whereas the HLA-DQ1 prevalence is higher in lepromatous and borderline leprosy (BB) (Worobec 2009).

Immunology

Granuloma formation around the bacillus, which occurs by having a good CMI response with production of Th1 cytokines (IL-2, IL-12, IFN-γ) and the CD4/CD8 ratio of two, restricts bacillus proliferation. The resulting disease is called tuberculoid leprosy (TT). During TT, there are few and asymmetric skin and nervous lesions and the bacillus is rarely found in these lesions. The lepromin skin test (see below) is positive in these patients but the serologic test is positive in only 40–50 % of these individuals.

If the CMI response is not adequate with a CD8/CD4 ratio of two, the Th2 cytokines (IL-4, IL-10) will be secreted. These processes do not form granulomas and prevent the bacillus proliferation. The resulting disease is called lepromatous leprosy (LL). In this type of leprosy, multiple symmetrical cutaneous and nervous lesions do exist, containing numerous Hansen (*M. leprae*) bacilli. Although the clinical signs are usually restricted to the skin and nerves, the bacillus disseminates to other organs. The lepromin skin test is negative for these patients but the serologic test is positive in more than 90 % of them. Lepromatous patients have polyclonal hypergammaglobulinemia and high antibody titers to *M. leprae* unique antigens and may have false-positive syphilis serology, rheumatoid factor, and antinuclear antibodies.

In the case of an intermediate CMI response, the patient will have the borderline leprosy (BB) with intermediate clinical symptoms and a negative lepromin test. In this group, there are borderline tuberculoid (BT) patients, who are more similar to the tuberculoid leprosy with a positive lepromin test, and borderline lepromatous (BL) patients who are more similar to LL with a negative lepromin test.

LL and TT patients are immunologically stable and will not convert to other disease forms. Borderline leprosy, on the other hand, may convert to other forms if the patient's CMI responses change (e.g., by treatment, pregnancy, immunosuppression, drug consumption, vaccination, HIV infection, etc.). This response can be associated with severe inflammation of skin and peripheral nerves which contain *M. leprae* antigens. This reaction is called a type-1 reaction and is a delayed-type hypersensitivity reaction.

In LL and BL patients numerous bacilli and high levels of serum antibody are found. Immune complexes precipitate in tissues that, in turn, activate the complement system, causing leukocytoclastic vasculitis in different body organs. This occurrence, which is called a type-2 leprosy reaction, is considered a type-III hypersensitivity reaction. The course of these immunologic reactions is not correlated to the course of the disease and can occur before, after, or during treatment.

The immune insufficiency in patients with leprosy is completely specific to the *M. leprae* antigens and all their other immune responses, even to other mycobacterium, are normal. Coinfection with HIV does not change the incidence or severity of leprosy and the treatment response is not be decreased as well; however, the risk of type-1 reaction might be increased after the antiretroviral treatment, which is called "immune reconstitution syndrome" (Lockwood and Lambert 2011).

Classification

Most commonly, patients with leprosy are classified by the use of two classifications: WHO classification, which is more simple and practical, and the Ridley-Jopling that is more specialized but less practical (Prasad 2005).

WHO Classification

In this classification, patients are classified in two groups based on their positive or negative skin smear test.

1. Paucibacillary patients with negative skin smear
2. Multibacillary patients with positive skin smear

Due to some reasons such as difficulty in preparation and interpretation of the skin smear and a false-negative result in some patients, WHO, recently, classified the patients based on the number of skin lesions.

Patients with five lesions or less are classified as paucibacillary group and those with more than five skin lesions are in the multibacillary leprosy group. Certainly, this type of classification has some drawbacks; for instance, some multibacillary patients will be classified in the paucibacillary group and will not receive the adequate treatment. In case of uncertainty regarding the exact classification, patients should be treated as a multibacillary leprosy case.

Ridely–Jopling Classification

During the 1960's Ridely and Jopling introduced their classification based on clinical signs and histopathological findings. It is now clear that the basis of this classification is CMI reaction against *M. leprae*.

1. Indeterminate (I)
2. Tuberculoid (TT)
3. Borderline tuberculoid (BT)
4. Mid-borderline (BB)
5. Borderline lepromatous (BL)
6. Lepromatous (LL)

The first three groups are almost similar to the paucibacillary group and the last three are roughly consistent with the multibacillary group of WHO. The first three groups have negative cutaneous smear and a limited number of skin and nerve lesions both with an asymmetrical distribution.

By moving toward the multibacillary pole, the number of lesions increases gradually and their distribution becomes more symmetrical. Furthermore, they are morphologically more diverse and the number of Hansen bacilli increases as well. However, the paucibacillary group has more pronounced neurological symptoms than the multibacillary group, while the latter shows these symptoms in later stages of their disease.

Clinical Features

There is a diverse range of clinical features in patients with leprosy. About 90 % of population are resistant to leprosy and their CMI system kill the bacilli after entering the body without developing any clinical symptoms. Susceptible individuals have a range of clinical findings that depends on their CMI response (Moschella 2004).

Indeterminate leprosy (I)

The most common outcome after exposure to *M. leprae* is spontaneous healing. If the disease occurs, the initial presentation is indeterminate leprosy. In the indeterminate type, usually there is a single hypopigmented macule or patch with an ill-defined border. Occasionally, there is a faded erythema that accompanies the hypopigmented macule. The common sites of involvement are face, upper arms, buttocks, and upper thighs. The scalp, axilla, and inguinal region are usually spared due to their higher temperature. Patient may complain from tingling, despite no sensory dysfunction during physical examination. This clinical manifestation is called indeterminate because the course of disease cannot be predicted. Although most of the patients recover spontaneously, some of them progress to the other forms of the disease.

Tuberculoid Leprosy (TT)

In TT leprosy, the skin and peripheral nerves are the only sites of involvement. There are one or a few skin lesions (less than five lesions). Lesions are well demarcated with elevated borders. The central part of the lesion may be hypopigmented or erythematous in darkly pigmented patients and in lightly pigmented skin, respectively. The lesion may be a macule/patch or an annular plaque (Fig. 1).

TT lesions are anesthetic. Loss of temperature sensation occurs early followed by loss of sensation of light touch, pain, and deep touch. Due to the extensive

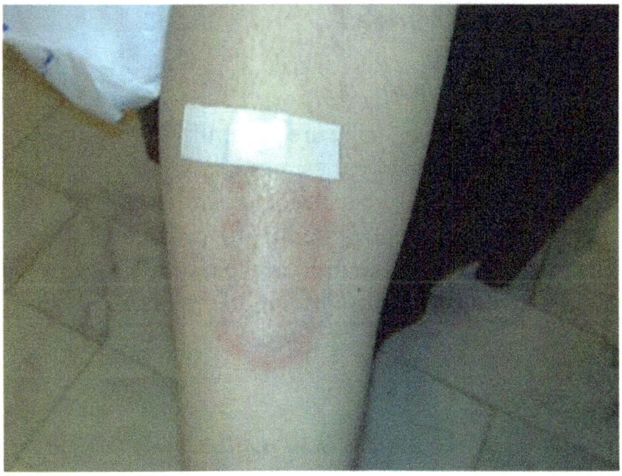

Fig. 1 Tuberculoid leprosy

nervous supply on the face, the loss of sensation of facial lesions may not be evident during the examination. Usually a peripheral nerve near the skin lesion is involved and becomes large and palpable and occasionally tender. Reduced hair growth and anhidrosis might be observed in a skin lesion due to autonomic nerve involvement. In addition, there may be atrophy of the muscle groups served by the involved nerve. Wasting of the thenar and hypothenar eminences, contracture of the fingers, paralysis of the facial muscles, and foot drop may also occur. Facial nerve damage dramatically increases the risk for ocular involvement and vision loss.

The TT course of the disease is prolonged. It may heal spontaneously in a few years (especially in children) or faster with treatment. Even without treatment, this manifestation never progresses to other forms of leprosy.

Borderline Tuberculoid Leprosy

BT lesions are well defined with a hypopigmented center; in lightly pigmented individuals, it may be erythematous. BT is defined by more lesions than TT leprosy and satellite lesions usually exist. The skin lesions are usually asymmetric and the borders are less infiltrative, however, sensation loss of the lesion is still an important feature.BT lesions have no scale or are slightly scaly. Peripheral nerve involvement is more prominent than TT leprosy and usually the large nerve trunks are involved asymmetrically. These nerves are large and tender or exhibit sensory, motor, and autonomic nerve fiber dysfunction (in the late stage) during neurologic examination.

Borderline Leprosy

BB leprosy usually has a short duration and will upgrade or downgrade to other forms. In this group, there are generalized and numerous lesions but not as many as in lepromatous leprosy. Lesion distribution is more symmetrical than BT lesions. The lesions may be macular, papular, or plaque type. In the typical form of the disease, lesions have a well-demarcated inner margin and an ill-defined outer margin, although these lesions are rare. Peripheral nerve involvement is variable and their manifestations depend on the involved nerve.

Borderline Lepromatous Leprosy

Skin lesions are almost symmetrical, multiple, and disseminated with no well-defined margin in BL leprosy. Initially, fade macules are present and gradually symmetric disseminated papules, plaques, and nodules will develop. In this type of leprosy, involvement of large nerve trunks without clinical symptoms is observed that is usually symmetrical. Although loss of sensation in a stocking and glove pattern may occur, sensation and sweating on individual skin lesions is normal.

The possibility of leprosy reactions is high in this type of leprosy. The signs and symptoms of oral and nasal involvement such as epistaxis, nasal septum perforation and saddle nose, and eyebrow loss usually are not observed in the early stages of this type of leprosy.

Lepromatous Leprosy

In this leprosy type, there is a huge proliferation of the bacteria and the bacilli are disseminated through skin, nerves, and other sites of the body. The skin lesions are diffuse and symmetric macules (with ill-defined borders in contrast to well-defined tuberculoid macules), papules, nodules (lepromas), and plaques are seen. The scalp and intertriginous areas are not involved. Diffuse infiltration of the bacillus in facial skin causes the typical feature of the leonine face (Fig. 2). Neuropathy might be observed as anesthesia in a bilateral symmetrical stocking and glove distribution during the late stage of the disease. But similar to BL, sensation and sweating on individual skin lesions are normal. Symptoms in the advanced stage include, saddle nose deformity, and nasal bridge damage due to bacillus infiltration, madarosis (decreased eyebrow hair), epistaxis, and ichthyosiform skin. Infiltration of the earlobe causes enlargement and swelling. Fusiform swelling of digits might be another sign of LL.

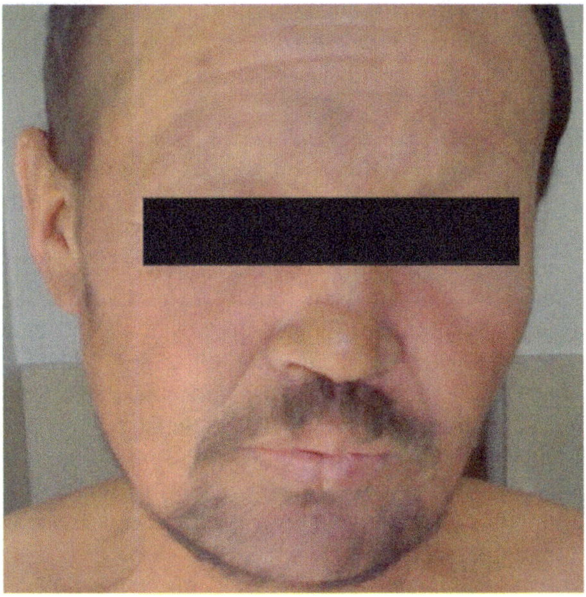

Fig. 2 Leonine facies in lepromatous leprosy

Histoid Leprosy

This uncommon form of multibacillary leprosy may occur de novo or more commonly in LL patients who become resistant to Dapsone. There are numerous yellow-red dermal or subcutaneous shiny papules and nodules appearing on a background of normal skin.

Nerve Involvement

Nerve involvement is characteristic of leprosy and due to neurotropism of *M. leprae*. Different types of peripheral nerves involvement might be observed in leprosy:

1. *Peripheral nerves enlargement*. This symptom is diagnosed through palpation of the peripheral nerves, which are asymmetric in the tuberculoid spectrum and symmetric in lepromatous pole of disease. The most important nerves that are involved in leprosy are the greater auricular, ulnar, and cutaneous branch of the radial nerve, lateral popliteal, sural, common peroneal, and posterior tibial nerves. The nerve trunks that are more superficial and are close to the skin such as the ulnar and greater auricular are more obvious in physical examination.
2. *Loss of skin sensation*. Temperature sensation is the first sense that is lost, followed by loss of sensation of light touch, pain, and deep touch. This impaired

sensation is in the skin lesion in the paucibacillary form. However, in the multibacillary type the anesthesia is in acral parts in a pattern of stocking and glove during the advanced stage of the disease with no loss of sensation in the skin lesions.

3. *Motor nerve paralysis*. This paralysis may have inflammatory symptoms (tenderness and swelling) or without symptoms (silent neuropathy). Gradually atrophy and wasting of small muscles of hands and feet and occasionally face served by involved nerves develops. Contracture and deformity of the extremities might be seen gradually. Ulnar and median nerve involvement can cause claw hand and thumb deformity, respectively. Wrist drop is the result of radial nerve involvement. Foot drop deformity and claw toe are due to common peroneal nerve and posterior tibial nerve involvement, respectively. Facial nerve involvement causes lagophthalmus (eye will not close during the sleep) and finally may lead to blindness due to trauma and secondary infection.
4. *Anhidrosis*. Loss of sweating is the result of autonomic nerve dysfunction. In the paucibacillary type, this symptom is limited to skin lesions but in the multibacillary form, it involves the acral areas and in late stage, the trunk becomes anhidrotic.
5. *Pure neural leprosy*. Although leprosy frequently involves both skin and peripheral nerves, in some patients the peripheral nerves are involved in the absence of cutaneous lesions. This type of leprosy is called pure neural leprosy. Some patients have undergone unnecessary orthopedic surgery and neurosurgical procedures due to the misdiagnosis. This type of leprosy is more commonly seen in paucibacillary patients and may represent up to 5 % of new cases of leprosy in some countries such as India and Nepal. Histological changes compatible with leprosy have been observed in normal appearing skin of some of these patients. (Menicucci et al. 2005)

Leprosy Reactions

In some leprosy patients, acute signs and symptoms of inflammation may occur. Physical examination may reveal erythema, tenderness, and swelling in skin lesions or nerves. New skin lesions with inflammation and erythema may occur abruptly. This phenomenon is called a leprosy reaction. Leprosy reactions are destructive inflammatory reactions due to immune system hyperactivity. Its occurrence is reported in 50 % of patients after starting therapy. In some instances, it is the first sign of leprosy and the cause of patient referral. In addition, this reaction can occur long after therapy cessation.

Leprosy reaction is an emergency with a high degree of importance that should be diagnosed and treated promptly. Otherwise, permanent sequelae and disability occur due to the fibrosis of the nerve fibers. Occasionally, the reaction may be subacute with the presence of progressive destructive non-inflammatory neuropathy. In case of reaction occurrence during therapy, both patient and physician

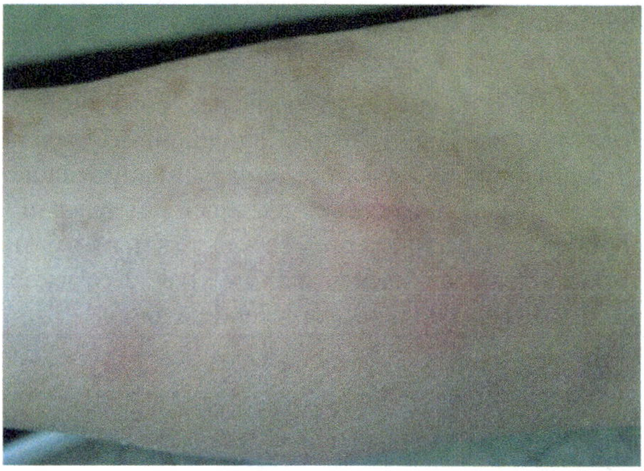

Fig. 3 Erythema nodosum leprosum

should be aware that these new symptoms are not the result of incompetency or side effects and that treatment should be continued.

There are two types of leprosy reactions. A type-1 reaction is due to a shift in delayed-type hypersensitivity (type IV reaction), which occurs in borderline leprosy (BT, BB, or BL). In these patients, the immune system is unstable and when it attenuates (for example during pregnancy, lactation, vaccination, immune suppressive drug consumption, and other infections), BT can shift to BB and even LL (reversal reaction or down grading). On the other hand, BL patients might shift to BT after reinforcement of the immune system especially after treatment (upgrading). In this reaction, the previous skin lesion becomes inflamed and erythematous, even new lesions might occur. In addition, the involved nerves might be swollen and inflamed. A type-1 reaction has no systemic symptoms and the manifestations are restricted to skin and peripheral nerves.

A type-2 reaction is a type-III hypersensitivity reaction and the result of antigen-antibody complex deposition in various tissues. It occurs in half of the LL and BL patients and is a multisystem disease due to leukocytoclastic vasculitis of various organs. The immune complex deposition in different organs causes inflammation: neuritis, arthritis, hepatitis, conjunctivitis, keratitis, uveitis, orchitis, and lymphadenopathy may develop. The systemic symptoms such as fever, chills, arthralgia, etc., are present.

The skin lesion is called erythema nodosum leprosum (ENL). Lesions are subcutaneous erythematous and painful nodules with shorter duration than the usual erythema nodosum. In addition to lower extremities, nodules might occur in upper extremities, face, and trunk (Fig. 3). This reaction may be mild to severe and may last from a few days to even years.

Systemic Involvement

Except for skin and peripheral nerves, other systemic manifestations are rare in leprosy (Klioze and Ramos-Karo 2000). However, in multibacillary leprosy, *M. leprae* might exist in most organs usually without any symptoms. The exceptions are the involvement of eyes (which is frequent in all types of leprosy due to the infiltration of bacilli, neuropathy, or reaction), upper respiratory tract, and testis (in multibacillary type) and the type-2 reaction which induces systemic leukocytoclastic vasculitis in several organs.

Diagnosis

Diagnosis of leprosy is based on clinical suspicion and accurate physical examination. The co-occurrence of peripheral neuropathy and skin lesions should lead to suspicion of leprosy, especially in patients with positive family history or background of living in endemic areas. Laboratory tests are not very sensitive and are not necessary if there are not adequate facilities.

Presence of two of the following clinical signs may confirm the diagnosis: (Prasad 2005)

1. Skin lesion
2. Sensory dysfunction in vicinity of skin lesions or in extremities
3. Enlargement of peripheral nerves

If there is any suspicion of leprosy, skin should be examined precisely and completely under proper lighting conditions. Any skin lesion might be due to leprosy because of the diversity of leprosy manifestations. But congenital, pruritic, hyperpigmented, or depigmented skin lesions and severe scaling are less suggestive of leprosy.

Leprosy is a chronic, long lasting disease that is progressive; repetitive courses of relapse and spontaneous remission are uncommon and are only observed in leprosy reactions.

After thorough skin examination, peripheral nerves should be palpated for thickness and tenderness, and then their function should be examined. Senses of temperature, superficial touch, pain, and deep touch should be examined by cold and hot water tubes, cotton ball or feather, needle tip, and pinching, respectively. This examination should be performed on skin lesion surfaces in case of paucibacillary leprosy and on skin of hands and feet in patients with multibacillary leprosy. The motor nerves should be evaluated by examination of the force of hand and foot muscles and presence of muscle atrophy or contracture. Hypohidrosis and hypotrichosis might be the signs of autonomic nerve dysfunction. In all patients, ophthalmologic examination is mandatory.

Although the diagnosis of leprosy is clinical and based on physical examination and WHO excludes the laboratory studies from the diagnostic criteria of leprosy, the following tests could be helpful.

Skin Smear

Skin slit smear samples can be obtained from the periphery of active skin lesions or cold body areas such as ear lobe, forehead, chin, and extensor surface of forearm, dorsal surface of fingers, buttocks, and trunk. Sometimes smears can be prepared from the nasal mucosa discharges. Nevertheless, the skin smear is more valuable.

Fite stain, which is the modified form of Ziehl–Neelson, is used for staining after the slide is dried, where the bacilli are seen as red rods against a blue background. Smear is positive in 100 % of multibacillary patients but is negative in all TT patients and most of BT patients. Bacteriological index (BI) indicates the density of bacilli in the smear on a 0–6 scale.

Although the bacilli are killed after a short duration after the onset of therapy, they may remain in skin tissue for many years and smears could remain positive. Therefore, the smear test is not a suitable way to monitor the disease treatment response. Live bacilli are solid, whereas killed bacilli are granular and crushed. The proportion of live solid bacilli in a smear is called morphological index (MI) which is a useful index for monitoring patients after treatment.

WHO has discontinued recommendation of taking skin smears in endemic areas for diagnosis or classification or in order to monitor progress with treatment (WHO 2003), because:

- Leprosy can be easily diagnosed and classified on the basis of clinical findings.
- MDT treatment regimens are standardized and usually do not require mid-course changes based on results of smear examinations.
- Cure of leprosy is based on completion of a full course of standard MDT regimen and skin smears from most leprosy patients will yield negative results.
- Unnecessary skin-piercing procedures are unethical, painful and carry the risk of serious infection (particularly HIV and hepatitis).
- Use of skin smears should be limited to referral centres, particularly for special investigations (suspicion of resistance, complex relapse cases) and research purposes.

Histopathology

Skin biopsy samples should contain the full thickness of the skin. Nerve biopsy is rarely used and might be performed from only sensory nerves such as great auricular, sural, and superficial peroneal nerves. Different types of leprosy have distinctive histologic patterns:

At the lepromatous pole, histological examination reveals diffuse dermal infiltration composed of lymphocytes and plasma cells. In addition, several bacilli are observed in the dermis, forming aggregates known as globi. The Grenz zone, a band of normal appearing dermis, separates the epidermis from the dermal infiltration. If treatment is successful, granular and fragmented bacilli will be seen.

At the tuberculoid pole, granulomatous infiltration in the dermis is observed that might have a linear pattern that follows the nerve course. Epithelioid cells and Langhans giant cells are surrounded by lymphocytes but there is no bacteria present in the tissue. The cutaneous nerves are edematous, a symptom that differs from other granulomatous reactions such as sarcoidosis.

The borderline spectrum of leprosy contains the histological features of both mentioned forms.

Serology

Serologic assays to detect antibodies against *M. leprae* specific antigens (anti-PGL) are only sensitive and specific in multibacillary disease because it is positive in nearly all of the patients. Serologic tests are only positive in half of the paucibacillary patients. Due to the presence of positive serology in 5–10 % of healthy people in endemic areas for leprosy (likely due to subclinical infection), this method is of no value in the diagnosis of leprosy and should only be utilized in epidemiologic surveys or to identify persons at risk of developing leprosy in endemic areas (Oskam et al. 2003).

Lepromin Skin Test

Lepromin test is an intradermal test in which a 0.1 ml suspension of heat-killed *M. leprae* (from lepromatous nodule or infected armadillo liver) is injected intradermally. Forty-eight hours later, the injection site is examined as in the tuberculin test. This reaction, called Fernandez reaction, has cross-reaction with other mycobacteria species. A reaction with induration of more than 7 mm developing 3–4 weeks following injection is more specific than the former reaction and is called Mitsuda reaction. This test is an indication of cell-mediated immune response against *M. leprae* antigen. It does not require the existence of bacilli in the body or the history of contact with it, so it is positive in healthy people that are resistant to leprosy and also in paucibacillary patients. However, it is negative in multibacillary patients. Thus this test cannot be used as a diagnostic method, but may be useful for the classification of leprosy.

Polymerase Chain Reaction

Polymerase chain reaction (PCR) can detect even very small numbers of organism in infected tissues and is universally positive in multibacillary disease, in which the diagnosis is not difficult. But it is negative in as many as half of the paucibacillary patients and cannot improve the diagnosis in these patients.

Differential Diagnosis

Hypopigmented lesions in leprosy may mimic pityriasis alba, pityriasis versicolor, the early stage of vitiligo, post inflammatory hypopigmentation, and achromic or anemic nevus. Erythematous lesions with well-demarcated borders in paucibacillary leprosy may be misdiagnosed as dermatophytosis, granuloma annulare, cutaneous TB, and allergic contact dermatitis. The differential diagnosis of diffuse erythematous lesions in multibacillary leprosy includes xanthomas, guttate psoriasis, pityriasis rosea, and sarcoidosis. Although leonine facies is a characteristic for lepromatous leprosy, it is not pathognomonic and it might be observed in other infiltrative diseases such as leishmaniasis, lymphoma, and pseudolymphoma.

Treatment

Despite thousands of years of history and recognition of causative agent of leprosy for many years, Dapsone as the first effective treatment was introduced in 1941. Dapsone is a bacteriostatic agent and it should be consumed throughout the patient's lifetime. Gradually, some of leprosy patients became resistant to Dapsone (secondary resistance). Transmission of Dapsone-resistant *M. leprae* to new patients caused primary resistance to Dapsone. The first bactericidal drug that had been discovered against *M. leprae* was Rifampin, which is still the drug of choice, and is the base of all the leprosy treatment regimens. Finally, the definite treatment for leprosy was introduced by WHO in 1982 as Multidrug Therapy (MDT): (WHO 1982).

Paucibacillary leprosy: Rifampin 600 mg once a month + Dapsone 100 mg daily for 6 months.

Multibacillary leprosy: Rifampin 600 mg once a month + Dapsone 100 mg daily + Clofazimine 300 mg once a month and 50 mg daily for 12 months.

Prior to 1998, the duration of the therapy was 24 months and then it was decreased to 12 months (Firooz 2006).

The most important bactericidal drug in leprosy and the base of all of regimens is Rifampin. Due to the bacillus proliferation period, increasing patient compliance, and decreasing the risk of resistance, Rifampin is used in a single monthly dose and under supervision.

If Rifampin is contraindicated or cannot be used for any other reasons, two drugs, mostly Minocycline and Ofloxacin can be substituted. Although there have not been any clinical trials comparing a placebo control to MDT, this treatment is very effective and the risk of recurrence is less than one per 1,000 cases per year. (Lockwood 2007; WHO 2003) The cases of recurrence are not due to drug resistance and patients should be retreated with the same MDT. Treatment in pregnancy and lactation is MDT and, in children, the dose should be adjusted concerning their age or weight (WHO 2003).

Side effects of MDT are rare. The most common side effects of Rifampin are gastrointestinal side effects and urinary discoloration. Hepatic toxicity is rare due to monthly consumption. Hemolytic anemia is not common during Dapsone consumption and it is not necessary to evaluate G6PD in all patients before starting treatment. The rare but important side effect of Dapsone is a hypersensitivity reaction with fever, generalized rash, pruritus, and lymphadenopathy. The most common side effect of Clofazimine is skin hyperpigmentation, which usually occurs a few months after starting the drug and will resolve spontaneously after drug discontinuation.

The new treatment regimen is ROM which consists of Rifampin 600 mg + Minocycline 100 mg + Ofloxacin 400 mg once a month for 6 months and 24 months in paucibacillary leprosy and multibacillary leprosy, respectively. Although this treatment appears to be as effective as MDT in randomized clinical trials, it is not useful in most of the endemic countries due to its high cost, and it is not recommended by WHO (Setia et al. 2011).

Treatment of leprosy reactions is an emergency and more important than the leprosy itself, because they will damage the nerves very quickly. In the case of a reaction, which may happen during MDT therapy, MDT should not be discontinued; rather it should be continued without any change.

For the mild reactions without nerve involvement, bed rest and NSAIDs are enough.

Nerve involvement requires prompt prednisolone therapy with the dose of 40–60 mg daily in an adult patient. It will be gradually tapered and then discontinued when the symptoms are controlled.

In type-2 reaction (ENL), thalidomide 100–200 mg daily with or without prednisolone is useful. Occasionally, differentiation between relapse and reaction is difficult. Usually reactions occur shortly after the therapy cessation while relapse happens after some years. Reaction is always acute but relapse has an insidious course. Reaction responds to oral corticosteroids rapidly, while relapse does not respond to corticosteroid therapy. If there is any doubt in the differentiation between reaction and relapse, it is recommended to start corticosteroid therapy. Therefore any possible nerve damage can be prevented.

Prevention

Long before discovering the cause of leprosy and its treatment, the only way for its prevention was to isolate patients and to reject them from society due to their deformed appearances. As most of the population are resistant to leprosy naturally, the paucibacillary leprosy is not contiguous, the possibility of transmission of the multibacillary form will disappear shortly after the treatment is started, and the only way for leprosy transmission is prolonged close contact, therefore, patient isolation is not required.

Those with household contact are more susceptible to catch the disease, and they should undergo precise examination: once for contact of paucibacillary leprosy patients and once a year for 5 years for contact of multibacillary leprosy patients. Contacts should be educated for leprosy symptoms and necessity of seeking medical care promptly in case of any suspicious lesion.

Extensive studies evaluating possible anti-leprosy vaccines were carried out before the advent of (Gupte et al 2003) different vaccines, such as BCG alone or with *M. leprae* or other mycobacterium antigens have been studied, but their efficacy was variable. There is some recommendation that leprosy patients' household contacts should be vaccinated with a booster dose of BCG for prevention (Merle et al. 2010).

Prevention of leprosy-induced disabilities is very important. Patients should moisturize skin daily. If there is any scratch or ulcer, lesions should be irrigated and antiseptics applied to prevent secondary infection and if there are any signs of infection, treatment with appropriate antibiotics is prescribed.

Also wearing fitted shoes, physiotherapy, and exercise for prevention of muscular atrophy and contracture, frequent use of artificial eye drops, and eye dressing by ointment during night for the prevention of xerosis and abrasion of cornea have significant importance.

Conclusion

The prevalence of leprosy has decreased after the introduction of MDT in 1981 and the goal of elimination (prevalence less than 1/10,000 population) has been achieved in most parts of the world including Middle East. Our most important task at present is early detection of new cases in order to both reduce the rate and severity of disabilities and also reduce the risk of transmission to achieve eradication in near future.

References

Browne SG (1985) The history of leprosy. In: Hastings RC (ed) Leprosy. Churchill Livingstone, Edinburgh, pp 1–14

Daumerie D (2003) Elimination of leprosy as a public health problem—current status and challenges ahead. Scientific Working Group Report on leprosy. WHO, Geneva. pp 57–72

Dowlati Y, Shizarpour M, Tabatabaei H, Firooz A, Nassiri-Kashani M, Mapar M (2012) Mycobacterial diseases. In: Mortazavi H, Firooz A, Shahidi M (eds) Text book of dermatology for medical students and general practitioners, 1st edn. Sepidbarg, Tehran, pp 118–137

Firooz A (2006) Multidrug therapy regimen for leprosy. J Am Acad Dermatol 55:1115

Ghorpade A (2009) Inoculable leprosy. Int J Dermatol 48:1267–1268

Gupte MD, Manickam P, Kishore Kumar B (2003) Vaccines for leprosy and other mycobacterial diseases—What do we know today? Scientific Working Group Report on leprosy. WHO, Geneva, pp 86–89

Han XY, Seo YH, Sizer KC et al (2008) A new Mycobacterium species causing diffuse lepromatous leprosy. Am J Clin Pathol 130:856–864

James WD, Berger TG, Elston DM (2011) Andrews' diseases of the skin, 11th edn. Saunders Elsevier, China, pp 334–344

Job CK, Jayakumar J, Kearney M, Gillis TP (2008) Transmission of leprosy: a study of skin and nasal secretions of household contacts of leprosypatients using PCR. Am J Trop Med Hyg 78:518–521

Klioze AM, Ramos-Karo FA (2000) Visceral leprosy. Int J Dermatol 39:641–658

Lockwood D (2007) Leprosy. BMJ Clin Evid 915:1–11

Lockwood DNJ, Lambert SM (2011) Human immunodeficiency virus and leprosy: an update. Dermatol Clin 29:125–128

Menicucci LA, Miranda A, Antunes SL et al (2005) Microscopic leprosy skin lesions in primary neuritic leprosy. J Am Acad Dermatol 52:648–652

Merle CS, Cunha SS, Rodrigues LC (2010) BCG vaccination and leprosy protection: review of current evidence and status of BCG in leprosycontrol. Expert Rev Vaccines 9:209–222

Mohammed AP, Ramanujam K (1966) Leprosy in twins. Int J Lepr 34:405–406

Monot M, Honoré N, Garnier T, Zidane N, Sherafi D, Paniz-Mondolfi A et al (2009) Comparative genomic and phylogeographic analysis of Mycobacterium leprae. Nat Genet 41:1282–1289

Moschella SL (2004) An update on the diagnosis and treatment of leprosy. J Am Acad Dermatol 51:417–426

Oskam L, Slim E, Bührer-Sékula S (2003) Serology: recent developments, strengths, limitations and prospects: a state of the art overview. Lepr Rev 74:196–205

Prasad PVS (2005) All about leprosy, 1st edn. Jaypee Brothers Medical Publishers, New Delhi, pp 45–60

Setia MS, Shinde SS, Jerajani H, Boivin JF (2011) Is there a role for rifampicin, ofloxacin and minocycline (ROM) therapy in the treatment of leprosy? Systematic review and meta-analysis. Trop Med Int Health 16:1541–1551

Truman RW, Singh P, Sharma R, Busso P, Rougemont J, Paniz-Mondolfi A et al (2011) Probable zoonotic leprosy in the southern United States. N Engl J Med 364:1626–1633

World Health Organization (1982) Chemotherapy of leprosy for control programmes: technical report series 675. WHO, Geneva

World Health Organization (2003) The final push strategy to eliminate leprosy as a public health problem: Questions and answers. WHO, Geneva, pp 25–27

World Health Organization (2013) Leprosy. Prevalence rates (per 10 000 population). Available at: http://apps.who.int/neglected_diseases/ntddata/leprosy/leprosy.html. Last accessed 10 June 2013

Worobec SM (2009) Treatment of leprosy/Hansen's disease in the early 21st century. Dermatol Ther 22:518–537

Dengue Fever in Asia and Africa

Sadegh Chinikar and Nariman Shah-Hosseini

Contents

S. Chinikar (✉) • N. Shah-Hosseini (✉)
Arboviruses and Viral Hemorrhagic Fevers Laboratory (National Ref. Lab), Department of Virology, Pasteur Institute of Iran, Tehran, Iran
e-mail: sadeghchinikar@yahoo.com; nariman.shahhosseini@yahoo.com

M.A. McDowell and S. Rafati (eds.), *Neglected Tropical Diseases - Middle East and North Africa*, Neglected Tropical Diseases, DOI 10.1007/978-3-7091-1613-5_8,

Abstract Dengue is one of the most significant viral hemorrhagic fever infections worldwide. In the past 50 years the incidence of dengue fever has increased 30-fold and has expanded to areas which were previously free from the virus. An estimated 2.5 billion people live in dengue-endemic countries. Dengue virus (DENV) belongs to the family of *Flaviviridae* and genus *Flavivirus* with a single stranded positive-sense RNA. The virus consists of four serotypes (DEN-1 to -4). The main route of transmission to humans is by the bites of infected *Aedes* mosquitoes, principally *A. aegypti*. Dengue has a wide spectrum of clinical symptoms, often with unpredictable clinical outcomes. After 4–10 days incubation period, infection by any of the four virus serotypes can produce a wide spectrum of illnesses, although most infections are asymptomatic or subclinical. While the infection is self-limiting in the majority of cases and patients fully recover, in a small proportion of patients the infection progresses to severe disease, mostly characterized by plasma leakage with or without hemorrhage. In its hemorrhagic form, it can be easily misdiagnosed as other hemorrhagic diseases.

Keywords Dengue Fever virus • Clinical symptoms • Diagnosis • Treatment • Vector control • Epidemiology

History

The first reported epidemic of dengue-like disease dates back to 1779–1780, when outbreaks occurred in Batavia (Jakarta), Cairo, and Philadelphia (Guzman and Isturiz 2010). The earliest known use of the word dengue to describe an illness was in Spain in 1801. However, the most likely origin of the word is from Swahili. In both the 1823 and 1870 epidemics of dengue-like illness in Zanzibar and the EastAfrican coast, the disease was called *Ki-Dingapepo*. From this came the name *dinga* or *denga*, which was used to describe the illness in both epidemics. The illness was first called *dunga* in Cuba during the 1828 epidemic, but later changed to *dengue*, the name by which it has been known ever since (Gubler 1998). However, reports of illnesses compatible with dengue fever occurred even earlier. The earliest record found to date was in a Chinese "encyclopedia of disease symptoms and remedies," first published during the Chin Dynasty (265–420 AD). The disease was called "waterpoison" by the Chinese and was thought to be somehow connected with flying insects associated with water (Gubler 2008).

Agent

Dengue and dengue hemorrhagic fever (DHF) are caused by infection with the Dengue Fever Virus (DFV) that is a member of the genus *Flavivirus*, family *Flaviviridae*. Dengue virus consists of four closely related, but antigenically distinct, virus serotypes (DENV-1, DENV-2, DENV-3, and DENV-4) (Fig. 1). Infection with one of these serotypes does not provide cross protective immunity, so persons living in a dengue endemic area can have four dengue infections during their lifetimes (Watterson et al. 2012; Thai et al. 2005).

Electron microscopic imaging revealed that dengue virus has a relatively smooth surface, with a diameter of approximately 500 A° and an electron-dense core surrounded by a lipid bilayer (Kuhn et al. 2002), which contains 180 identical copies of the envelope protein. They are attached to the surface of the membrane by a short transmembrane segment. The envelope protein is responsible for attachment to a cell surface and triggers the process of infection (Zhang et al. 2003), (Fig. 2).

The *flavivirus* genome is approximately 11,000 bases and is made up of three structural and seven nonstructural proteins. There are three major subgroups within this family: tick-borne, mosquito-borne, and viruses with no known arthropod vector (Zhao et al. 1986; Roehrig et al. 2008), (Fig. 3).

Clinical Manifestations

After a person is bitten by an infective mosquito, the virus undergoes an incubation period of 3–14 days (average, 4–7 days). Generally, younger children under age 15 years are asymptomatic and have a milder illness than older children and adults. Infants and young children may have an undifferentiated febrile disease with a maculopapular rash (Rigau-Perez et al. 1998). Several population-based studies have demonstrated that the severity of DF clinical features can be highlighted with increasing age of the patient and with repeated infections (Rigau-Perez et al. 1998). Leucopenia and mild thrombocytopenia are among the main paraclinical findings for DF. The principal clinical symptoms of dengue are high fever (may be as short as 2 days and as long as 10 days), severe headache, backache, joint pains, nausea and vomiting, eye pain, and rash. A test of capillary fragility caused by an abnormality in the capillary wall or thrombocytopenia can be used to diagnose DF. This test, called a tourniquet test, is administered by applying a blood pressure cuff to a person's arm for 5 min and inflating to a pressure halfway between the diastolic and systolic blood pressure. The number of petechiae within a circumscribed area of the skin is then counted or the results reported in a range from negative (no petechiae) to +4 positive (confluent petechiae). A positive tourniquet test may be found in over one-third of patients with DF. Clinical findings alone are not very helpful to distinguish DF from other febrile illnesses such as the chikungunya, measles, leptospirosis, typhoid, or malaria. If symptoms start more

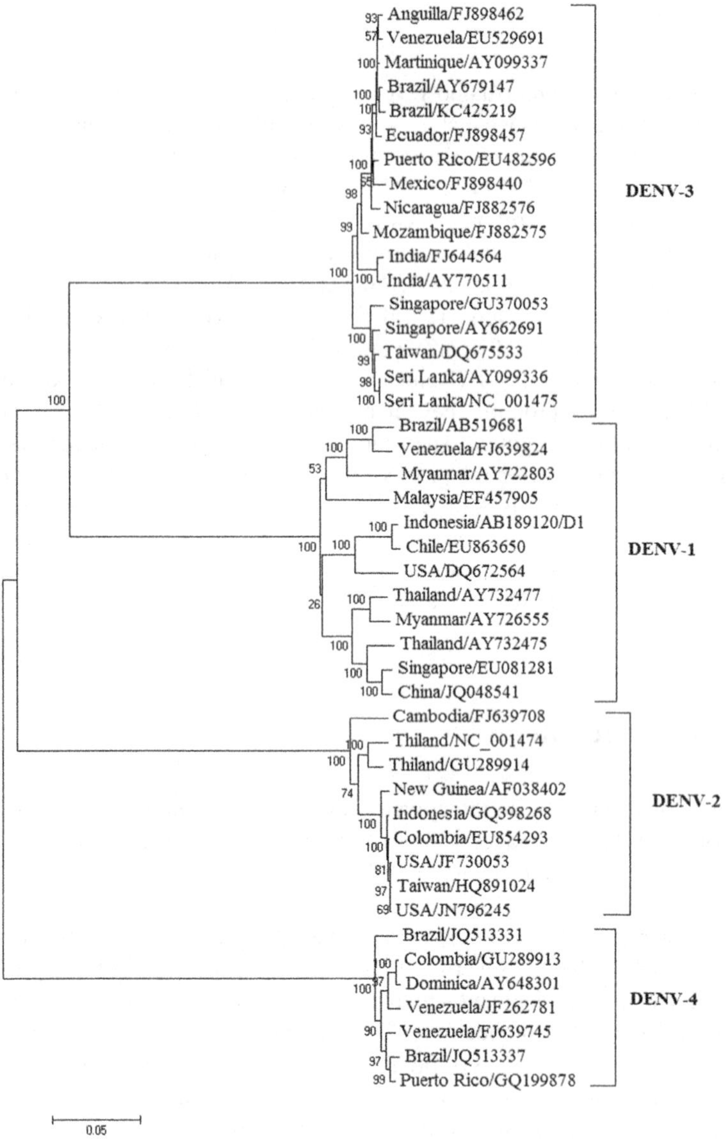

Fig. 1 Phylogeny tree illustrating four serotypes of dengue virus. The sequence alignment was performed by ClustalW and a phylogenetic tree generated by the Neighbor-Joining method (NJ) with Kimura 2-parameter distance using Mega 5 software. Bootstrap values were based on 1,000 replicates

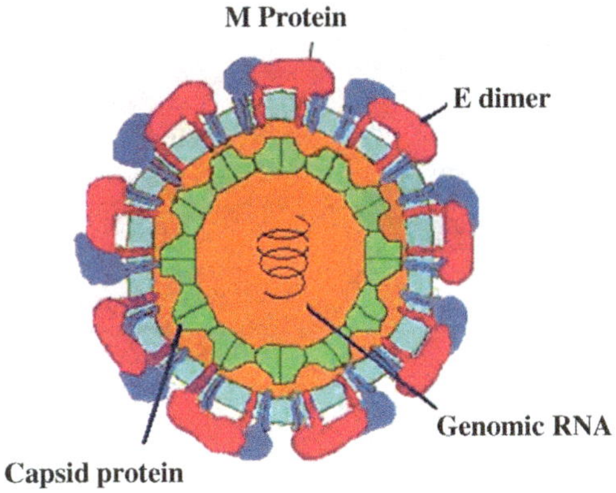

Fig. 2 Schematic graph of dengue virus

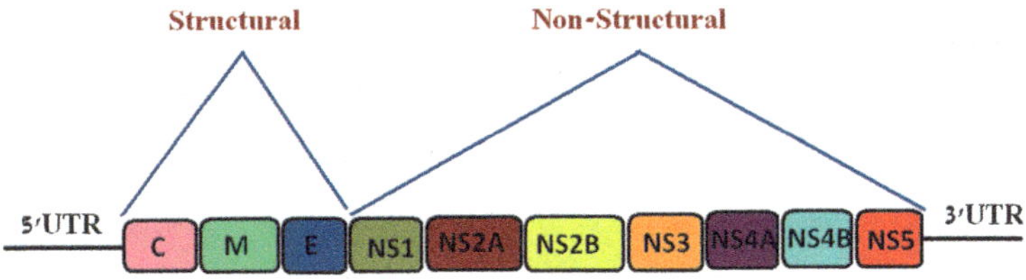

Fig. 3 Genome of dengue fever virus

than 2 weeks after the patient has left a dengue-endemic area, or if the fever lasts more than 2 weeks, dengue can be effectively ruled out (Rigau-Perez et al. 1998).

The WHO clinical case definition for DHF is the presence of all four criteria: fever that lasts from 2 to 7 days (typically high 38–40 °C), with general signs and symptoms that could occur with many other illnesses (e.g., nausea, vomiting, abdominal pain, and headache). The major pathophysiological change that can be used for determination of DHF severity and differentiation from DF is the leakage of plasma. The most common hemorrhagic feature is a positive tourniquet test (over 50 % of patients). Petechiae, easily bruised skin, and subcutaneous bleeding at venepuncture sites are present in most cases (Murgue et al. 1999).

DHF can be classified into four severity grades (I–IV), with grades III and IV being defined as dengue shock syndrome (DSS). DSS symptoms include thrombocytopenia (platelet count <100,000 cells/mm3), hemorrhagic tendency, hypotension, frank shock, and evidence of capillary leakage (i.e., hematocrit increase of ≥ 20 % from baseline, pleural effusion, ascites, or hypoproteinemia). Hemorrhagic tendency is defined as spontaneous bleeding or signs of capillary fragility. Spontaneous bleeding is evidenced as bleeding from the nose, gums, or gastrointestinal (GI) tract or hypermenorrhea or ecchymosis (Wichmann et al. 2007; Research SPf et al. 2009). The liver enzymes are usually mildly abnormal but jaundice is rare. The four warning signs for impending shock are intense, sustained

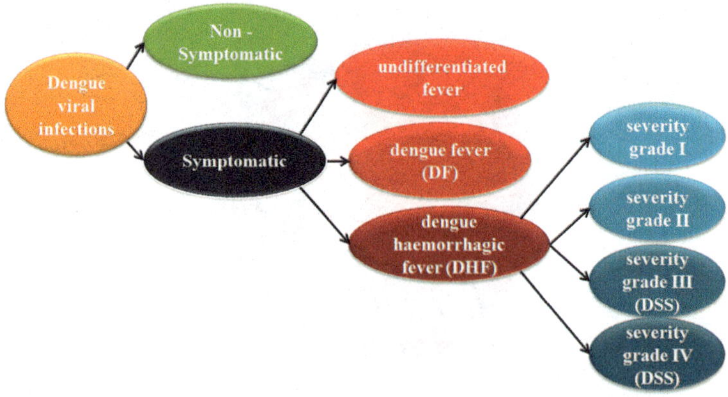

Fig. 4 WHO classification for dengue virus infections

abdominal pain; persistent vomiting; restlessness or lethargy; and a sudden change from fever to hypothermia with sweating and prostration (Rigau-Perez et al. 1998), (Fig. 4).

Laboratory Diagnostic Methods

For serological and virological detection of dengue fever, serum is the specimen of choice. Circulating virus remains detectable in the blood during the febrile period (for an average of 5 days after onset of symptoms) and is then rapidly cleared with the appearance of specific antibody. The virus is detectable in diagnostic samples up to 5 days at 4 °C. For longer storage, samples should be frozen at 60 °C or lower. Most laboratories attempting virus isolation utilize an established cell culture line of *Aedes albopictus* cells. After a week of incubation, the cells are stained with fluorescein-conjugated polyclonal antibodies to detect virus isolates (Rigau-Perez et al. 1998).

Based on the World Health Organization (WHO) guidelines, reported patients are classified into 2 categories: patients with probable and those with confirmed dengue virus infection (Research SPf et al. 2009). A probable diagnosis of dengue virus infection is supported by a single positive IgM antibody test on a late acute- or convalescent-phase serum specimen (i.e., 4–7 or 8–45 days, respectively, after the onset of symptoms). For confirmation, virus is detected by reverse-transcriptase polymerase chain reaction (RT-PCR), shortening the time required for detection, or by a considerable change in IgM or IgG antibody titers in paired serum samples (Wichmann et al. 2007).

Serological Assays

Although Enzyme-Linked Immunosorbent Assay (ELISA) is suitable for detection of dengue antibodies and has shown high sensitivity and specificity (Vaughn et al. 1999). Interpretation of ELISA can be complicated, however, as dengue virus along with other members of *Flaviviridae*, including those causing yellow fever and several encephalitis (e.g., Japanese, St. Louis, West Nile, and tick borne), share group antigens that can crossreact in serological tests (Calisher et al. 1989). This method relies on the presence of IgM antibodies or a rise in IgG antibody titre in paired acute and convalescent phase sera (Thai et al. 2005; Rigau-Perez et al. 1998).

IgM: To detect IgM antibody, serum samples are provided during the acute phase of the illness and 90 % of patients are IgM positive by the sixth day after onset of symptoms. Generally, this antibody class may be detectable for 60 days. A routine test for measurement of IgM is an ELISA (Rigau-Perez et al. 1998; Matheus et al. 2005; Falconar et al. 2006).

IgG:Patients who are infected by dengue virus for the first time, IgG antibody starts to rise by the fifth day after onset of symptoms. Titers increase gradually for some weeks and then remain detectable for many years. Patients who are infected by dengue virus for the second time, IgG antibody is generally already present in early acute serum samples and titers rise rapidly over a few days. IgG antibody is commonly measured by the hemagglutination inhibition test or ELISA(Rigau-Perez et al. 1998; Matheus et al. 2005; Falconar et al. 2006).

Molecular Detection

Reverse transcription polymerase chain reaction (RT-PCR) is a rapid test for molecular detection of genetic elements of DFV, however, the RT-PCR test is experimental and no commercial products are available. For this purpose, viral RNA is extracted from serum, reverse transcribed and amplified by Dengue Virus-specific primers. For DENV serotyping, nested PCR may be performed from the initial amplification reaction. The PCR products are visualized by agarose gel electrophoresis stained with ethidium bromide or by quantitative methods (Rigau-Perez et al. 1998; Grobusch et al. 2006).

Treatment

Treatment strategies vary from patient to patient. General prescriptions for patients with DF include rest, oral fluids, antipyretics (acetaminophen but not aspirin). To compensate for lost fluids an intravenous line should be placed. Monitoring of

Fig. 5 *Aedes albopictus* (*left*) and *Aedes aegypti* (*right*). Pictures by James Gathany, CDC

blood pressure, hematocrit, platelet count, hemorrhagic manifestations, urinary output, and level of consciousness can play a vital role in the management of DF patients. In case of severe plasma leakage in DHF patients, hematocrit may continue to rise even while intravenous fluids are being administered; however, the "leaky capillary" period is short and intravenous fluids are usually required for only 1–2 days. If fluids replacement has not been adequately supplied, worsening shock, acidosis, and disseminated intravascular coagulation is expected, while fluid overload causes massive effusions, respiratory compromise, and congestive heart failure (Hung et al. 2006). Isotonic solutions and plasma expanders, such as Ringer's acetate or Ringer's lactate, plasma protein fraction, and dextran 40 are helpful and usually given to patients (Waterman and Gubler 1989; Wills et al. 2005).

Insect Vectors

The *Aedes aegypti* mosquito is the most important transmitter or vector of dengue viruses, although a 2001 outbreak in Hawaii was transmitted by *Aedes albopictus. A. aegypti*is a small, black-and-white mosquito that is highly adapted to humans and urban environments and was spread throughout the tropics of the world by the shipping trade (Hopp and Foley 2003), (Fig. 5).

A. aegypti prefers to lay its eggs in artificial containers commonly found in and around homes in the tropics, for instance, flower vases, old automobile tires, buckets that collect rainwater, and trash in general. Containers used for water storage have a significant role in producing large numbers of adult mosquitoes in close proximity to dwellings where people live and work. The adult mosquitoes prefer to rest indoors, are unobtrusive, and prefer to feed on humans during daylight hours. Female *A. aegypti* mosquitoes are very nervous feeders, disrupting the feeding process at the slightest movement, only to return to the same or a different person to continue feeding moments later. Due to this behavior, females will usually feed on several persons during a single blood meal and, if infective, may transmit dengue virus to multiple persons in a short period of time even if they only probe without taking blood. It is commonly observed that several members of the same household become ill with dengue fever within a 24- to 36-h time frame, suggesting transmission by a single infective mosquito. These characteristics make *A. aegypti* an efficient epidemic vector (Gubler 1998).

Vector Control

As no vaccine is currently available, the primary means of controlling dengue is by controlling the mosquitoes. In contrast to other major vector-borne diseases, such as malaria, leishmaniasis, and Chagas disease, in which most vector control and prevention activities target the adult stages of the vector, the prevention of *Aedes* sp. infestation has typically been directed against the immature stages of the mosquito. Larval control decreases infestation levels throughout the year or during the rainy season (Vanlerberghe et al. 2013).

In Asia and the Americas, *A. aegypti* mosquitoes breed in human-made containers used for water storage or in other items that collect water. In Africa, they also breed extensively in natural habitats such as tree holes and leaf axils. Eggs of these mosquitoes are laid singly or in rafts, and they may stick to the surface or they may sink if the water is disturbed. *A. albopictus*, a secondary dengue vector in Asia, recently has become established in the USA, several Latin American and Caribbean countries, parts of Europe, and at least one country in Africa (Chow et al. 1998; Moore and Mitchell 1997).

Vector control is implemented by environmental management and application of insecticides, along with social awareness regarding proper solid waste disposal and improved water storage practices (Ligon 2005). As weather conditions have a significant effect on mosquito populations and thus dengue cases, developing a dengue forecasting model to provide early warning of a dengue outbreak to allow adequate time for effective vector control to be implemented has a critical role for surveillance systems. In this regard, in the late 1990s, a study that investigated the relationship between dengue cases and *Aedes* mosquito population as well as weather conditions in Singapore demonstrated that increasing temperatures precede rising dengue incidence (Heng et al. 1998).

Another investigation on the association between weather variables and dengue cases in Singapore suggested that minimum and maximum temperature are strong weather predictors for the increase of dengue cases, whereas rainfall and relative humidity have little correlation with dengue cases (Hii et al. 2012).

Environmental Management

Global warming, ozone depletion, and devastating rain are some of the most outstanding dire repercussions of the modern world. Consequently, climate change and discharging waste carelessly have led to increasing in insect population (Barclay 2008). To fight against dengue fever, covering water containers, changing stored water, introduction of *Mesocyclops* into water tanks as predators of *A. aegypti*, removal of discarded containers through community and running cleanup campaigns are some strategies to control the population of *A. aegypti* as the main vector for dengue virus in environment (Nam et al. 2005; Nam et al. 2000).

Well-screened door and windows or air conditioning are also among practical methods of vector control. Window curtains and domestic water container covers treated with insecticide can reduce densities of dengue vectors to low levels and potentially affect dengue transmission (Reddy and Zembower 2008).

Insecticides

Chemical

DDT was the first organ chlorine insecticide. Operations against adult mosquitoes involve spraying organophosphates or pyrethroids in outbreak areas where a suspected dengue case is reported. Using larvicides such as temephos against container breeding *Aedes* mosquitoes in clean water is also an effective method (Seleena et al. 2001; Itrat et al. 2008) at controlling vector populations. Carbamates are particularly useful against adult mosquitoes and synthetic pyrethroids are effective against both larvae and adult mosquitoes (Priest 1992).

Biological

Microbial control agents such as the bacteria *Bacillus thuringiensis* serovar *israelensis* (*B.t.i.*) are known for their efficacy and selectivity against mosquito larvae. *B.t.i.* has been formulated into hydrophilic powders, corn-cob granules, suspension concentrates, briquettes, pellets, and tablets. *B.t.i.* has been used for at least a decade in many countries; its success in controlling mosquitoes in Europe and Asia is well established (Azirun 2010). The second bacterium, *B. sphaertcus*, has recently been registered for use in the USA (Seleena et al. 2001).

One of the major disadvantages in the use of *B.t.i.* is its rapid inactivation (24-48 h) in the environment. Thus larval populations of stagnant water mosquitoes recover within 5–7 days following treatment.

Nevertheless, using biological control methods preponderate to chemical types, since firstly, biological control programs are more environmentally friendly, and secondly, mosquitoes can acquire resistance to chemicals (Priest 1992).

Social Awareness

To achieve sustainable and cost-effective *A. aegypti* control in long term, community-based, grass roots efforts are a good strategy. Thanks to social awareness, dengue fever control can be changed from an emergency response to a preventive approach. Training of health professionals down to community volunteers are an essential element to success (Teng and Singh 2001).

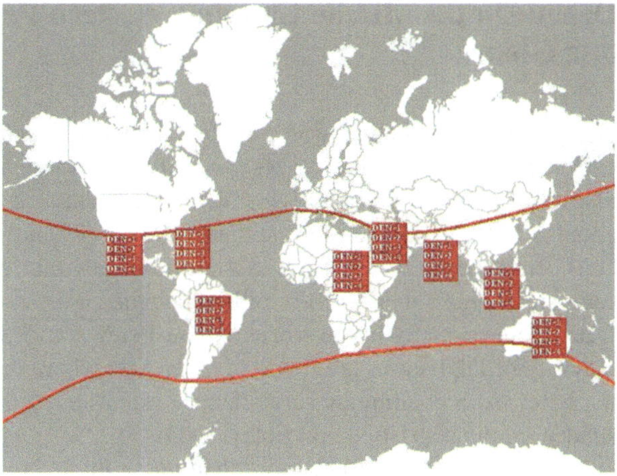

Fig. 6 Worldwide distribution of dengue virus serotypes

Education campaigns are targeted medical community as well as community members, especially in endemic regions. These campaigns are often supported by schools and the Youth, Women's and Farmers' Unions and include open-air community meetings, radio and television broadcasts, loudspeaker announcements, travelling school drama performances, church meetings, football competitions, as well as posters, leaflets, and billboards (Spiegel et al. 2005).

Education of school children is a primary factor that triggers the engagement of the entire community in mosquito control. Likewise, the participation of the local population can be helpful to control the transmission of dengue fever. Education constitutes an effective manner to modify behavior in relation to the formation of mosquito larval habitats, potentially has a long effect, and should form part of a strategy to include the community (Spiegel et al. 2005).

It must be emphasized, however, that community ownership and sustainability can only be achieved if government officials are committed to this type of partnership and implement policies to facilitate community involvement in *A. aegypti* control. Based on studies television, teachers, relatives, Internet, newspaper, radio, and personal physician are the most effective ways to education people about dengue fever. Thus, surveillance systems should heed special attention to these communication methods (Van Benthem et al. 2002).

Epidemiology of Dengue Fever

Major epidemics of dengue-like illnesses have been reported in the Americas, southern Europe, North Africa, the eastern Mediterranean, Asia, and Australia, as well as the islands in the Indian Ocean, the south and central Pacific Ocean, and the Caribbean as far back as the latter part of the eighteenth century (Ligon 2005), (Fig. 6).

Dengue in the Middle East and Western Pacific Region at a Glance

Fifty years ago, dengue was reported in just a handful of countries, now it is endemic in over 100 and causes more illness and death than any other arboviral disease of humans (Gubler 1998). WHO has estimated that dengue fever threatens about 2.5 billion people in more than 100 tropical and subtropical countries and is considered the most important mosquito-borne viral disease in the world. Current estimates suggest that 50–100 million dengue cases and 250,000–500,000 cases of DHF occur annually in the world (Rigau-Perez et al. 1998; Wills et al. 2004). The majority of DHF cases are reported from Asia where the disease has affected most countries, so a continuous surveillance system is required to monitor and prevent sudden outbreaks (Hopp and Foley 2003).

Between 2001 and 2008, more than one million cases were reported in Cambodia, Malaysia, Philippines, and Vietnam, the four countries in the Western Pacific Region with the highest numbers of cases and deaths (Arima 2011).

Dengue in Iran

Dengue is a serious public health concern in the Western Pacific Region, which are interested destinations among Iranian travelers for business and education (Kumaria 2010). In 2008, the first case of imported dengue fever reported in Iran, where the patient had previously travelled to Malaysia (Kuala Lumpur) with a history of going to forest in this journey. Thereafter, no other cases were officially reported in the Iranian population until 2012 (Chinikar et al. 2010). After first dengue case in Iran, further investigations on human suspected cases and vectors led to new findings. The existence of *A. aegypti* as the main vector of the disease still has not been fully proven in Iran.

Patients with dengue fever are mainly male in Iran. Gender differences in Iranian dengue fever patients can be explained by employment, as males participate in more business travel to endemic areas than females. Overall, it can be concluded that the majority of cases of dengue fever in Iran occur among returning travelers (Chinikar et al. 2013).

Dengue in Yemen

The historic record of dengue infection in Yemen goes back to the nineteenth century when a severe outbreak was reported in 1870–1873. In 1954 a dengue epidemic was also reported in Yemen, when 98 % of the population in Alhudaydah was affected (Madani et al. 2013). However, more frequent outbreaks of dengue have emerged since 2000, but some of these outbreaks were not well-documented or published, which was the case in Shabwah governorate in 2001, 2002, and 2005 and the outbreaks in Aden and Taiz (2010). Documented outbreaks were in Shabwah governorate (1994), Hadramout/Mukalla (2005), and Al-Hudidah governorate (1994, 2000, 2004, and 2005). Furthermore, travel-associated dengue has been reported among travelers from Yemen to the USA and Italy. These travel-associated cases were caused by dengue virus (DENV) 2 and 3. In 2010, several areas in Yemen experienced dengue outbreaks with approximately 100,000 infections and 200 deaths, the most devastating of which occurred in Hadramout coastal districts in southeastern Yemen (Ghouth et al. 2012).

Dengue in Saudi Arabia

In 1994, the first outbreak of dengue fever coincided with the period of the Haj (annual pilgrim), consequently the virus was isolated, and it is possible that virus was introduced with the pilgrims traveling from Southeast Asia (Al-Azraqi et al. 2013). The majority of dengue fever patients are male in Saudi Arabia (Al-Azraqi et al. 2013). In Saudi Arabia, young adults are the most common age group involved for DF. The high rates of infection among those aged 15–30 years reflects the fact that younger adults are more likely to spend time outdoors doing work or recreational activities and hence be exposed to mosquito bites (Alzahrani et al. 2013).

In terms of seasonal pattern of dengue occurrence in Jeddah, data shows the highest number of cases occurring in the first quarter of the year. This could be attributed to the rain that occasionally occurs at that time. Recent entomological studies have shown that adult female *A. aegypti* has been detected in Jizan, Abha (where monkeys are also found and hence the potential of a sylvatic cycle for DENV) and also from various other parts of the Kingdom of Saudi Arabia. The majority of documents have demonstrated that dengue virus type 1, 2, and 3 are common in Saudi Arabia (Ayyub et al. 2006; Fakeeh and Zaki 2001; Khan et al. 2008). According to Saudi authorities' report in 2007, the case fatality rate for dengue was 4.6 per thousand (6/1308) in Jeddah. By 2008, 2,500 confirmed infections, with 59 cases of dengue hemorrhagic fever, 18 of DSS, and 12 deaths, were documented in three major outbreaks in Saudi Arabia (Shibl et al. 2012). Afterwards, in 2009, a total of 3,350 dengue cases in Saudi Arabia were detected (Al-Azraqi et al. 2013).

Dengue in Pakistan

In 1934, Barraud collected *A. aegypti* from Peshawar, Dera Ismail Khan, Lahore, Larkana, and Karachi. In 1949, Qutubuddin reported *A. aegypti* from Kohat-Hangu valley in northern Pakistan. The distribution of *A. aegypti* decreased significantly after 1950 as a byproduct of a malaria vector eradication program. In the southern metropolis of Karachi, *A. aegypti* was reported in a survey conducted in 1983 to determine the species of mosquitoes acting as vectors of different diseases. However, at that time, the distribution of *A. aegypti* was limited to the port city of Karachi, and it was not collected from the neighboring district of Thatta. The second re-emergence of *A. aegypti* occurred in Landi Kotal (Rasheed et al. 2013).

The presence of dengue virus has been detected using neutralization and hemagglutination inhibition antibodies in local populations in Pakistan since the 1960s. However, the first epidemic was not reported until 1994 in Karachi. This was followed by some cases in 1995, but the disease was confined to the port city of Karachi. Since 2006, dengue epidemics have occurred every year and the range has extended to most cities in Pakistan. Dengue now affects thousands of people and has caused hundreds of deaths (Rasheed et al. 2013).

Dengue fever is a challenging issue in young patients. Changing lifestyles, urbanization, explosive population growth, destruction of city water supplies, migration, and increased air travel are some of the reasons that cause increases in the prevalence of dengue infection. The peak incidence of dengue is coinciding with hot summer and monsoon season (Chan et al. 1995; Khan et al. 2010).

Dengue in India

Dengue virus was first isolated in India in 1945, while the first recorded outbreak of dengue fever in India dates from 1812. Also, the first dengue epidemic in India occurred in Kolkata in 1963–1964, and since that time, the epidemiology of dengue virus has been evolving. All four dengue virus serotypes co-circulate in the India. The circulation of multiple dengue virus serotypes can cause concurrent infections like dengue outbreak in Davangere from June 2011–March 2012 (VinodKumar et al. 2013). Studies have depicted that rain, temperature, and relative humidity play a significant role in outbreak of dengue fever as climatic factors. Young adults are among high risk groups for dengue fever in India. In terms of sex distribution, dengue fever is more common in Indian men than women (Dar et al. 1999; Singh et al. 2005; Chakravarti and Kumaria 2005; Raheel et al. 2010). The numbers of dengue confirmed cases have steadily increased in Indian territories from 2007 onwards. In 2012 the number of dengue cases was almost three times higher than in 2011. The epidemiology of dengue in India shows a classic epidemic pattern of transmission with sporadic outbreaks, with low to moderate numbers of cases, usually localized to urban centers and neighboring regions, but occasionally spreading and causing larger epidemics (Mariappan 2013).

Dengue in Malaysia

Dengue is endemic in Malaysia. The first reported cases of dengue were in 1902. The disease was notified in 1973 and the first outbreak of DHF was reported in 1962. During the decade of 1973–1982, there were 12,077 dengue cases with a case fatality rate of 3.38 %. In the following decade of 1983–1992, the number of reported cases increased to 26,361 but the case fatality rate dropped to 0.55 %. From 2000 to 2010, a total of 385,758 dengue cases (including DHF) and 997 deaths due to dengue were reported in Malaysia. There is an overall escalating trend in annual incidences and death of dengue in Malaysia between 2000 and 2010 (Mia et al. 2013). The reason for raising the prevalence of DF in recent years could be related to boosting the economy, rapid industrialization and creating many manmade opportunities for *Aedes* mosquito breeding. This coupled with rural–urban migration and pockets of illegal settlements, indiscriminate solid-waste disposal and a tropical rainfall, provide fertile grounds for *Aedes* breeding, and the rise of dengue transmission in the country (Teng and Singh 2001). Epidemiology data has shown that dengue fever is most seen in men than women. All age group are affected with the most vulnerable among the school-going children and young adults (Mia et al. 2013). All four serotypes of dengue exit in Malaysia. The case fatality rate for dengue 3 outbreak is highest (0.77 %), followed by dengue 2 (0.54 %), and dengue 1 (0.35 %) (Lam 1994; George 1987).

Dengue in Thailand

The incidence of DHF in Thailand shows some fluctuations from year to year. Dengue is being observed mostly among children in Thailand. All four serotypes of dengue virus are circulating in Thailand. Several factors have significant role in pattern of DHF, such as environmental and climate factors and viral factors (Cummings et al. 2004).

Dengue in Indonesia

Dengue fever was first reported from the Indonesia during outbreaks on the island of Java in 1968. All 4 dengue serotypes exist in Indonesia, but the dengue three serotypes is the most common. Regarding age and sex incidence, it can be claimed that the majority of patients are children. Dengue fever outbreaks have been occurring mainly in the rainy season (Sumarmo et al. 1983; Corwin et al. 2001).

Dengue in Singapore

DHF appeared in Singapore in the 1960s and quickly became a major cause of childhood death. Public health response to dengue began in 1966. DHF and DF were made a reportable disease in 1972 and 1977, respectively. From 1966 to 1968, following a series of entomologic surveys and a pilot project to control the *Aedes* vectors in an area with high incidence of DHF, a vector control system based on entomologic surveillance and larval source reduction was developed (Ooi et al. 2006; Ooi 2001). With the reduced *A.aegypti* population, Singapore experienced a 15-year period of low dengue incidence. However, Singapore has faced a resurgence of dengue/DHF since 1990, with high incidence in adults and few cases in children (Gubler and Clark 1995; Ooi et al. 2001).

Dengue in Vietnam

In Vietnam, epidemics of dengue often occur from June to November. Peak transmission occurs from July to September and is closely associated with the rainy season, the prime breeding period for the mosquito vector (Cuong 2013). Dengue is endemic throughout Vietnam, but transmission is highest in the south where very large epidemics occur regularly and all four serotypes of dengue virus have already been identified (Thai et al. 2005).

Since 1963, the incidence of DHF has constantly increased in Vietnam. From 1995, a preventive approach of dengue fever in Vietnam was developed with the assistance of the World Health Organization. Three years later, the eradication of *A. aegypti* was achieved at PhanBoi Village, 31 km east of Hanoi (Nam et al. 2005; Kay et al. 2002; Phuong et al. 2006). In 1998, a widespread DHF epidemic affected 19 provinces in southern Vietnam; 119,429 cases of DHF and 342 deaths were reported (Ha et al. 2000; Coudeville and Garnett 2012).

Dengue in Africa Continent at a Glance

The existence of dengue dates back to 1956, when a retrospective serosurvey confirmed that a dengue epidemic occurred in Durban, South Africa, in 1926–1927. Additional evidence was obtained in the 1960s, when DENV-1 and DENV-2 were isolated for the first time from human samples in Nigeria (Diallo et al. 2003). Since 1980, epidemic dengue fever caused by all four serotypes has increased dramatically in Africa. Most dengue cases has reported in East Africa, and major epidemics were reported for the first time in the South Africa and Senegal (in the early 1900s), Yemen (1920), Seychelles (1977), Kenya (1982, DEN-2), Mozambique (1985, DEN-3), Djibouti (1991–1992, DEN-2), Somalia (1982, 1993,

DEN-2), and Saudi Arabia (1994, DEN-2) (Gubler and Clark 1995). The first DEN viruses isolated in Africa were from humans in Nigeria (Gubler 2004).

Dengue in Sudan

Dengue has been reported in different regions of the Sudan (Adam et al. 2010). DEN-2 was first reported in Port Sudan in 1986. An outbreak of acute febrile illness occurred later in 1989 in the Northern Province of Sudan and the prevalence of DEN-2 antibody was 24 %. Additional serological evidence of DEN-2 infection in Sudan was reported 1995. *Aedes* mosquitoes have been reported from many different regions of Sudan (Malik et al. 2011).

Dengue in Somalia

In 1993, dengue fever was reported among United States troops in Somalia. It was found that 2 % of cases were caused by DEN-3 and 41 % by DEN-2. During the outbreak, only cases of classical dengue were seen, and there were no cases of severe dengue (Kanesa-thasan et al. 1998; Sharp et al. 1995).

Dengue in Kenya

In 1982, an outbreak of dengue fever caused by DEN-2 was reported in the Kenyan coastal towns of Malindi and Kilifi; clinical presentation was consistent with classical dengue fever, with no severe dengue reported. Since then there have been sporadic cases of dengue reported in Kenya (unpublished observations), and a serology survey carried out in 2005 (unpublished) revealed the occurrence of dengue transmission in coastal and inland parts of Kenya (Sang 2007; Johnson et al. 1982).

Dengue in Mozambique

In 1984–1985, an outbreak of dengue was reported in Pemba, Mozambique and led to detection of DENV-3 for the first time. During this epidemic, two deaths were reported and most patients appeared to be experiencing secondary infection with *flavivirus* (Sang (2007); Messer et al. 2003).

Dengue in Senegal

In 1970, evidence of dengue virus circulation in Senegal was obtained when DENV-2 was isolated for the first time from human blood. Different studies identified several DENV-2 epizootics through periodic amplification of the sylvatic cycle. Although DENV 1–4 were isolated incidentally in Senegal from humans, only DENV-2 have been repeatedly shown to be circulating in mosquito, human, and monkey populations with a sylvatic focus in Kedougou in southeastern Senegal (Diallo et al. 2008). DENV-2 reemerged in Senegal, 1999, after 8 years of silence (Diallo et al. 2003).

Dengue in Nigeria

There are handful documented reports about the epidemiology of dengue fever in Nigeria. In the 1960s, DEN-1, -2 and -3 were isolated for the first time from samples taken from humans in Nigeria (Vasilakis et al. 2008; Carey et al. 1971). Based on a serological survey for dengue immunity in 1977, neutralization tests performed on 1,816 human sera from different geographical locations revealed that 45 % of Nigerians were immune to dengue type-2 virus. The percentage of immunity in adults aged 20 years and older (51 %) was considerably higher than in children (37 %) ($P < 0.01$). The highest percentage of dengue N antibody was observed in the derived savannah zone (63 %), followed by the rainforest zone (42 %). The Southern Guinea savannah and plateau zones had lower percentages of dengue-immune persons. There was a higher prevalence of antibodies in urban (48 %) than in rural communities (37 %) (Fagbami et al. 1977).

Gender Differences

Understanding male–female differences in infection rates and severity of disease is important for public health control programs. The incidence of dengue fever in males and females varies in different studies. For instance, some studies have shown a similar incidence rate for males and females (Kaplan et al. 1983), while other reports have demonstrated higher incidence rates in females only (San Martin et al. 2010), and others in males only (Shekhar and Huat 1992; Agarwal et al. 1999). Three independent studies from epidemics in India and Singapore have shown nearly twice the number of male patients compared to females. In contrast, Kaplan (Kaplan et al. 1983), in a rare study testing for significance, found a higher proportion of women in all of his four Mexican samples. Surprisingly, some studies have demonstrated that in spite of higher incidence of DF in males, severe illness seen among females (Chinikar et al. 2013; Guha-Sapir and Schimmer 2005).

Age Distribution

The age distribution of dengue fever has important implications for control and prevention. In general, dengue fever is usually seen in children and middle age and is an important cause of pediatric hospital admission. In general, children are less symptomatic than adults in the febrile phase of uncomplicated dengue (Thu et al. 2012). These are people who are active outdoors, whether working, schooling, or playing outside their homes. However, the number of dengue fever infection cases and death cases among older age groups has been increasing in recent years (Ooi et al. 2006; Guha-Sapir and Schimmer 2005).

Race Susceptibility

Race-related susceptibility to dengue has been seen in a handful studies. Take the Cuban epidemics study, 1977 and 1981, as an example, results illustrated that blacks and whites were equally infected with DEN-1 and DEN-2 viruses, while severe dengue disease was observed less frequently in dengue-infected black persons than whites. The reason behind protecting person of African origin may be genetic polymorphism in cytokine profiles and coagulation proteins. Other studies in Asia have shown that the higher incidence of DHF among Chinese compared to Malaysian males. Although there is no solid statement regarding race susceptibility, conducting this kind of studies can help for better understanding of virus pathology (Guha-Sapir and Schimmer 2005).

References

Adam I, Jumaa AM, Elbashir HM, Karsany MS (2010) Research maternal and perinatal outcomes of dengue in PortSudan, Eastern Sudan. Parity 2:2.3

Agarwal R, Kapoor S, Nagar R, Misra A, Tandon R, Mathur A et al (1999) A clinical study of the patients with dengue hemorrhagic fever during the epidemic of 1996 at Lucknow, India. Southeast Asian J Trop Med Public Health 30(4):735–740

Al-Azraqi TA, El Mekki AA, Mahfouz AA (2013) Seroprevalence of dengue virus infection in Aseer and Jizan regions, Southwestern Saudi Arabia. Trans R Soc Trop Med Hyg 107(6):368–371

Alzahrani AG, Al Mazroa MA, Alrabeah AM, Ibrahim AM, Mokdad AH, Memish ZA (2013) Geographical distribution and spatio-temporal patterns of dengue cases in Jeddah Governorate from 2006-2008. Trans R Soc Trop Med Hyg 107(1):23–29

Arima Y (2011) Epidemiologic update on the dengue situation in the Western Pacific Region, 2010. Western Pac Surveill Response J 2(2):1–5

Ayyub M, Khazindar AM, Lubbad EH, Barlas S, Alfi AY, Al-Ukayli S (2006) Characteristics of dengue fever in a large public hospital, Jeddah, Saudi Arabia. J Ayub Med Coll Abbottabad 18 (2):9–13

Azirun MSB (2010) Susceptibility of field-collected Aedes aegypti (L.)(Diptera: Culicidae) to Bacillus thuringiensis israelensis and temephos. Trop Biomed 27(3):493–503

Barclay E (2008) Is climate change affecting dengue in the Americas? Lancet 371(9617):973–974
Calisher CH, Karabatsos N, Dalrymple JM, Shope RE, Porterfield JS, Westaway EG et al (1989) Antigenic relationships between flaviviruses as determined by cross-neutralization tests with polyclonal antisera. J Gen Virol 70(1):37
Carey D, Causey O, Reddy S, Cooke A (1971) Dengue viruses from febrile patients in Nigeria, 1964-68. Lancet 297(7690):105–106
Chakravarti A, Kumaria R (2005) Eco-epidemiological analysis of dengue infection during an outbreak of dengue fever. India Virol J 14:2–32
Chan Y, Tan H, Seah C, Li J, Chow V, Salahuddin N et al (1995) Dengue haemorrhagic fever outbreak in Karachi, Pakistan, 1994. Trans R Soc Trop Med Hyg 89(6):619–620
Chinikar S, Ghiasi S, Moradi M, Madihi S (2010) Laboratory detection facility of dengue fever (DF) in Iran: the first imported case. Int J Infect Dis 8(1):18
Chinikar S, Ghiasi SM, Shah-Hosseini N, Mostafavi E, Moradi M, Khakifirouz S et al (2013) Preliminary study of dengue virus infection in Iran. Travel Med Infect Dis 11:166–169
Chow V, Chan Y, Yong R, Lee K, Lim L, Chung Y et al (1998) Monitoring of dengue viruses in field-caught Aedes aegypti and Aedes albopictus mosquitoes by a type-specific polymerase chain reaction and cycle sequencing. Am J Trop Med Hyg 58(5):578–586
Corwin AL, Larasati RP, Bangs MJ, Wuryadi S, Arjoso S, Sukri N et al (2001) Epidemic dengue transmission in southern Sumatra, Indonesia. Trans R Soc Trop Med Hyg 95(3):257–265
Coudeville L, Garnett GP (2012) Transmission dynamics of the four dengue serotypes in southern Vietnam and the potential impact of vaccination. PLoS One 7(12):e51244
Cummings DAT, Irizarry RA, Huang NE, Endy TP, Nisalak A, Ungchusak K et al (2004) Travelling waves in the occurrence of dengue haemorrhagic fever in Thailand. Nature 427 (6972):344–347
Cuong HQ (2013) Spatiotemporal dynamics of dengue epidemics, southern Vietnam. Emerg Infect Dis 19(6):945–953
Dar L, Broor S, Sengupta S, Xess I, Seth P (1999) The first major outbreak of dengue hemorrhagic fever in Delhi, India. Emerg Infect Dis 5(4):589
Diallo M, Ba Y, Sall AA, Diop OM, Ndione JA, Mondo M et al (2003) Amplification of the sylvatic cycle of dengue virus type 2, Senegal, 1999-2000: entomologic findings and epidemiologic considerations. Emerg Infect Dis 9(3):362
Diallo M, Ba Y, Faye O, Soumare ML, Dia I (2008) Vector competence of Aedes aegypti populations from Senegal for sylvatic and epidemic dengue 2 virus isolated in West Africa. Trans R Soc Trop Med Hyg 102(5):493–498
Fagbami A, Monath T, Fabiyi A (1977) Dengue virus infections in Nigeria: a survey for antibodies in monkeys and humans. Trans R Soc Trop Med Hyg 71(1):60–65
Fakeeh M, Zaki A (2001) Virologic and serologic surveillance for dengue fever in Jeddah, Saudi Arabia, 1994-1999. Am J Trop Med Hyg 65(6):764–767
Falconar AKI, de Plata E, Romero-Vivas CME (2006) Altered enzyme-linked immunosorbent assay immunoglobulin M (IgM)/IgG optical density ratios can correctly classify all primary or secondary dengue virus infections 1 day after the onset of symptoms, when all of the viruses can be isolated. Clin Vaccine Immunol 13(9):1044
George R (1987) Dengue haemorrhagic fever in Malaysia: a review. Southeast Asian J Trop Med Public Health 18(3):278
Ghouth ASB, Amarasinghe A, Letson GW (2012) Dengue outbreak in hadramout, Yemen, 2010: an epidemiological perspective. Am J Trop Med Hyg 86(6):1072
Grobusch M, Niedrig M, Gobels K, Klipstein-Grobusch K, Teichmann D (2006) Evaluation of the use of RT-PCR for the early diagnosis of dengue fever. Clin Microbiol Infect 12(4):395–397
Gubler DJ (1998) Dengue and dengue hemorrhagic fever. Clin Microbiol Rev 11(3):480–496
Gubler DJ (2004) The changing epidemiology of yellow fever and dengue, 1900 to 2003: full circle? Comp Immunol Microbiol Infect Dis 27(5):319–330

Gubler DJ (ed) (2008) Dengue/dengue haemorrhagic fever: history and current status. New treatment strategies for dengue and other flaviviral diseases: Novartis Foundation Symposium 277. Wiley Online Library
Gubler DJ, Clark GG (1995) Dengue/dengue hemorrhagic fever: the emergence of a global health problem. Emerg Infect Dis 1(2):55
Guha-Sapir D, Schimmer B (2005) Dengue fever: new paradigms for a changing epidemiology. Emerg Themes Epidemiol 2(1):1–10
Guzman A, Isturiz RE (2010) Update on the global spread of dengue. Int J Antimicrob Agents 36 (1):S40–S42
Ha D, Tien N, Huong V, Loan H, Thang C (2000) Dengue epidemic in southern Vietnam, 1998. Emerg Infect Dis 6(4):422
Heng B, Goh K, Neo K (1998) Environmental temperature, Aedes aegypti house index and rainfall as predictors of annual epidemics of dengue fever and dengue haemorrhagic fever in Singapore. Dengue in Singapore. Ministry of Environment, Singapore, pp 138–149
Hii YL, Zhu H, Ng N, Ng LC, Rocklov J (2012) Forecast of dengue incidence using temperature and rainfall. PLoS Negl Trop Dis 6(11):e1908
Hopp J, Foley JA (2003) Worldwide fluctuations in dengue fever case related to climate variability. Climate Res 25(1):85–94
Hung NT, Lan NT, Lei H-Y, Lin Y-S, Lien LB, Huang K-J et al (2006) Volume replacement in infants with dengue hemorrhagic fever/dengue shock syndrome. Am J Trop Med Hyg 74 (4):684–691
Itrat A, Khan A, Javaid S, Kamal M, Khan H, Javed S et al (2008) Knowledge, awareness and practices regarding dengue fever among the adult population of dengue hit cosmopolitan. PLoS One 3(7):e2620
Johnson B, Ocheng D, Gichogo A, Okiro M, Libondo D, Kinyanjui P et al (1982) Epidemic dengue fever caused by dengue type 2 virus in Kenya: preliminary results of human virological and serological studies. East Afr Med J 59(12):781
Kanesa-thasan N, Chang G, Smoak B, Magill A, Burrous M, Hoke C Jr (1998) Molecular and epidemiologic analysis of dengue virus isolates from Somalia. Emerg Infect Dis 4(2):299
Kaplan JE, Eliason DA, Moore M, Sather GE, Schonberger LB, Cabrera-Coello L et al (1983) Epidemiologic investigations of dengue infection in Mexico, 1980. Am J Epidemiol 117 (3):335–343
Kay BH, Nam VS, Van Tien T, Yen NT, Phong TV, Diep VTB et al (2002) Control of aedes vectors of dengue in three provinces of Vietnam by use of Mesocyclops (Copepoda) and community-based methods validated by entomologic, clinical, and serological surveillance. Am J Trop Med Hyg 66(1):40–48
Khan NA, Azhar EI, El-Fiky S, Madani HH, Abuljadial MA, Ashshi AM et al (2008) Clinical profile and outcome of hospitalized patients during first outbreak of dengue in Makkah. Saudi Arabia Acta tropica 105(1):39–44
Khan E, Kisat M, Khan N, Nasir A, Ayub S, Hasan R (2010) Demographic and clinical features of dengue fever in Pakistan from 2003-2007: a retrospective cross-sectional study. PLoS One 5 (9):e12505
Kuhn RJ, Zhang W, Rossmann MG, Pletnev SV, Corver J, Lenches E et al (2002) Structure of dengue virus: implications for flavivirus organization, maturation, and fusion. Cell 108 (5):717–725
Kumaria R (2010) Correlation of disease spectrum among four Dengue serotypes: a five years hospital based study from India. Braz J Infect Dis 14(2):141–146
Lam S (1994) Two decades of dengue in Malaysia. Trop Med 35(4):195–200
Ligon BL (ed) (2005) Dengue fever and dengue hemorrhagic fever: a review of the history, transmission, treatment, and prevention. Seminars in pediatric infectious diseases, vol 16. pp 60–65
Madani TA, Abuelzein E-TM, Al-Bar HM, Azhar EI, Kao M, Alshoeb HO et al (2013) Outbreak of viral hemorrhagic fever caused by dengue virus type 3 in Al-Mukalla, Yemen. BMC Infect Dis 13(1):1–6

Malik A, Earhart K, Mohareb E, Saad M, Saeed M, Ageep A et al (2011) Dengue hemorrhagic fever outbreak in children in Port Sudan. J Infect Public Health 4(1):1–6
Mariappan T (2013) Current emerging situation of dengue in India. Trop Doct 43(3):116–119
Matheus S, Deparis X, Labeau B, Lelarge J, Morvan J, Dussart P (2005) Use of four dengue virus antigens for determination of dengue immune status by enzyme-linked immunosorbent assay of immunoglobulin G avidity. J Clin Microbiol 43(11):5784
Messer WB, Gubler DJ, Harris E, Sivananthan K, de Silva AM (2003) Emergence and global spread of a dengue serotype 3, subtype III virus. Emerg Infect Dis 9(7):800
Mia MS, Begum RA, Er A, Abidin RDZRZ, Pereira JJ (2013) Trends of dengue infections in Malaysia, 2000–2010. Asian Pac J Trop Med 6(6):462–466
Moore CG, Mitchell CJ (1997) Aedes albopictus in the United States: ten-year presence and public health implications. Emerg Infect Dis 3(3):329
Murgue B, Deparis X, Chungue E, Cassar O, Roche C (1999) Dengue: an evaluation of dengue severity in French Polynesia based on an analysis of 403 laboratory-confirmed cases. Trop Med Int Health 4(11):765–773
Nam VS, Yen NT, Holynska M, Reid JW, Kay BH (2000) National progress in dengue vector control in Vietnam: survey for Mesocyclops (Copepoda), Micronecta (Corixidae), and fish as biological control agents. Am J Trop Med Hyg 62(1):5–10
Nam VS, Yen NT, Phong TV, Ninh TU, Mai LQ, Lo LV et al (2005) Elimination of dengue by community programs using Mesocyclops (Copepoda) against Aedes aegypti in central Vietnam. Am J Trop Med Hyg 72(1):67–73
Ooi EE (2001) Changing pattern of dengue transmission in Singapore. Dengue Bull 25:40–44
Ooi EE, Hart TJ, Tan HC, Chan SH (2001) Dengue seroepidemiology in Singapore. Lancet 357 (9257):685–686
Ooi E-E, Goh K-T, Gubler DJ (2006) Dengue prevention and 35 years of vector control in Singapore. Emerg Infect Dis 12(6):887
Phuong HL, de Vries PJ, Nga TT, Giao PT, Hung LQ, Binh TQ et al (2006) Dengue as a cause of acute undifferentiated fever in Vietnam. BMC Infect Dis 6(1):123
Priest F (1992) Biological control of mosquitoes and other biting flies by Bacillus sphaericus and Bacillus thuringiensis. J Appl Microbiol 72(5):357–369
Raheel U, Faheem M, Riaz MN, Kanwal N, Javed F, Qadri I (2010) Dengue fever in the Indian subcontinent: an overview. J Infect Dev Ctries 5(04):239–247
Rasheed S, Butlin R, Boots M (2013) A review of dengue as an emerging disease in Pakistan. Public Health 127:11–17
Reddy P, Zembower T (2008) Emerging pathogens and new recommendations in travel medicine. Infect Med 25(8):352–365
Research SPf, Diseases TiT, Diseases WHODoCoNT, Epidemic WHO, Alert P (2009) Dengue: guidelines for diagnosis, treatment, prevention and control. World Health Organization, Geneva
Rigau-Perez JG, Clark GG, Gubler DJ, Reiter P, Sanders EJ, Vance VAXEA (1998) Dengue and dengue haemorrhagic fever. Lancet 352(9132):971–977
Roehrig JT, Hombach J, Barrett AD (2008) Guidelines for plaque-reduction neutralization testing of human antibodies to dengue viruses. Viral Immunol 21(2):123–132
San Martin JL, Brathwaite O, Zambrano B, Solorzano JO, Bouckenooghe A, Dayan GH et al (2010) The epidemiology of dengue in the Americas over the last three decades: a worrisome reality. Am J Trop Med Hyg 82(1):128
Sang RC (2007) Dengue in Africa. TropIKA net Available online only: http://www.tropikanet/svc/review/061001-Dengue_in_Africa
Seleena P, Lee H, Chiang Y (2001) Thermal application of Bacillus thuringiensis serovar israelensis for dengue vector control. J Vector Ecol 26:110–113
Sharp TW, Wallace MR, Hayes CG, Sanchez JL, DeFraites RF, Arthur RR et al (1995) Dengue fever in US troops during operation restore hope, Somalia, 1992-1993. Am J Trop Med Hyg 53 (1):89–94

Shekhar KC, Huat OL (1992) Epidemiology of Dengue/Dengue Hemorrhagic Fever in Malaysia-"A Retrospective Epidemiological Study 1973–1987. Part I: Dengue hemorrhagic fever (DHF). Asia Pac J Public Health 6(2):15–25

Shibl A, Senok A, Memish Z (2012) Infectious diseases in the Arabian Peninsula and Egypt. Clin Microbiol Infect 18(11):1068–1080

Singh N, Jhamb R, Agarwal S, Gaiha M, Dewan R, Daga M et al (2005) The 2003 outbreak of Dengue fever in Delhi, India

Spiegel J, Bennett S, Hattersley L, Hayden MH, Kittayapong P, Nalim S et al (2005) Barriers and bridges to prevention and control of dengue: the need for a social-ecological approach. Ecohealth 2(4):273–290

Sumarmo HW, Jahja E, Gubler D, Suharyono W, Sorensen K (1983) Clinical observations on virologically confirmed fatal dengue infections in Jakarta, Indonesia. Bull World Health Organ 61(4):693

Teng AK, Singh S (2001) Epidemiology and new initiatives in the prevention and control of dengue in Malaysia. Dengue Bull 25:7–12

Thai KT, Binh TQ, Giao PT, Phuong HL, Hung LQ, Nam NV et al (2005) Seroprevalence of dengue antibodies, annual incidence and risk factors among children in southern Vietnam . Trop Med Int Health 10(4):379–386

Thu TLT, Minh DN, Van NT, Tinh HT, Van Vinh CN, Wolbers M et al (2012) Clinical features of dengue in a large Vietnamese cohort: Intrinsically lower platelet counts and greater risk for bleeding in adults than children. PLoS Negl Trop Dis 6(6):e1679

Van Benthem B, Khantikul N, Panart K, Kessels P, Somboon P, Oskam L (2002) Knowledge and use of prevention measures related to dengue in northern Thailand. Trop Med Int Health 7 (11):993–1000

Vanlerberghe V, Trongtokit Y, Jirarojwatana S, Jirarojwatana R, Lenhart A, Apiwathnasorn C et al (2013) Coverage-dependent effect of insecticide-treated curtains for dengue control in Thailand. Am J Trop Med Hyg 89(1):93–98

Vasilakis N, Tesh RB, Weaver SC (2008) Sylvatic dengue virus type 2 activity in humans, Nigeria, 1966. Emerg Infect Dis 14(3):502

Vaughn DW, Nisalak A, Solomon T, Kalayanarooj S, Kneen R, Cuzzubho A et al (1999) Rapid serologic diagnosis of dengue virus infection using a commercial capture ELISA that distinguishes primary and secondary infections. Am J Trop Med Hyg 60:693–698

VinodKumar C, Kalapannavar N, Basavarajappa K, Sanjay D, Gowli C, Nadig NG et al (2013) Episode of coexisting infections with multiple dengue virus serotypes in central Karnataka, India. J Infect Public Health 6:302–306

Waterman SH, Gubler DJ (1989) Dengue fever. Clin Dermatol 7(1):117–122

Watterson D, Kobe B, Young PR (2012) Residues in domain III of the dengue virus envelope glycoprotein involved in cell-surface glycosaminoglycan binding. J Gen Virol 93(1):72–82

Wichmann O, Gascon J, Schunk M, Puente S, Siikamaki H, Gjorup I et al (2007) Severe dengue virus infection in travelers: risk factors and laboratory indicators. J Infect Dis 195(8):1089–1096

Wills BA, Oragui EE, Dung NM, Loan HT, Chau NV, Farrar JJ et al (2004) Size and charge characteristics of the protein leak in dengue shock syndrome. J Infect Dis 190(4):810–818

Wills BA, Dung NM, Loan HT, Tam DT, Thuy TT, Minh LT et al (2005) Comparison of three fluid solutions for resuscitation in dengue shock syndrome. N Engl J Med 353(9):877–889

Zhang W, Chipman PR, Corver J, Johnson PR, Zhang Y, Mukhopadhyay S et al (2003) Visualization of membrane protein domains by cryo-electron microscopy of dengue virus. Nat Struct Mol Biol 10(11):907–912

Zhao B, Mackow E, Buckler-White A, Markoff L, Chanock RM, Lai C-J et al (1986) Cloning full-length dengue type 4 viral DNA sequences: analysis of genes coding for structural proteins. Virology 155(1):77–88

Brucellosis

Anna Dean, Esther Schelling, and Jakob Zinsstag

Contents

Abstract Brucellosis is one of the most common zoonotic diseases globally, with many of the world's high burden countries found in the MENA region. The main species threatening human health in the region are those associated with livestock, namely *Brucella melitensis* and *Brucella abortus*. The main route of infection is through consumption of unpasteurised dairy products or direct contact with infected

A. Dean • E. Schelling • J. Zinsstag (✉)
Department of Epidemiology and Public Health, Swiss Tropical and Public Health Institute, Basel, Switzerland

University of Basel, Basel, Switzerland
e-mail: deanan@who.int; jakob.zinsstag@unibas.ch

M.A. McDowell and S. Rafati (eds.), *Neglected Tropical Diseases - Middle East and North Africa*, Neglected Tropical Diseases, DOI 10.1007/978-3-7091-1613-5_9,

livestock goats, sheep, cattle or camels. Brucellosis often affects poor, marginalised communities who do not have the financial resources to seek treatment. Chronic disease is common, including osteoarticular complications and infertility. The burden of brucellosis is grossly under-estimated in the region, due to diagnostic challenges, weak public health systems and a lack of prioritisation by decision-makers. However, the impact of brucellosis extends far beyond its clinical scope, threatening livelihoods, national economies and animal welfare. Reduced milk and meat production can have major socioeconomic implications for livestock owners. At the national level, brucellosis is a major barrier to the international trade of animals. In order to reduce the burden of human disease, brucellosis must first be controlled in its animal reservoir, usually through a combination of vaccination and test-and-slaughter strategies. Collaboration between human health and veterinary sectors is essential, using a One Health approach. There is a need for high quality epidemiological and economic data from the region, in order to advocate to governments and international funding bodies for greater investment in the prevention and control of brucellosis.

Keywords Brucellosis • Brucella • Zoonotic • Zoonosis • Livestock • Cattle • Sheep • Goat • Camel • Milk • Livelihoods • Economy • Health

Introduction

The health of people and animals are intimately linked. In the Middle East and North Africa (MENA) region, people rely on livestock for nutrition, income, draught power and religious and ceremonial purposes. Zoonotic diseases account for 61 % of pathogens known to infect humans and 75 % of all emerging infectious diseases (Taylor et al. 2001; Jones et al. 2008). Brucellosis has afflicted human health and livestock productivity for millennia. Evidence of infection has been demonstrated in human bones excavated from civilisations in ancient Egypt and Pompeii (Pappas and Papadimitriou 2007). *Brucella melitensis* was first isolated by David Bruce in 1887 from the spleen of a British soldier who died of Malta Fever, a common illness of British soldiers stationed in Malta. However, it was not until 1905 that Themistocles Zammit isolated the organism from goat's milk. The discovery that apparently healthy goats could carry disease revolutionised thinking regarding animal vectors of disease (Wyatt 2005). In the twenty-first century, brucellosis remains one of the most common zoonotic diseases globally. Many of the world's high burden countries are found in the MENA region.

Table 1 *Brucella* species and their animal hosts

Species	Animal hosts	Major zoonosis
Classical species		
B. melitensis	Sheep, goats	✓
B. abortus	Cattle	✓
B. suis	Pigs, hares, reindeers, rodents	✓
B. canis	Dogs	✓
B. ovis	Sheep	
B. neotomae	Rodents	
Newly described species		
B. pinnipedialis	Seals	
B. ceti	Dolphins, porpoises	
B. microti	Voles	
B. inopinata	Not known	

This table is adapted from Whatmore (2009). The only non-zoonotic classical species are *B. ovis* and *B. neotomae*. Although *Brucella* spp. of marine origin have been isolated from human cases, the major species presenting a risk to human health are *B. melitensis*, *B. abortus*, *B. suis* and to a lesser extent, *B. canis*

Pathogen

Classification

The causative agents of brucellosis are bacteria belonging to the genus *Brucella* of the order of *Rhizobiales* of the 2-alpha subgroup of *Proteobacteria*. The *Brucella* genus consists of ten species of gram negative, aerobic, non-spore forming, facultatively intracellular coccobacilli (Whatmore 2009). *Brucella* spp. are highly related, displaying a similarity of around 98 % in most coding sequences. Despite this, they display wide variation in phenotypic characteristics, host predilection and pathogenicity (Al Dahouk et al. 2012).

Mammalian hosts are the natural reservoir of *Brucella* spp. This was previously believed to include only terrestrial mammals until, in 1994, *Brucella* was isolated from marine mammals. The names and preferred hosts of the ten *Brucella* species are given in Table 1.

Pathogenesis

By hiding intracellularly, *Brucella* can multiply for prolonged periods in the reticuloendothelial system, protected from the host's immune system. Disease relapse can occur months or even years after apparently disease-free periods. Compared to many other bacterial pathogens, the factors contributing to persistence and reproduction of *Brucella* in the host are not well understood. The organism does not display classic virulence factors such as capsules or secreted proteases, exotoxins, endotoxins or pili/fimbrae (Franco et al. 2007; Seleem et al. 2010). Lipopolysaccharide (LPS) is an important component of the outer membrane, although it has a non-classical structure compared to other Gram-negative organisms. Consequently, this LPS is several hundred times less active and toxic than that of *Escherichia coli*, resulting in only a low activation of the alternative complement pathway of the host's immune response. This protects the organism from destruction by the host. Another key virulence factor is it type-IV secretion system, coded by the *virB* region of the genome. This is essential for intracellular survival and replication (Fugier et al. 2007).

Growth Requirements

Culture and isolation of *Brucella* should be performed in a laboratory of biosafety level 3 (WHO et al. 2006). Selective media include Farrell's medium and the modified Thayer-Martin medium, which contain combinations of different antibiotics to prevent growth of contaminants. As Farrell's may have an inhibitory effect on the growth of some strains, the two should be used in parallel (Zinsstag et al. 2011). Several biovars of *B. abortus* require atmospheric conditions of 5–10 % CO_2 (OIE 2009a).

Phenotypic Characterisation

Colony morphology can be smooth or rough, depending on the LPS structure. The phage susceptibility pattern is important for species identification, whereas agglutination with monospecific antisera, growth on dye-containing media and the production of hydrogen sulphide allow discrimination of the biovars of the major three species, *B. melitensis* (three biotypes), *B. abortus* (nine biotypes) and *B. suis* (five biotypes). These biotypes were, for some time, the only means of epidemiologically investigating disease transmission. However, biotyping is time consuming, requires the handling of dangerous live organisms and is somewhat subjective in its performance and interpretation (Whatmore 2009).

Genotypic Characterisation

Tandemly repeated sequences of DNA show variability in their numbers of repeats. An analysis of Variable Number of Tandem Repeats (VNTR) allows discrimination of *Brucella* strains both at the taxonomic and sub-species level. This approach has been used to demonstrate common sources of infection, confirm relapse of illness (Le Flèche et al. 2006; Al Dahouk et al. 2007) and trace laboratory-acquired infection (Marianelli et al. 2008). The existence of an online database, hosted by University Paris Sud, Orsay, France, of all published VNTR profiles allows for harmonisation and global comparison (Grissa et al. 2008).

Transmission

Brucella can enter a host through ingestion, penetration of skin abrasions or conjunctiva or by inhalation. The main route of infection of people is through consumption of unpasteurised dairy products or direct contact with infected bodily fluids of livestock (WHO et al. 2006; Franco et al. 2007; Seleem et al. 2010). Assisting the delivery of an aborted animal foetus or performing animal slaughters are recognised risk factors (Earhart et al. 2009; John et al. 2010). There is a high risk of laboratory-acquired infection through inhalation (Staszkiewicz et al. 1991; Fiori et al. 2000). *Brucella* spp. are classified by the US Centers for Disease Control and Prevention (CDC) in Atlanta as a potential Category B bioweapon (CDC 2013).

Clinical Manifestations, Diagnosis and Treatment

Animals

Clinical Manifestations

In livestock, the most common manifestation of brucellosis infection is abortion during the second half of gestation, reduced milk yield and arthritis. Orchitis may also occur and is characteristic of ram infection with the non-zoonotic organism, *B. ovis*. Infected animals may continue to secrete organisms in their milk for their rest of the lives or in the placenta of apparently normal births (Seleem et al. 2010).

Diagnosis

Due to the absence of a perfect test, serology is notoriously difficult to interpret. All serological tests have limitations when used as screening tools (OIE 2009a). Both the Rose Bengal Test (RBT) and enzyme-linked immunosorbent assay (ELISA) are suitable screening tests for the control of brucellosis at the local or national level. The sensitivity of the RBT in small ruminants is increased by the use of a serum: reagent ratio of 3:1, instead of the 1:1 ratio used in cattle. Positive reactions should be tested by a confirmatory strategy, such as the Complement Fixation Test (CFT). The ELISA and milk ring test can also be used for bulk milk samples from cows (OIE 2009a, b). *Brucella* organisms can be isolated from aborted foetal tissues

(Leyla et al. 2003; Muendo et al. 2012), vaginal fluid (Muendo et al. 2012), hygroma fluid (Bankole et al. 2010; Dean et al. 2014) and milk (Schelling et al. 2003; Muendo et al. 2012). However, given the high risk of laboratory-acquired infection, laboratories in developing countries often do not have an adequate level of biosafety for safely performing microbiological culture. Molecular typing by Multiple Locus VNTR Analysis (MLVA) can be used as a molecular epidemiology tool to understand transmission patterns and monitor surveillance and control programs (Le Flèche et al. 2006; Al Dahouk et al. 2007).

Humans

Symptoms

Human disease is predominantly caused by *B. melitensis, B. abortus, B. suis* and *B. canis*. In the MENA region, the main species threatening human health are those associated with livestock, namely *B. melitensis and B. abortus*. After a period of acute fever and lymphadenopathy, the organism can be seeded anywhere in the body, causing a chronic, granulomatous infection (Franco et al. 2007).

The symptoms associated with acute brucellosis are non-specific and thus present a diagnostic challenge. Table 2 presents the proportions of patients with different symptoms, based on a systematic review and meta-analysis of literature published between 1990 and 2010 (Dean et al. 2012). Fever is a common feature, accompanied by a combination of sweats, chills, fatigue and headache. On physical examination, abdominal pain, splenomegaly and hepatomegaly may also be detected. Localisation of the organism in bones and joints is common, with more than half of patients experiencing joint and/or back pain, due to arthritis, spondylitis and sacroilitis. One in ten men will develop epidiymo-orchitis, which can have serious repercussions such as abscessation and infertility. Severe outcomes are not rare, with neurological events occurring in 4 in 100 cases and endocarditis in 1 in 100 cases. The frequency of self-reported symptoms such as sweats, chills, fatigue, headache and malaise is lower in children, possibly reflecting difficulty in obtaining accurate case histories. Brucellosis is likely to be under-diagnosed in children.

Diagnosis

Seroconversion is rarely observed, given that the incubation time of the disease can be long, with patients diagnosed once infection has already been established (Zinsstag et al. 2011). The RBT detects the presence of IgM, IgG and IgA. It is rapid and simple to perform in the field but only indicates previous exposure to *Brucella*, not active infection. However, serial testing of serum dilutions can improve the discriminatory power, making the RBT the test of choice in remote, resource-poor settings (Diaz et al. 2011). Wherever possible, additional

Table 2 Clinical manifestations of brucellosis as a proportion of cases (%)

	Age category		
Manifestation	Children	Adults	All ages
General			
Fever	82 (69–91)	73 (59–85)	79 (49–97)
Sweats	23 (11–37)	55 (35–74)	73 (60–85)
Chills	18 (9–29)	47 (34–60)	60 (34–83)
Fatigue	19 (13–23)	33 (13–100)	51 (27–75)
Headache	9 (5–15)	34 (19–50)	52 (32–72)
Malaise	24 (16–34)	81 (71–89)	74 (48–93)
Nausea/vomiting	–	16 (5–31)	26 (15–38)
Weight loss	13 (8–18)	31 (15–50)	29 (15–47)
Abdominal			
Abdominal pain	14 (1–38)	9 (1–22)	26 (13–41)
Splenomegaly	31 (19–43)	24 (18–31)	25 (17–34)
Hepatomegaly	27 (15–41)	22 (16–26)	22 (15–29)
Hepatitis	1 (0–5)	8 (1–38)	3 (1–6)
Musculoskeletal			
Arthralgia	71 (56–84)	65 (49–79)	62 (52–70)
Arthritis	41 (18–65)	13 (3–28)	25 (17–34)
Myalgia	18 (11–26)	56 (38–75)	49 (36–63)
Back pain	10 (3–21)	49 (31–67)	45 (31–60)
Sacroiliitis	6 (3–10)	32 (20–46)	14 (7–22)
Spondylitis	18 (1–28)	12 (7–19)	11 (6–18)
Specific organs			
Epididymo-orchitis	10 (1–32)	10 (7–15)	9 (6–13)
Endocarditis	3 (1–6)	2 (1–3)	1 (1–2)
Neurological	2 (1–4)	5 (3–7)	4 (2–6)
Respiratory	5 (1–14)	2 (1–5)	9 (4–14)
Cutaneous	5 (2–10)	4 (1–11)	8 (4–14)

This table is adapted from the systematic review and meta-analysis of 57 clinical studies published between 1990 and 2010 of Dean et al. (2012). Pooled proportions of patients with each manifestation are presented as percentages with the 95 % confidence intervals given in parentheses. Study populations were classified as children only (<15 years), adults only (≥15 years) or all ages (studies of both children and adults)

confirmatory tests should be used and the results interpreted in light of the clinical status of the patient. Lateral flow immunochromatography requires no special training or equipment and can distinguish IgM and IgG to detect acute, persistent and relapsing disease (Smits et al. 2003; Irmak et al. 2004). Other commonly used tests include the serum agglutination test, Coomb's test, ELISA and immunocapture (Orduña et al. 2000; Serra and Viñas 2004; Araj et al. 2005; Gómez et al. 2008; Casanova et al. 2009). Their sensitivity and specificity for the diagnosis of active infection vary according to the background level of exposure of a given population. The cut-off thresholds of these tests may not, therefore, be appropriate for every epidemiological setting (Franco et al. 2007; Araj 2010).

The gold standard for the diagnosis of active brucellosis is isolation of the organism from bodily fluids or tissues, such as blood, joint fluid or cerebrospinal fluid (Franco et al. 2007). However, laboratories in developing countries often do not have the capacity to safely perform microbiological culture. MLVA of human-isolated strains can assist trace-back analysis for identification of the origin of an infection (Le Flèche et al. 2006; Al Dahouk et al. 2007; Marianelli et al. 2007; Kattar et al. 2008; Tiller et al. 2009; Nöckler et al. 2009; Valdezate et al. 2010; Kiliç et al. 2011; Ferreira et al. 2012).

Treatment

Disease relapses occur in approximately 5–15 % of uncomplicated cases and are more likely when monotherapy is used. Combination antimicrobial therapy is essential. For uncomplicated adult brucellosis, this is based on 6 weeks of oral doxycycline therapy at 100 mg twice per day for 6 weeks. This can be combined with either streptomycin intramuscularly at 15 mg/kg daily for 2–3 weeks or 600–900 mg rifampicin daily for 6 weeks. Parenteral gentamicin may be considered as an acceptable alternative at 5 mg/kg once per day for 7 days (Ariza et al. 2007).

Disease Burden and Impact

Brucellosis is characterised by a dual burden of disease, affecting both human and animal populations. The impact of brucellosis extends far beyond its clinical scope, threatening livelihoods, national economies and animal welfare.

Human Health and Wellbeing

Although brucellosis can present as a severe, life-threatening neurological or cardiac event, the chronic form of the disease is much more common and can last for years if untreated. In an endemic area of Russia prior to the availability of effective antibiotics, approximately 40 % of 1,000 brucellosis cases followed over a 20 year period continued to suffer from clinical manifestations two years after the onset of disease (Wundt 1968). A study in Tunisia showed that patients with localised osteoarticular brucellosis experienced a greater diagnostic delay than other cases (Zribi et al. 2009). Chronic pain affects the social and working life of the sufferer (Breivik et al. 2006), and brucellosis certainly impacts on wellbeing and quality of life. Male infertility, which can result from epididymo-orchitis, has been shown to be associated with social stigma in Egypt and Lebanon (Inhorn 2004).

Table 3 Brucellosis incidence by country (cases per 100,000 person-years)

	Scientific literature		Other sources
Country	Study level	Incidence per 100,000 per year	National incidence per 100,000 per year
Algeria	–	–	8.43
Egypt	Sub-national	0.28–70.00	0.30
Iran	Sub-national	0.73–141.60	23.86
Iraq	Sub-national	52.29–268.81	27.84
Jordan	National	25.70–130.00	2.34
Kuwait	–	–	3.39
Lebanon	–	–	4.95
Oman	Sub-national	11.01	3.56
Palestine	Sub-national	8.00	–
Saudi Arabia	National	137.61	21.44
	Sub-national	6.00–149.54	
Syria	–	–	160.34
Tunisia	–	–	3.54
Turkey	Sub-national	11.93–49.54	26.22
United Arab Emirates	–	–	4.1

The incidence rates reported in the scientific literature were identified by a systematic review conducted by Dean et al. of high quality studies published between 1990 and 2010 (Dean et al. 2012). The level at which the study was conducted is given. National level incidence rates derived from other sources were identified by a non-systematic review conducted by Pappas et al. and are predominantly based on reports from government authorities and international organisations (Pappas et al. 2006). These incidence rates were lower when compared to the scientific literature, highlighting under-reporting by national health systems. Data were not available for all countries in the MENA region

The burden of brucellosis is grossly under-estimated worldwide and in the MENA region. Brucellosis often affects poor, marginalised communities who do not have the financial resources to seek treatment, thus delaying diagnosis. Barriers to accessing health care, case mismanagement and misdiagnosis and a lack of political commitment are important contributors to this under-estimation. A systematic review of scientific publications between 1990 and 2010 identified high quality data from only Egypt, Iran, Iraq, Jordan, Oman, Saudi Arabia and Turkey (Table 3). Brucellosis incidence varied widely not only between countries but also within countries. Fivefold differences in one study in Iraq (Yacoub et al. 2006) and four fold differences in similarly designed studies in Egypt (Crump et al. 2003; Jennings et al. 2007) suggest that demographic, occupational and socioeconomic factors may play a role. Passively acquired national data are likely to underestimate the true disease burden. In an Egyptian study based on an acute febrile illness surveillance system to identify and confirm suspected cases, brucellosis incidence in the study area was determined to be 70 cases per 100,000 person-years. However, just 5.7 % of these cases were identified through hospital-based surveillance, from which the incidence rate would appear to be only 3.8 cases per 100,000 person-years for laboratory confirmed cases or 6 cases per 100,000 person-years for clinical cases. Routine hospital-based incidence data therefore underestimated incidence by 12–18 times (Jennings et al. 2007). Official data from the Ministry of Health gave an even lower incidence rate of only 0.3 cases per 100,000 person-years (Pappas et al. 2006).

Livelihoods and National Economies

Human brucellosis brings both direct costs of treatment, including consultation fees and medications, as well as indirect costs such as transportation to the health centre or loss of income due to an inability to work (Roth and Zinsstag 2001). Livestock are an important route out of poverty (WHO 2006) and, in the MENA region, many smallholder farmers rely on livestock for the livelihoods. Reduced milk and meat production can have major socioeconomic implications for livestock owners. At the national level, brucellosis is a major barrier to the international trade of animals (Thiermann 2004), which may slow the economic development of the country. Economic assessments of the cost of brucellosis and the profitability of interventions are urgently needed for the MENA region.

Animal Health and Welfare

Brucellosis is endemic in livestock across the MENA region, representing a constant reservoir of infection for people. Ovine and caprine brucellosis caused by *B. melitensis* is the predominant form of the disease (Refai 2002). Although animals may clinically recover from acute brucellosis, chronic arthritis with large fluid-filled hygromas can also result (WHO et al. 2006). These can be a source of discomfort and lameness and represent an animal welfare concern.

Disease Control Options

In order to reduce the burden of human disease, brucellosis must first be controlled in its animal reservoir. Some industrialised countries have successfully eliminated brucellosis. Vaccination, either implemented on mass or targeting only replacement stock, is a key component of a brucellosis control strategy. Deciding whether disease control or disease elimination should be the goal in a given setting depends on the disease prevalence, structure of the livestock production system, mixing and mobility of the animal population, availability of economic and other resources, potential ramifications on international trade, and economic and public health impact (Benkirane 2006).

When more than 10 % of herds or flocks are infected, mass vaccination is appropriate, although it should be remembered that this will interfere with serological testing and can cause abortions in pregnant animals (see below). If 5–10 % of herds/flocks are infected, vaccination of young replacement stock should be combined with the test-and-slaughter of adult animals. When disease prevalence is very low (1–2 % of herds/flocks infected) and there is no mixing between herds, an exclusive test-and-slaughter or stamping-out policy will lead to disease elimination if rigorously implemented (Zinsstag et al. 2011).

The *B. melitensis* Rev 1 vaccine induces strong immunity in sheep and goats. Under experimental conditions, it provides 80–100 % protection, compared to infection of 100 % of control animals (Jacques et al. 2007; Barrio et al. 2009). However, the vaccine does have some drawbacks. Subconjunctival administration is the preferred route, as it provides strong immunity but a reduced serological response compared to subcutaneous administration, causing less interference with serological testing as part of a combined elimination campaign. The vaccine can cause abortions in pregnant animals and should therefore either be administered to only young replacement animals at 3–4 months of age or to the whole population outside of the breeding season at a time when it can be assured that ewes are not pregnant. As the vaccine also poses a health risk to humans, people administering the vaccine must wear gloves and eye protection (Zinsstag et al. 2011).

The attenuated S19 *B. abortus* strain provides protection in cattle to both *B. abortus* and *B. melitensis*. The risk of abortion is much lower than that induced by Rev 1 in small ruminants, at a proportion of less than 1 % when administered subcutaneously in cows that were 7–8 months pregnant (Nicoletti 1990). The risk is even lower when administered subcutaneously. Mass vaccination of the whole female population over 4 months of age is recommended, using a single reduced dose (5×10^9 UFC/animal) subconjunctivally, either annually or every 2 years. Bulls should not be vaccinated, due to possible adverse effects. Although *B. abortus* S19 is not as dangerous as *B. melitensis* Rev1, biosafety precautions are still necessary when vaccinating animals (Zinsstag et al. 2011).

Epidemiological Considerations

The MENA region is a large geographical area comprised of countries with different political, economic, cultural and environmental characteristics. A major challenge to livestock control in the region is political instability and conflict, with associated break-downs in public health infrastructure (Gwida et al. 2010; Hotez et al. 2012).

Livestock owners commonly use mobile and semi-mobile herd management practices, moving seasonally between different pastoral zones. For certain ethnic groups, such as the Bedouin, mobility is central to cultural identity and is important for maximising productivity and accessing trade routes and markets. However, the frequent, unrestricted movement of livestock across borders poses a challenge to the control of brucellosis. In recent years, oil wealth has led to modernisation of these movements, incorporating the use of modern communication systems to identify ideal grazing areas as well as vehicles for transporting food and water for livestock. As a result, livestock movements in the region have become more extensive and opportunistic (Di Nardo et al. 2011). There may now be a greater opportunity to communicate with owners regarding disease control interventions than previously.

The trade of live animals is much more common in the region than the trade of animal products, linked to religious slaughtering practices and Muslim festivals such as Hajj, Ramadan, Eid ul-Fitr and Eid ul-Adha (Di Nardo et al. 2011). Mass livestock movements and slaughtering during this period result in increased direct contact with animal bodily fluids including aerosolisation, posing a risk for transmission of zoonotic diseases to people (Davies 2006).

There are, however, natural barriers to disease spread in the MENA region. In Oman, vast differences in the incidence of human cases between southern and northern regions may be due to their separation by a desert which prevents the trafficking of livestock. The island state of Bahrain has not reported human cases for several years (Pappas and Memish 2007), disease surveillance and quarantine activities being easier to implement when surrounded by sea.

Although goats are the most commonly reared livestock species in the region, camels are also an important source of milk, meat and wool, particularly in arid regions. They are traditionally grazed under nomadic conditions together with small ruminant herds, but many are also kept in backyards or within intensive dairy systems (Radwan et al. 1992). Given their mixing with other ruminants and their proximity to humans, camels are also an important reservoir of brucellosis.

Urgent Needs

Brucellosis is an ancient disease, yet it remains a major cause of human illness, reduced livestock productivity and economic losses in the MENA region. The under-reporting of human brucellosis reflects a weakness in health systems. Strengthening of disease surveillance systems is essential, including training of health workers and increased laboratory capacity. Given the zoonotic nature of the disease, improved case identification and a reduction in the burden of human disease will only be achieved through collaboration with the veterinary services. A One Health approach is needed, with closer cooperation of human and animal health sectors including joint surveillance systems and combined implementation of control strategies (Bonfoh et al. 2011).

Community education is important for reducing the risk of exposure to disease, through measures such as boiling milk before consumption or wearing protective equipment when handling aborted materials or slaughtering animals. It will also be essential for gaining support from farmers for vaccination campaigns and other interventions. These strategies must be carefully designed to include groups who may be at a higher risk of disease, such as nomadic communities or abattoir workers.

In order to advocate to governments and international agencies for greater investment in brucellosis, a sound evidence base is required. There is a need for high quality epidemiological and economic data from the region. Such information could be used as an advocacy tool to seek political commitment and ensure the allocation of necessary resources. Control interventions must be considered in terms of the overall benefits that they bring to society as a whole, rather than to the health or veterinary sectors alone. The Disability-Adjusted Life Year (DALY) is an economic measure used to rank diseases globally in terms of their impact. It is defined as a "time-based measure that combines years of life lost due to premature mortality and years of life lost due to time lived in states of less than full health" (WHO 2013). Although not without its criticisms (Mont 2007), the DALY provides policy-makers, funding bodies and researchers with a comparative view of population health, guiding the decision-making process about where global health efforts should be directed (Horton 2012). The calculation of a DALY estimate would place brucellosis on the global agenda. The first informed estimate of a disability weight for brucellosis has recently been published and is an important first step towards achieving this (Dean et al. 2012).

Cross-border movements are a central feature of livestock production in the region of both cultural and economic importance. Consequently, the burden and impact of brucellosis will only be reduced by adopting a regional approach to disease control. Neighbouring countries should synchronise interventions, such as mass vaccination campaigns, and work together to strengthen border security.

Acknowledgements This work was funded by the National Center of Competence in Research North–South (NCCR North–South) in Switzerland.

References

Al Dahouk S, Le Flèche P, Nöckler K et al (2007) Evaluation of Brucella MLVA typing for human brucellosis. J Microbiol Methods 69:137–145. doi:10.1016/j.mimet.2006.12.015

Al Dahouk S, Hofer E, Tomaso H et al (2012) Intraspecies biodiversity of the genetically homologous species Brucella microti. Appl Environ Microbiol 78:1534–1543. doi:10.1128/AEM.06351-11

Araj G (2010) Update on laboratory diagnosis of human brucellosis. Int J Antimicrob Agents 36(Suppl 1):S12–S17. doi:10.1016/j.ijantimicag.2010.06.014

Araj G, Kattar MM, Fattouh LG et al (2005) Evaluation of the PANBIO Brucella immunoglobulin G (IgG) and IgM enzyme-linked immunosorbent assays for diagnosis of human brucellosis. Clin Diagn Lab Immunol 12:1334–1335. doi:10.1128/CDLI.12.11.1334-1335.2005

Ariza J, Cascio A, Colmenero JD et al (2007) Perspectives for the treatment of brucellosis in the 21st century: the Ioannina recommendations. PLoS Med 4:e317. doi:10.1371/journal.pmed.0040317

Bankole AA, Saegerman C, Berkvens D et al (2010) Phenotypic and genotypic characterisation of Brucella strains isolated from cattle in the Gambia. Vet Rec 166:753–756. doi:10.1136/vr.b4862

Barrio MB, Grilló MJ, Muñoz PM et al (2009) Rough mutants defective in core and O-polysaccharide synthesis and export induce antibodies reacting in an indirect ELISA with smooth lipopolysaccharide and are less effective than Rev 1 vaccine against Brucella melitensis infection of sheep. Vaccine 27:1741–1749. doi:10.1016/j.vaccine.2009.01.025

Benkirane A (2006) Ovine and caprine brucellosis: world distribution and control/eradication strategies in West Asia/North Africa region. Small Rumin Res 62:19–25. doi:10.1016/j.smallrumres.2005.07.032

Bonfoh B, Kasymbekov J, Dürr S et al (2011) Representative seroprevalences of brucellosis in humans and livestock in Kyrgyzstan. Ecohealth. doi:10.1007/s10393-011-0722-x

Breivik H, Collett B, Ventrafridda V et al (2006) Survey of chronic pain in Europe: prevalence, impact on daily life, and treatment. Eur J Pain 10:287–333. doi:10.1016/j.ejpain.2005.06.009

Casanova A, Ariza J, Rubio M et al (2009) BrucellaCapt versus classical tests in the serological diagnosis and management of human brucellosis. Clin Vaccine Immunol 16:844–851. doi:10.1128/CVI.00348-08

CDC (2013) Bioterrorism Agents/Diseases. Centers for Disease Control and Prevention. http://emergency.cdc.gov/agent/agentlist-category.asp. Accessed 15 Jan 2013

Crump JA, Youssef FG, Luby SP et al (2003) Estimating the incidence of typhoid fever and other febrile illnesses in developing countries. Emerg Infect Dis 9:539–544

Davies FG (2006) Risk of a rift valley fever epidemic at the haj in Mecca, Saudi Arabia. Rev Sci Tech 25:137–147

Dean A, Crump L, Greter H et al (2012) Clinical manifestations of human brucellosis: a systematic review and meta-analysis. PLoS Negl Trop Dis 6:e1929. doi:10.1371/journal.pntd.0001929

Dean AS, Schelling E, Bonfoh B, Kulo AE, Boukaya GA, Pilo P (2014) Deletion in the gene BruAb2_0168 of Brucella abortus strains: diagnostic challenges. Clin Microbiol Infect. doi:10.1111/1469-0691.12554

Di Nardo A, Knowles NJ, Paton DJ (2011) Combining livestock trade patterns with phylogenetics to help understand the spread of foot and mouth disease in sub-Saharan Africa, the Middle East and Southeast Asia. Rev Sci Tech 30:63–85

Diaz R, Casanova A, Ariza J, Moriyon I (2011) The Rose Bengal Test in human brucellosis: a neglected test for the diagnosis of a neglected disease. PLoS Negl Trop Dis 5:e950. doi:10.1371/journal.pntd.0000950

Earhart K, Vafakolov S, Yarmohamedova N et al (2009) Risk factors for brucellosis in Samarqand Oblast, Uzbekistan. Int J Infect Dis 13:749–753. doi:10.1016/j.ijid.2009.02.014

Ferreira AC, Chambel L, Tenreiro T et al (2012) MLVA16 typing of Portuguese human and animal Brucella melitensis and Brucella abortus isolates. PloS One 7:e42514. doi:10.1371/journal.pone.0042514

Fiori PL, Mastrandrea S, Rappelli P, Cappuccinelli P (2000) Brucella abortus infection acquired in microbiology laboratories. J Clin Microbiol 38:2005–2006

Franco M, Mulder M, Gilman R, Smits H (2007) Human brucellosis. Lancet Infect Dis 7:775–786. doi:10.1016/S1473-3099(07)70286-4

Fugier E, Pappas G, Gorvel J-P (2007) Virulence factors in brucellosis: implications for aetiopathogenesis and treatment. Expert Rev Mol Med 9:1–10. doi:10.1017/S1462399407000543

Gómez MC, Nieto JA, Rosa C et al (2008) Evaluation of seven tests for diagnosis of human brucellosis in an area where the disease is endemic. Clin Vaccine Immunol 15:1031–1033. doi:10.1128/CVI.00424-07

Grissa I, Bouchon P, Pourcel C, Vergnaud G (2008) On-line resources for bacterial micro-evaluatioin studies using MLVA or CRISPR typing. Biochimie 90:660–668. doi:10.1016/j.biochi.2007.07.014

Gwida M, Al Dahouk S, Melzer F et al (2010) Brucellosis—regionally emerging zoonotic disease? Croat Med J 51:289–295. doi:10.3325/cmj.2010.51.289

Horton R (2012) GBD 2010: understanding disease, injury, and risk. Lancet 380:2053–2054. doi:10.1016/S0140-6736(12)62133-3

Hotez PJ, Savioli L, Fenwick A (2012) Neglected tropical diseases of the Middle East and North Africa: review of their prevalence, distribution, and opportunities for control. PLoS Negl Trop Dis 6:e1475. doi:10.1371/journal.pntd.0001475

Inhorn MC (2004) Middle Eastern masculinities in the age of new reproductive technologies: male infertility and stigma in Egypt and Lebanon. Med Anthropol Q 18:162–182

Irmak H, Buzgan T, Evirgen O et al (2004) Use of the Brucella IgM and IgG flow assays in the serodiagnosis of human brucellosis in an area endemic for brucellosis. Am J Trop Med Hyg 70:688–694

Jacques I, Verger J-M, Laroucau K et al (2007) Immunological responses and protective efficacy against Brucella melitensis induced by bp26 and omp31 B. melitensis Rev.1 deletion mutants in sheep. Vaccine 25:794–805. doi:10.1016/j.vaccine.2006.09.051

Jennings GJ, Hajjeh RA, Girgis FY et al (2007) Brucellosis as a cause of acute febrile illness in Egypt. Trans R Soc Trop Med Hyg 101:707–713

John K, Fitzpatrick J, French N et al (2010) Quantifying risk factors for human brucellosis in rural northern Tanzania. PloS One 5:e9968. doi:10.1371/journal.pone.0009968

Jones K, Patel N, Levy M et al (2008) Global trends in emerging infectious diseases. Nature 451: 990–993. doi:10.1038/nature06536

Kattar MM, Jaafar RF, Araj G et al (2008) Evaluation of a multilocus variable-number tandem-repeat analysis scheme for typing human Brucella isolates in a region of brucellosis endemicity. J Clin Microbiol 46:3935–3940. doi:10.1128/JCM.00464-08

Kiliç S, Ivanov IN, Durmaz R et al (2011) Multiple-locus variable-number tandem-repeat analysis genotyping of human Brucella isolates from Turkey. J Clin Microbiol 49:3276–3283. doi:10.1128/JCM.02538-10

Le Flèche P, Jacques I, Grayon M et al (2006) Evaluation and selection of tandem repeat loci for a Brucella MLVA typing assay. BMC Microbiol 6:9. doi:10.1186/1471-2180-6-9

Leyla G, Kadri G, Ümran O (2003) Comparison of polymerase chain reaction and bacteriological culture for the diagnosis of sheep brucellosis using aborted fetus samples. Vet Microbiol 93:53–61. doi:10.1016/S0378-1135(02)00442-X

Marianelli C, Graziani C, Santangelo C et al (2007) Molecular epidemiological and antibiotic susceptibility characterization of Brucella isolates from humans in Sicily, Italy. J Clin Microbiol 45:2923–2928. doi:10.1128/JCM.00822-07

Marianelli C, Petrucca A, Pasquali P et al (2008) Use of MLVA-16 typing to trace the source of a laboratory-acquired Brucella infection. J Hosp Infect 68:274–276. doi:10.1016/j.jhin.2008.01.003

Mont D (2007) Measuring health and disability. Lancet 369:1658–1663. doi:10.1016/S0140-6736(07)60752-1

Muendo EN, Mbatha PM, Macharia J et al (2012) Infection of cattle in Kenya with Brucella abortus biovar 3 and Brucella melitensis biovar 1 genotypes. Trop Anim Health Prod 44:17–20. doi:10.1007/s11250-011-9899-9

Nicoletti P (1990) Vaccination. In: Neilsen K, Duncan J (eds) Animal brucellosis. CRC, Boca Raton, pp 283–299

Nöckler K, Maves R, Cepeda D et al (2009) Molecular epidemiology of Brucella genotypes in patients at a major hospital in central Peru. J Clin Microbiol 47:3147–3155. doi:10.1128/JCM.00900-09

OIE (2009a) Chapter 2.4.3. Bovine brucellosis. Manual of diagnostic tests and vaccines. Available via OIE: http://www.oie.int/fileadmin/home/eng/health_standards/tahm/2.04.03_bovine_brucell.pdf. Accessed 21 July 2013

OIE (2009b) Chapter 2.7.2. Caprine and ovine brucellosis. Manual of diagnostic tests and vaccines. Available via OIE: http://www.oie.int/fileadmin/home/eng/health_standards/tahm/2.07.02_caprine_ovine_bruc.pdf. Accessed 21 July 2013

Orduña A, Almaraz A, Prado A et al (2000) Evaluation of an immunocapture-agglutination test (Brucellacapt) for serodiagnosis of human brucellosis. J Clin Microbiol 38:4000–4005. doi:0095-1137/00/$04.0010

Pappas G, Memish ZA (2007) Brucellosis in the middle East: a persistent medical, socioeconomic and political issue. J Chemother 19:243–248

Pappas G, Papadimitriou P (2007) Challenges in Brucella bacteraemia. Int J Antimicrob Agents 30(Suppl 1):29–31. doi:10.1016/j.ijantimicag.2007.06.011

Pappas G, Papadimitriou P, Akritidis N et al (2006) The new global map of human brucellosis. Lancet Infect Dis 6:91–99

Radwan AI, Bekairi SI, Prasad PV (1992) Serological and bacteriological study of brucellosis in camels in central Saudi Arabia. Rev Sci Tech 11:837–844

Refai M (2002) Incidence and control of brucellosis in the Near East region. Vet Microbiol 90: 81–110

Roth F, Zinsstag J (2001) Human health benefits from livestock vaccination for brucellosis: case study. Bull World Health Organ 81:867–876. doi:10.1590/S0042-96862003001200005

Schelling E, Diguimbaye C, Daoud S et al (2003) Brucellosis and Q-fever seroprevalences of nomadic pastoralists and their livestock in Chad. Prev Vet Med 61:279–293. doi:10.1016/j.prevetmed.2003.08.004

Seleem MN, Boyle SM, Sriranganathan N (2010) Brucellosis: a re-emerging zoonosis. Vet Microbiol 140:392–398. doi:10.1016/j.vetmic.2009.06.021

Serra J, Viñas M (2004) Laboratory diagnosis of brucellosis in a rural endemic area in northeastern Spain. Int Microbiol 7:53–58

Smits HL, Abdoel TH, Solera J et al (2003) Immunochromatographic Brucella-specific immunoglobulin M and G lateral flow assays for rapid serodiagnosis of human brucellosis. Clin Diagn Lab Immunol 10:1141–1146. doi:10.1128/CDLI.10.6.1141-1146.2003

Staszkiewicz J, Lewis CM, Colville J et al (1991) Outbreak of Brucella melitensis among microbiology laboratory workers in a community hospital. J Clin Microbiol 29:287–290

Taylor L, Latham S, Woolhouse M (2001) Risk factors for human disease emergence. Philos Trans R Soc Lond 356:983–989. doi:10.1038/nature06536

Thiermann A (2004) Emerging diseases and implications for global trade. Rev Sci Tech 23: 701–707

Tiller RV, De BK, Boshra M et al (2009) Comparison of two multiple-locus variable-number tandem-repeat analysis methods for molecular strain typing of human Brucella melitensis isolates from the Middle East. J Clin Microbiol 47:2226–2231. doi:10.1128/JCM.02362-08

Valdezate S, Navarro A, Villalón P et al (2010) Epidemiological and phylogenetic analysis of Spanish human Brucella melitensis strains by multiple-locus variable-number tandem-repeat typing, hypervariable octameric oligonucleotide fingerprinting, and rpoB typing. J Clin Microbiol 48:2734–2740. doi:10.1128/JCM.00533-10

Whatmore AM (2009) Current understanding of the genetic diversity of Brucella, an expanding genus of zoonotic pathogens. Infect Genet Evol 9:1168–1184. doi:10.1016/j.meegid.2009.07.001

WHO (2006) The control of neglected zoonotic diseases: a route to poverty alleviation. World Health Organization. Available via WHO: http://www.who.int/zoonoses/Report_Sept06.pdf. Accessed 21 Jul 2013

WHO (2013) Global burden of disease (GBD). http://www.who.int/topics/global_burden_of_disease/en/. Accessed 21 Jul 2013

WHO, FAO, OIE (2006) Brucellosis in humans and animals. World Health Organization. Available via WHO: https://extranet.who.int/iris/restricted/bitstream/10665/43597/1/WHO_CDS_EPR_2006.7_eng.pdf. Accessed 21 Jul 2013

Wundt W (1968) Krankheiten durch Brucellen. Infekt. Band Ii Krankh. Durch Bakterien, 1st ed. Springer, Berlin, pp 31–554

Wyatt HV (2005) How Themistocles Zammit found Malta Fever (brucellosis) to be transmitted by the milk of goats. J R Soc Med 98:451–454

Yacoub A, Bakr S, Hameed AM et al (2006) Seroepidemiology of selected zoonotic infections in Basra region of Iraq. East Mediterr Health J 12:112–118

Zinsstag J, Schelling E, Solera J et al (2011) Brucellosis. In: Palmer SR, Soulsby L, Torgerson PR, Brown D WG (eds) Oxford textbook of zoonoses: biology, clinical practice and public health control, 2nd ed. Oxford University Press, Oxford, pp 54–62

Zribi M, Ammari L, Masmoudi A et al (2009) Clinical manifestations, complications and treatment of brucellosis: 45 patients study. Pathol Biol 57:349–352. doi:10.1016/j.patbio.2008.02.003

Toxoplasmosis in the Middle East and North Africa

Aïda Bouratbine and Karim Aoun

Contents

Abstract In this chapter, we summarized current knowledge on the prevalence, distribution, and major clinical manifestations of toxoplasmosis in the MENA. Risk factors for *Toxoplasma* infection and approaches to control were also discussed focusing on particular aspects to the region.

Keywords MENA • Neglected tropical disease • Toxoplasmosis • *Toxoplasma* infection • Seroprevalence • Acquired toxoplasmosis • Congenital toxoplasmosis • Risk factors • Prevention

Introduction

Toxoplasmosis is a worldwide disease caused by an obligate intracellular protozoan parasite, *Toxoplasma (T.) gondii*. This parasite can infect any warm-blooded vertebrate and is a pathogen of medical and veterinary significance. Since its first

A. Bouratbine (✉) • K. Aoun
Department of Parasitology, LR 11-IPT-06, Pasteur Institute of Tunis, Tunis, Tunisia
e-mail: aida.bouratbine@pasteur.rns.tn; karim.aoun@pasteur.rns.tn

M.A. McDowell and S. Rafati (eds.), *Neglected Tropical Diseases - Middle East and North Africa*, Neglected Tropical Diseases, DOI 10.1007/978-3-7091-1613-5_10,

description by Nicolle and Manceaux in 1908 in Tunisia in a North African rodent, the gondi (Nicolle and Manceaux 1908), the parasite was progressively recognized as the agent of a widespread zoonosis. However, its entire life cycle was not definitively understood until the late 1960s, with the discovery of the central role of felids acting as definitive hosts, harboring the sexual parasitic cycle and spreading oocysts through feces (Frenkel et al. 1970). Oocysts are remarkably stable in the environment and are transmitted to other animals through inadvertent ingestion. The prevalence of *Toxoplasma* infection in intermediate hosts depends on the presence of felids, stray and domestic cats, or wild felid species, in their surroundings. However, the processes promoting infection are highly intricate, involving (1) climate characteristics conditioning oocyst survival, (2) host species susceptibility to *Toxoplasma* infection, and (3) host species diet and feeding behavior. In fact, the prevalence of infection is often higher in omnivores and carnivores than herbivores due to the cumulative efficacy of the predator–prey cycle of *T. gondii*. The majority of horizontal transmissions to humans are caused either by the ingestion of parasite tissue cysts in infected meat or by the ingestion of soil, water, or food contaminated with sporulated oocysts derived from the environment or, less often, directly from feline feces. When primary infection is acquired by a pregnant woman, the parasite can colonize placental tissues during the dissemination process and from there can reach the fetal compartment causing congenital toxoplasmosis (Robert-Gangneux and Darde 2012).

Human toxoplasmosis is usually subclinical or appears like a minor viral illness (Montoya and Liesenfeld 2004). Visual impairment may also reveal primary infection (Delair et al. 2008; Delair et al. 2011). On the other hand, parasite's bradyzoites can persist inside human cells for protracted periods leading to a reactivation of latent infection, typically during acquired immunodeficiency syndrome (AIDS) where *T. gondii* reactivation causes severe encephalitis (Porter and Sande 1992). Furthermore, vertical transmission of active infection to the fetus may result in a large spectrum of clinical presentations, ranging from a clinically silent form to serious affections, namely, major ocular and neurological impairment, mental retardation, and spontaneous abortions or stillbirth (Villena et al. 2010). It is now recognized that the severity of infection may depend on the immune status of the host but also on the genotype of the *T. gondii* strain. Indeed, the severity of acquired and congenital infections is low in Western European countries and North America, where type II strains are prevailing, but much higher in other parts of the world, such as South America or Africa, where other genotypes circulate (Robert-Gangneux and Darde 2012).

Toxoplasma gondii has a worldwide distribution, with an estimated one-third of the entire global population infected (Montoya and Liesenfeld 2004). In North America and Europe, toxoplasmosis is one of the leading illnesses associated with food-borne hospitalizations and deaths (Mead et al. 1999; Jones and Dubey 2012) Also, the disease burden of congenital toxoplasmosis, as represented by disability-adjusted life years (DALY), is the highest among all food-borne pathogens (Havelaar et al. 2007). In 2012, the TDR disease reference group on zoonoses and marginalized infectious diseases of poverty highlighted that (1) prevalence of toxoplasmosis is strongly correlated with socioeconomic status, (2) burden of toxoplasmosis is borne disproportionately by poor communities in which food safety is inadequate, and (3) limited access to routine prenatal evaluation for toxoplasmosis adds to the burden (Anonymous 2012). Furthermore, in low- and middle-income countries, toxoplasmosis may trap the poor in a vicious cycle of poverty because of the tendency of this infection to impair child development, pregnancy outcome, and worker productivity. These distinguishing features of the disease incorporate toxoplasmosis in the groups of "neglected infections of poverty" and "neglected tropical diseases" (NTDs) (Hotez and Gurwith 2011; Hotez et al. 2012).

The Middle East and North Africa (MENA) region is made up by approximately 20 countries where almost 400 million people, i.e., 5 % of the world's population, live. Toxoplasmosis is present throughout the MENA. However, there is a dearth of available information about this disease (Hotez et al. 2012). Estimates from 2005 indicate that 3.6 % of the MENA population lives below the World Bank poverty figure of US$1.25 per day, while 16.9 % lives below US$2 per day. These "bottom 14 million" and "bottom 65 million," respectively, are probably the groups of people with the greatest vulnerability to toxoplasmosis (Hotez et al. 2012). Here, we summarize current knowledge on the prevalence, distribution, and major clinical manifestations of toxoplasmosis in the MENA. Risk factors for *Toxoplasma* infection and approaches to control will be discussed focusing on particular aspects to the region.

Seroprevalence of *Toxoplasma* Infection

Prevalence and distribution of toxoplasmosis in MENA were assessed by the review of the literature using the online database PubMed with Medical Subject Headings, toxoplasmosis, and countries of MENA as defined by the World Bank. Several local studies about *T. gondii* seroprevalence in MENA were recorded. These studies targeted general population, pregnant women, and women at childbearing. Several of these studies offer extended and reliable information about some countries, being nationwide or using large and representative samples. However, for some countries the available information may be limited and of questionable reliability (Table 1).

Globally, no obvious gradient in toxoplasmosis seroprevalence was observed in MENA. High prevalence foci (40–60 %) exist in North Africa and the Middle East including Morocco, Tunisia, Egypt, Lebanon, Jordan, Iraq, and Iran whereas lower prevalence (20–40 %) was reported in Saudi Arabia, in Bahrain and Qatar (Table 1, Fig. 1). No information was available about current *T. gondii* seroprevalence in Algeria, Libya, Syria, and Yemen (Fig. 1). It is important to note that Yemen has the highest percentage of people living in poverty in MENA region, whereas Egypt represents the nation with the largest total number of people living in poverty. Significant numbers of impoverished people also live in Morocco, Algeria, Tunisia,

Table 1 Toxoplasmosis seroprevalence in MENA countries

Country	Region	Sample (size)	Seroprevalence	References
Bahrain	Manama	Childbearing age (3,499)	22.3 %	Tabbara and Saleh (2005)
Egypt	Qalyubia	General pop (152)	57.9 %	Hussein et al. (2001)
	Mansoura	Blood donors (260)	59.6 %	Elsheikha et al. (2009)
	Menoufia	Pregnant (323)	67.5 %	El Deeb et al. (2012)
Iran	Babol	Childbearing age (241)	63.9 %	Youssefi et al. (2007a, b)
	Gorgan	Pregnant (300)	48.3 %	Saeedi et al. (2007)
	Isfahan	General pop (599)	41.1 %	Mostafavi et al. (2011)
	Isfahan	Childbearing age (217)	47.5 %	Mostafavi et al. (2012)
Iraq	Basra	General pop	41.1–52.1 %	Yacoub et al. (2006)
Israel	Jerusalem	General pop (2,794)	19.9–60.4 %	Markovich et al. (2013)
Jordan	Amman	Pregnant (280)	47.1 %	Jumaian (2005)
Kuwait	Kuwait	Pregnant (224)	53.1 %	Iqbal and Khalid (2007)
Lebanon	Beirut	Childbearing age (3516)	55–67 %	Bouhamdan et al. (2010)
Morocco	Rabat	Pregnant (2,456)	50.6 %	El Mansouri et al. (2007)
Qatar	Doha	General pop (1,625)	29.8 %	Abu-Madi et al. (2008)
Saudi Arabia	East	General pop (1,400)	25–26 %	Al-Qurashi (2004)
	Riyadh	Pregnant (2,176)	38 %	Almogren (2011)
	Najran	General pop (210)	31.9 %	Alqahtani and Hassan (2012)
Tunisia	Nord	General pop (1,421)	58.4 %	Bouratbine et al. (2001)
	Sfax	Pregnant (15,952)	39.3 %	Sellami et al. (2010)
	Tunis	Pregnant (2,351)	47.7 %	Fakhfakh et al. (2013)
	Tunis	Pregnant (2,070)	45.6 %	Ben Abdallah et al. (2013)

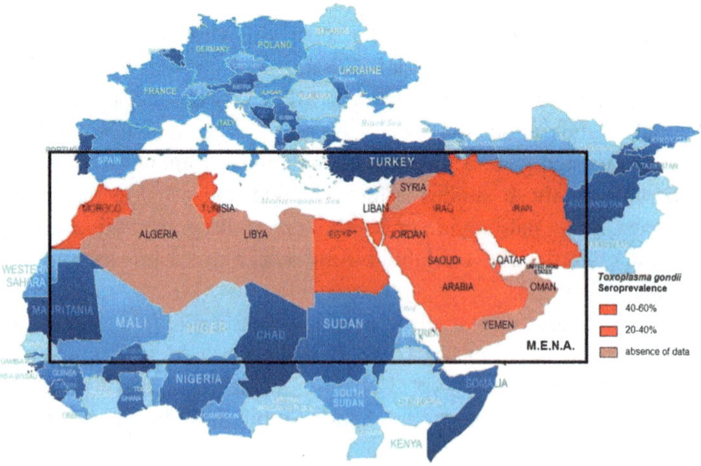

Fig. 1 *Toxoplasma gondii* seroprevalence in MENA countries (Map of the MENA region from Hotez et al. 2012)

Iraq, and Iran. People originating from these countries are probably the groups with the greatest vulnerability to toxoplasmosis.

Seroprevalence varied significantly between different regions in the same country and among different population groups of the same region (Table 1). These differences may be related to weather, cultural, and especially, socioeconomic differences of the population. These findings were particularly highlighted in a recent study undertaken in an Israeli population, where the highest age-adjusted seroprevalence rate was observed in Arabs (non-Bedouins) (60.4 %) and was significantly higher compared to the rate in Jews (19.9 %) and Bedouins (27.5 %). For Jews, seropositivity was associated with place of birth and socioeconomic status. The low seroprevalence rate in Bedouins (despite their poor living conditions and close contact with livestock) was attributed to the dry and hot climate conditions in their area of residence (Markovich et al. 2013).

Toxoplasma gondii seroprevalence in general population measures the accumulated exposure to the parasite during a person's lifetime in a particular social setting. Thus, the seropositivity rate could be treated as a quantitative measure of the relative protection for an individual of this population (particularly women at childbearing age) against primary infection. Hence, in settings with low general population seroprevalence, the potential for a particular woman to be infected is consequently low, but if she is infected during pregnancy it will most likely be her primary infection, with the risk of abortion or congenital toxoplasmosis. In settings with high general population seroprevalence, the chances of a particular woman to acquire a primary infection during pregnancy are low because she has most likely already been exposed to *T. gondii*. However if she has not been exposed to the parasite until childbearing age, chances of contracting a primary infection during pregnancy are high (Pappas et al. 2009). Conversely, seropositivity rate may be used as a quantitative measure of relative risk for disease reactivation in HIV-infected individuals. In fact, HIV-positive patients are more commonly at risk for disease reactivation resulting from cyst rupture than for a newly acquired infection. In the settings with high general population seroprevalence, the relatively high prevalence of *T. gondii* infection in asymptomatic HIV-infected individuals suggests that toxoplasmosis may represent a frequent opportunistic parasitic disease (Mohraz et al. 2011; Addebbous et al. 2012).

More accurate information about epidemiological features of toxoplasmosis can be obtained by studying seroprevalence according to age groups. In fact, seroprevalence increases with age, but the rate of acquisition of infection in relation to age may vary according to the region and socioeconomic level. Hence, near-maximal seroprevalence may be reached during childhood in populations living under poor-hygiene conditions, probably linked to telluric or waterborne contamination by oocyst ingestion. A study undertaken in Northern Tunisia showed that *T. gondii* seroprevalence rose from 24.5 % at 10 years to 52.1 % at 20 years of age. A maximum level, around 70 %, was reached by about 30 years. The risk of acute infection after this age seemed low as judging by additional serological data. This epidemiological profile suggested that even if infection is a frequent event in childhood, additional infections are acquired between 15 and 30 years.

Moreover, a significantly higher prevalence was detected in urban residents, which presented a different epidemiological profile than individuals from rural origin (Bouratbine et al. 2001).

Further epidemiological information can also be obtained by studying serological markers of recent infection, namely, specific Immunoglobin (Ig) M anti-*T. gondii*. Several studies from Iran reported a high percentage of specific IgM identification in women at childbearing age (12.4–17.7 %) (Saeedi et al. 2007, Youssefi et al. 2007a) which suggests a high level of primo infection in this specific group. However, primo infection seems to be a rare event during pregnancy in North Africa. In fact the percentage of specific IgM identification in seropositive pregnant women was around 1.3–2.8 % in Tunisia, Morocco, and Egypt (Boughattas et al. 2010; Sellami et al. 2010; El Deeb et al. 2012). In Saudi Arabia where *T. gondii* seroprevalence is low, IgM antibodies were rarely identified (0.57 %) (Al-Mulhim and Al-Qurashi 2001).

Clinical Aspects of Toxoplasmosis

Acquired Toxoplasmosis

In immunocompetent subjects, primary acquired infection is asymptomatic in the majority of cases. However, rare patients may experience fever or cervical lymphadenopathy, sometimes associated with myalgia, asthenia, or other nonspecific clinical signs (Sellami et al. 2010). Visual impairment may also reveal primary infection. In fact, it was previously thought that ocular toxoplasmosis was only the result of congenital infection. However, in a recent retrospective study undertaken in France about 425 consecutive cases of ocular toxoplasmosis, 100 (23.5 %) were attributed to acquired toxoplasmosis, 62 (14.6 %) to congenital transmission, and 263 (61.9 %) were of unknown origin. The mean age of the patients with congenital ocular toxoplasmosis was 9.1 ± 8.8 years and was 21.7 ± 12.6 years in the patients with acquired ocular toxoplasmosis ($P < 0.0001$). Bilateral chorioretinitis was observed in 4 % of acquired cases and in 43.5 % of congenital cases ($P < 0.0001$). Cases of congenital ocular toxoplasmosis were more severe than acquired cases (Delair et al. 2008). There are several published data about ocular toxoplasmosis in MENA countries (Khairallah et al. 2011; Tabatabaei et al. 2011; Kianersi et al. 2012). Nevertheless, there is no information about the origin (congenital versus acquired) of the infection.

In immunocompromised patients, toxoplasmosis is always a life-threatening disease. In North African HIV-infected individuals, toxoplasmosis represents one of the main causes of death (Sodqi et al. 2012). Toxoplasmosis is also one of the most frequent opportunistic infection in patients originating from MENA with CD4 + T-cell counts under 100 cells/µl (Naba et al. 2010; Sellami et al. 2010). Toxoplasmic encephalitis is the most predominant manifestation of the disease. Less frequently, toxoplasmic retinochoroiditis may occur independently of any other signs of evolutive infection and must be distinguished from other infectious etiologies.

Congenital Toxoplasmosis

Congenital infection is the most important part of the disease burden due to *Toxoplasma* infection in humans. Classically, congenital infection results from primary acquired maternal infection during gestation. Mother *Toxoplasma* infection is often asymptomatic and diagnosis of primary infection in pregnant women relies on serology. Currently, most clinical laboratories use an ELISA method for the routine screening of specific *anti-T. gondii* IgG and IgM, whereas other techniques are mostly reserved for reference laboratories. Detection of specific IgG is indicative of exposure to the parasite, and identification of specific IgM is indicative of recent infection. In case of IgM presence a more accurate serological tool is needed to establish if maternal infection is acquired during pregnancy. Determination of IgG avidity is currently the means of dating infection according to the age of gestation (Siala et al. 2006). When a primary maternal infection is highly suspected, the current practice is to propose a prenatal diagnosis in order to evaluate fetus infection. This latter method mostly relies on ultrasound and polymerase chain reaction (PCR)-based detection of parasite DNA in amniotic fluid (Siala et al. 2007). At birth, neonatal diagnosis is critical to diagnose infection as well as to compensate for the few false-negative results of antenatal diagnoses (Ben Abdallah et al. 2009)

The frequency of vertical transmission and the severity of fetal damage depend on the stage of pregnancy when maternal infection occurs. The placenta plays a main role in the process, as it constitutes a natural barrier which is supposed to protect the fetus and conversely a target tissue for parasite multiplication. In fact, the placental barrier is more efficient at the beginning of gestation, leading to the passage of parasites in less than 10 % of cases during the first trimester, but becomes more permeable throughout pregnancy, allowing parasite transmission in around 30 % of cases in the second trimester and 60–70 % of cases in the third trimester and even more close to the time of delivery (Dunn et al. 1999). The severity of fetal infection is inversely correlated, since neonates are asymptomatic in more than 80 % of cases when infected during the third trimester of gestation (Desmonts and Couvreur 1974). However, when transplacental transmission occurs during the first trimester, the consequences for fetal development are heavy, often leading to severe abnormalities or to abortion. A newborn exposed to *T. gondii* in utero may develop congenital toxoplasmosis with major ocular and neurological consequences. Indeed, parasite multiplication may induce necrosis foci and strong inflammation, leading to major abnormalities in brain and eye tissues. It can induce the destruction or profound remodeling of the white substance. Infected necrotized foci may block the aqueduct of Sylvius, resulting in hydrocephalus of lateral ventricles. These foci further calcify and can be detected by transfontanellar echography or cranial X ray. Major sequelae include mental retardation, seizures, microcephalus, hydrocephalus, deafness, and psychomotor deficiency (Olariu et al. 2011). Retinochoroiditis is a common feature that can be observed whatever the time of maternal infection. Its particularity resides in its frequently delayed clinical expression after birth (Delair et al. 2008; Delair et al. 2011).

Little is known about prevalence and clinical aspects of congenital toxoplasmosis in MENA. Results of a neonatal screening performed in a Tunisian hospital between 2003 and 2004 estimated the prevalence of congenital toxoplasmosis, 7.9 per 10,000 live births (Ben Hamida Nouaili et al. 2009). Among the data from 11 congenital toxoplasmosis cases collected in Pasteur Institute of Tunis, one case resulted in neonatal death, 3 cases (27 %) had retinochoroiditis of variable severity, and 7 (63 %) live-born infants were asymptomatic (Ben Abdallah et al. 2009).

The possible relationship between congenital toxoplasmosis and *T. gondii* genotype has been approached in several studies. In France, where a systematic diagnosis of congenital toxoplasmosis is performed, more than 80 % of infections are caused by the archetypal genotype II. In South America, where non-archetypal strains predominate, the few reports about isolates from congenital cases indicate the role of type I, atypical, or recombinant I/III strains in severe cases (Robert-Gangneux and Darde 2012). Few data are available about genetics of *Toxoplasma* strains from MENA. One report about *T. gondii* genotypes in Egyptian female patients with abortion and intrauterine fetal death found a predominance of type II genotypes, but the analysis was restricted to the single *SAG2* marker which was not able to detect recombinant or exotic strains (Abdel-Hameed and Hassanein 2008). More recent study determined the *Toxoplasma gondii* genotypes in amniotic fluid, placenta, and cerebrospinal fluid samples from 11 congenital toxoplasmosis cases in Tunisia. Direct genotypic characterization of *T. gondii* strains was performed by polymerase chain reaction amplification of six genetic markers. Multilocus analysis revealed that all Tunisian isolates, except one, harbored recombinant I/II and/or I/III strain which was in concordance with the very few results reported in patients of African origin. On the other hand, the presence of strains with close homology to the natural recombinant I/III P-Br strain, incriminated in severe cases in South America, was also noted (Boughattas et al. 2010). One fatal congenital toxoplasmosis case associated with I/III recombinant genotype was described (Boughattas et al. 2011).

Risk Factors for *Toxoplasma* Infection

In MENA, many factors can explain *T. gondii* seroprevalence differences between countries and between regions within the same country.

Higher prevalences were observed in MENA Mediterranean countries with a humid and warm climate, and conversely, lower prevalences were noted in arid countries. In fact, the risk of infection in cats increases when the weather is warm

Table 2 Toxoplasmosis seroprevalence in livestock according to the country

Country	Sheep	Goat	Cattle	References
Iran	22.8 % ($n = 105$)	14.2 % ($n = 35$)	–	Zia-Ali et al. (2007)
	31.2 % ($n = 285$)	–	–	Youssefi et al. (2007b)
	35 % ($n = 588$)	30 % ($n = 400$)	0 % ($n = 290$)	Sharif et al. (2007)
	21.1 % ($n = 156$)	–	1.6 % ($n = 125$)	Raeghi et al. (2011)
	29.1 % ($n = 1000$		–	Bonyadian et al. (2007)
Morocco	27.6 % ($n = 261$)			
Saudi Arabia	23–52.2 % ($n = 397$)	19–51.7 % ($n = 290$)		Sanad and Al-Ghabban (2007)
Syria	44.5 % ($n = 800$)			el-Moukdad (2002)
Tunisia	10.8 % ($n = 350$)			Gharbi et al. (2013)

and moist, or moderate and less moist, reflecting the influence of climatic conditions on oocyst survival and prey population. Despite the short duration of oocyst shedding by cats, the burden in the environment may be very high. A single cat may shed more than 100 million non-sporulated oocysts. These oocysts need between 1 and 5 days to mature and become infective for other hosts. Once sporulated, oocysts are resistant to harsh environmental conditions and remain viable in a moist environment for more than a year and also in water for long periods of time (Robert-Gangneux and Darde 2012).

Climatic factors affect the survival of oocysts in the environment and, consequently, the infection rates in livestock which is a significant source of human infection. In MENA, high seroprevalences were observed in sheep and goats (Table 2). Also viable *T. gondii* organisms have been recovered from these animals which may represent the main source of infected meat in the region (Asgari et al. 2011; Abu-Dalbouh et al. 2012; Gharbi et al. 2013). The presence of cats (OR = 4.74), a large flock size (OR = 2.76), and the method of disposing the aborted fetuses (OR = 3.77) were all statistically significant ($P < 0.05$) risk factors that were positively associated with *Toxoplasma* positivity in goat and sheep flocks (Abu-Dalbouh et al. 2012).

Besides climatic factors, anthropogenic factors may explain a large part of the variations in human seroprevalence. Probably, one of the main risk factors for toxoplasmosis in MENA countries is linked to dietary habits. In fact, seroprevalence associated with the type of meat consumed (lamb, pork, and beef, etc.) varies among different countries according to local eating habits and according to the prevalence in livestock. In MENA, sheep and goats which are highly infected represent the main meat-producing animals. Moreover, the sheep is often eaten grilled as "mechoui" or "shawarma." This way of cooking may not achieve the temperatures that are required to kill all tissue cysts of *Toxoplasma* in all parts of the meat. Several risk factor studies identified eating undercooked meat and "shawarma," as associated with an increased risk of *Toxoplasma* infection (Elsheikha et al. 2009; Fakhfakh et al. 2013). Also eating unwashed raw vegetables or fruits was incriminated as risk factor associated with seropositivity in pregnant women (Fakhfakh et al. 2013). Drinking goat milk which was incriminated in toxoplasmosis human transmission by some authors (Robert-Gangneux and Darde 2012) was not explored in MENA region.

Other anthropogenic factors that may explain variations in human seroprevalence are represented by economic, social, or cultural habits. Relative high prevalence identified in children in MENA is probably linked to telluric contamination by oocyst ingestion in populations living under poor-hygiene conditions (Bouratbine et al. 2001; Sharif et al. 2010). Also, socioeconomic status, level of education, and urban or rural origin were incriminated as significant factors influencing human seroprevalence (Bouratbine et al. 2001; Elsheikha et al. 2009).

Prevention and Control Measures

Priority research needs for toxoplasmosis have been recently summarized by the TDR disease reference group on zoonoses and marginalized infectious diseases of poverty (Anonymous 2012). Experts' technical report highlighted the need:

- To globally quantify the proportion of chronic abortions that are attributable to toxoplasmosis and to estimate the impact of improved water quality and sanitation on toxoplasmosis infection.
- To reduce parasite transmission by (1) the development of animal vaccines and (2) the promotion of culturally acceptable health education programs to improve food hygiene at home, especially for pregnant women.
- To improve case management by (1) the cost-effectiveness assessment of the integration of existing serological test regimes into antenatal care programs in low-income settings; (2) the development of new economic and safe diagnostic techniques for acute infection during pregnancy to detect toxoplasmosis in the mother and fetus, and (3) the development of cost-effective diagnostic and management protocols for central nervous system toxoplasmosis in high-risk HIV-seropositive patients.

In fact, there is a lack of knowledge about epidemiology and risk factors of toxoplasmosis in developing countries. More researches are needed to estimate the burden of disease and to better understand modalities of transmission. Among the key policy recommendations to consider for this neglected infection in MENA include more efforts for active surveillance and the emergency for the assessment of the full extent of *Toxoplasma* infection in terms of prevalence, incidence, geographic distribution, disease burden, and adverse economic impact. There is also an urgent need in MENA countries to determine the main ecological factors and mechanisms that account for transmission of these diseases. Moreover, what should also be explored is the risk associated with the consumption of particularly untreated or unfiltered water in countries having as the main source of drinking water the surface one. On the other hand, *Toxoplasma* genotypes should be more widely studied to assess strain virulence in the MENA region.

There is evidence to suggest that health education approaches may help to reduce the risk of *Toxoplasma* infection (Robert-Gangneux and Darde 2012). Hygienic measures are the keystone of toxoplasmosis prevention and should be adapted to the local settings of transmission and the local cultural context. They can be summarized in specific guidelines and largely disseminated to population at risk, especially to pregnant women and seronegative immunocompromised patients. Particularly, individuals ought to be advised that they should wash their hands after contact with raw meat, gardening or other external activity with soil contact, and after having close contact with cats. In addition, they should thoroughly wash fruits and vegetables (especially those growing in contact with soil) if eaten raw and consume well-done or stew meat. From a general point of view considering the potentially huge contamination of the environment by spread oocysts, one must keep in mind all hygienic measures in relation with external activities.

Serological screening of pregnant women is not the rule and differs among countries according to healthcare policies. The aim of serological screening and repeated testing of seronegative women is the diagnosis of maternal primo-infection and the mother treatment to avoid vertical transmission or/and to limit fetal damage in case of transplacental transmission (Robert-Gangneux and Darde 2012). However, over the last decade, contradictory results on treatment efficacy have opened a large debate questioning the pertinence of survey (Robert-Gangneux and Darde 2012). In settings with high general population seroprevalence, the probability for a seronegative woman to acquire a primary infection during pregnancy is high, and clearly, surveillance with a simple blood test early in pregnancy would benefit most nonimmune females. Thus, decisions about implementation of prenatal screening in a given country should base on seroprevalence data, disease burden, technical resources, and diagnostic costs. On the other hand, implementing newborn screening may represent an appropriate intervention to promote early treatment. However, there is a need to better understand the disease burden for this condition in MENA.

There was no evidence for a sustainable HIV epidemic in the general population in any of the MENA countries, except possibly for southern Sudan. The general pattern in different countries in MENA points toward emerging epidemics in high-risk populations including injecting drug users, men who have sex with men, and to a lesser extent female sex workers, with heterogeneity between countries on the relative role of each of these high-risk groups (Abu-Raddad et al. 2010). In this population at risk, screening for HIV infection is of great importance. Public health efforts should focus on early identification of HIV infection, initiation, and compliance of highly active antiretroviral therapy (HAART) (Kiderlen et al. 2011). In fact, HAART has reduced the rate of toxoplasmosis and other opportunistic infections. On the other hand, *T. gondii*-specific T-cell response is restored after successful combined antiretroviral therapy in patients with AIDS and current or previous toxoplasmic encephalitis (Lejeune et al. 2011).

Finally, research and development about toxoplasmosis should be more promoted. Better diagnostic methods are needed to detect parasites in slaughtered animals, water, and foods, and new control tools, such as animal vaccines, are required to reduce the transmission of this zoonose to human. New diagnosis methods should be developed to detect toxoplasmosis in humans and to improve case management. The use of noninvasive sampling methods and the development of point of care techniques would be more practical for *Toxoplasma* screening, especially under field conditions (Chahed Bel-Ochi et al. 2013).

Conclusion

Knowledge about *Toxoplasma* infection in MENA remains limited. More investigations are needed to quantify the burden of disease especially in neonates and immunocompromised individuals, to identify the main transmission routes in the different population groups across the countries of the region, and to assess *Toxoplasma* strain virulence in MENA. Particularly, newborn screening for toxoplasmosis is critical for determining the full extent of this condition in MENA and identifying infants at risk. Decisions about implementation of prenatal screening in pregnant women, as a part of primary prevention, should be based on seroprevalence data and disease burden. In settings with high general population seroprevalence, surveillance with a simple blood test early in pregnancy would benefit most nonimmune females, for whom hygienic advices can be provided with more insistency. Assessment of the cost – effectiveness of integration of existing serological test regimes into antenatal care programs in low-income settings and development of new economic and safe diagnostic techniques will contribute to a better case management in MENA countries. In high-risk populations of HIV infection, public health efforts should focus on early identification of HIV infection and a rapid initiation of HAART to avoid any opportunistic infection, including toxoplasmosis.

References

Abdel-Hameed DM, Hassanein OM (2008) Genotyping of Toxoplasma gondii strains from female patients with toxoplasmosis. J Egypt Soc Parasitol 38(2):511–520

Abu-Dalbouh MA, Ababneh MM, Giadinis ND, Lafi SQ (2012) Ovine and caprine toxoplasmosis (Toxoplasma gondii) in aborted animals in Jordanian goat and sheep flocks. Trop Anim Health Prod 44(1):49–54

Abu-Madi MA, Al-Molawi N, Behnke JM (2008) Seroprevalence and epidemiological correlates of Toxoplasma gondii infections among patients referred for hospital-based serological testing in Doha, Qatar. Parasit Vectors 1(1):39

Abu-Raddad LJ, Hilmi N, Mumtaz G, Benkirane M, Akala FA, Riedner G, Tawil O, Wilson D (2010) Epidemiology of HIV infection in the Middle East and North Africa. AIDS 24(Suppl 2): S5–S23

Addebbous A, Adarmouch L, Tali A, Laboudi M, Amine M, Aajly L, Rhajaoui M, Chabaa L, Zougaghi L (2012) IgG anti-toxoplasma antibodies among asymptomatic HIV-infected patients in Marrakesh-Morocco. Acta Trop 123(1):49–52

Al-Mulhim AA, Al-Qurashi AM (2001) Sero-prevalence of toxoplasmosis in pregnant mothers and new born infants in eastern province, Saudi Arabia. J Family Community Med 8(1):45–48

Al-Qurashi AR (2004) Seroepidemiological study of toxoplasmosis in rural areas in the eastern region of Saudi Arabia. J Egypt Soc Parasitol 34(1):23–34

Almogren A (2011) Antenatal screening for Toxoplasma gondii infection at a tertiary care hospital in Riyadh, Saudi Arabia. Ann Saudi Med 31(6):569–572

Alqahtani J, Hassan MM (2012) Incidence of Toxoplasmosis gondii in Najran region, KSA. J Egypt Soc Parasitol 42(2):253–260

Anonymous (2012) Research priorities for zoonoses and marginalized infections. World Health Organ Tech Rep Ser (971): ix-xi, 1-119, 112 p following 119

Asgari Q, Sarnevesht J, Kalantari M, Sadat SJ, Motazedian MH, Sarkari B (2011) Molecular survey of Toxoplasma infection in sheep and goat from Fars province, Southern Iran. Trop Anim Health Prod 43(2):389–392

Ben Abdallah R, Aoun K, Siala E, Souissi O, Maatoug R, Hlioui S, Bouratbine A (2009) Congenital toxoplasmosis: clinical and biological analysis of 11 cases in Tunisia. Arch Pediatr 16(2):118–121

Ben Abdallah R, Siala E, Bouafsoun A, Maatoug R, Souissi O, Aoun K, Bouratbine A (2013) Toxoplasmosis mother-to-child screening: study of cases followed in the Pasteur Institute of Tunis (2007-2010). Bull Soc Pathol Exot 106(2):108–112

Ben Hamida Nouaili E, Chaouachi S, Meftah N, Bardi R, Sfar R, Marrakchi Z (2009) Neonatal screening of congenital toxoplasmosis. Prospective study. Tunis Med 87(3):196–199

Bonyadian M, Hematzade F, Manuchehri K (2007) Seroprevalence of antibodies to Toxoplasma gondii in sheep in center of Iran. Pak J Biol Sci 10(18):3228–3230

Boughattas S, Ben-Abdallah R, Siala E, Souissi O, Aoun K, Bouratbine A (2010) Direct genotypic characterization of Toxoplasma gondii strains associated with congenital toxoplasmosis in Tunisia (North Africa). Am J Trop Med Hyg 82(6):1041

Boughattas S, Ben-Abdallah R, Siala E, Souissi O, Maatoug R, Aoun K, Bouratbine A (2011) Case of fatal congenital toxoplasmosis associated with I/III recombinant genotype. Trop Biomed 28 (3):615–619

Bouhamdan SF, Bitar LK, Saghir HJ, Bayan A, Araj GF (2010) Seroprevalence of Toxoplasma antibodies among individuals tested at hospitals and private laboratories in Beirut. J Med Liban 58(1):8–11

Bouratbine A, Siala E, Chahed M, Aoun K, Ben IR (2001) Sero-epidemiologic profile of toxoplasmosis in northern Tunisia. Parasite (Paris, France) 8(1):61

Chahed Bel-Ochi N, Bouratbine A, Mousli M (2013) Enzyme-linked immunosorbent assay using recombinant SAG1 antigen to detect Toxoplasma gondii-specific immunoglobulin G antibodies in human sera and saliva. Clin Vaccine Immunol 20(4):468–473

Delair E, Latkany P, Noble AG, Rabiah P, McLeod R, Brezin A (2011) Clinical manifestations of ocular toxoplasmosis. Ocul Immunol Inflamm 19(2):91–102

Delair E, Monnet D, Grabar S, Dupouy-Camet J, Yera H, Brezin AP (2008) Respective roles of acquired and congenital infections in presumed ocular toxoplasmosis. Am J Ophthalmol 146 (6):851–855

Desmonts G, Couvreur J (1974) Congenital toxoplasmosis. A prospective study of 378 pregnancies. N Engl J Med 290(20):1110–1116

Dunn D, Wallon M, Peyron F, Petersen E, Peckham C, Gilbert R (1999) Mother-to-child transmission of toxoplasmosis: risk estimates for clinical counselling. Lancet 353 (9167):1829–1833

el-Moukdad AR (2002) Serological studies on prevalence of Toxoplasma gondii in Awassi sheep in Syria. Berl Munch Tierarztl Wochenschr 115(5–6):186–188

El Deeb HK, Salah-Eldin H, Khodeer S, Allah AA (2012) Prevalence of Toxoplasma gondii infection in antenatal population in Menoufia governorate, Egypt. Acta Trop 124(3):185–191

El Mansouri B, Rhajaoui M, Sebti F, Amarir F, Laboudi M, Bchitou R, Hamad M, Lyagoubi M (2007) Seroprevalence of toxoplasmosis in pregnant women in Rabat, Morocco. Bull Soc Pathol Exot 100(4):289–290

Elsheikha HM, Azab MS, Abousamra NK, Rahbar MH, Elghannam DM, Raafat D (2009) Seroprevalence of and risk factors for Toxoplasma gondii antibodies among asymptomatic blood donors in Egypt. Parasitol Res 104(6):1471–1476

Fakhfakh N, Kallel K, Ennigro S, Kaouech E, Belhadj S, Chaker E (2013) Risk factors for Toxoplasma gondii and immune status of pregnant women: cause and effect? Tunis Med 91 (3):188–190

Frenkel JK, Dubey JP, Miller NL (1970) Toxoplasma gondii in cats: fecal stages identified as coccidian oocysts. Science 167(3919):893–896

Gharbi M, Zribi L, Jedidi M, Chakkhari H, Hamdi S, R'Hayem S, Zribi N, Souli M, Darghouth MA (2013) Prevalence of Toxoplasma gondii infection in Tunisian sheep. Bull Soc Pathol Exot 106(3):184–187

Havelaar AH, Kemmeren JM, Kortbeek LM (2007) Disease burden of congenital toxoplasmosis. Clin Infect Dis 44(11):1467–1474

Hotez PJ, Gurwith M (2011) Europe's neglected infections of poverty. Int J Infect Dis 15(9):e611–e619

Hotez PJ, Savioli L, Fenwick A (2012) Neglected tropical diseases of the Middle East and North Africa: review of their prevalence, distribution, and opportunities for control. PLoS Negl Trop Dis 6(2):e1475

Hussein AH, Ali AE, Saleh MH, Nagaty IM, Rezk AY (2001) Prevalence of toxoplasma infection in Qualyobia governorate, Egypt. J Egypt Soc Parasitol 31(2):355–363

Iqbal J, Khalid N (2007) Detection of acute Toxoplasma gondii infection in early pregnancy by IgG avidity and PCR analysis. J Med Microbiol 56(Pt 11):1495–1499

Jones JL, Dubey JP (2012) Foodborne toxoplasmosis. Clin Infect Dis 55(6):845–851

Jumaian NF (2005) Seroprevalence and risk factors for Toxoplasma infection in pregnant women in Jordan. East Mediterr Health J 11(1–2):45–51

Khairallah M, Kahloun R, Ben Yahia S, Jelliti B (2011) Clinical, tomographic, and angiographic findings in patients with acute toxoplasmic retinochoroiditis and associated serous retinal detachment. Ocul Immunol Inflamm 19(5):307–310

Kianersi F, Naderi Beni A, Naderi Beni Z (2012) Clinical manifestation and prognosis of active ocular toxoplasmosis in Iran. Int Ophthalmol 32(6):539–545

Kiderlen TR, Liesenfeld O, Schurmann D, Schneider T (2011) Toxoplasmic encephalitis in AIDS-patients before and after the introduction of highly active antiretroviral therapy (HAART). Eur J Clin Microbiol Infect Dis 30(12):1521–1525

Lejeune M, Miro JM, De Lazzari E, Garcia F, Claramonte X, Martinez E, Ribera E, Arrizabalaga J, Arribas JR, Domingo P, Ferrer E, Plana M, Valls ME, Podzamczer D, Pumarola T, Jacquet A, Mallolas J, Gatell JM, Gallart T (2011) Restoration of T cell responses to toxoplasma gondii after successful combined antiretroviral therapy in patients with AIDS with previous toxoplasmic encephalitis. Clin Infect Dis 52(5):662–670

Markovich MP, Shohat T, Riklis I, Avni R, Yujelevski-Rozenblit D, Bassal R, Cohen D, Rorman E (2013) Seroepidemiology of Toxoplasma gondii infection in the Israeli population. Epidemiol Infect 1–7

Mead PS, Slutsker L, Dietz V, McCaig LF, Bresee JS, Shapiro C, Griffin PM, Tauxe RV (1999) Food-related illness and death in the United States. Emerg Infect Dis 5(5):607–625

Mohraz M, Mehrkhani F, Jam S, SeyedAlinaghi S, Sabzvari D, Fattahi F, Jabbari H, Hajiabdolbaghi M (2011) Seroprevalence of toxoplasmosis in HIV(+)/AIDS patients in Iran. Acta Med Iran 49(4):213–218

Montoya JG, Liesenfeld O (2004) Toxoplasmosis. Lancet 363(9425):1965–1976

Mostafavi N, Ataei B, Nokhodian Z, Monfared LJ, Yaran M, Ataie M, Babak A (2012) Toxoplasma gondii infection in women of childbearing age of Isfahan, Iran: a population-based study. Adv Biomed Res 1:60

Mostafavi SN, Ataei B, Nokhodian Z, Yaran M, Babak A (2011) Seroepidemiology of Toxoplasma gondii infection in Isfahan province, central Iran: a population based study. J Res Med Sci 16(4):496–501

Naba MR, Kanafani ZA, Awar GN, Kanj SS (2010) Profile of opportunistic infections in HIV-infected patients at a tertiary care center in Lebanon. J Infect Public Health 3(3):130–133

Nicolle C, Manceaux L (1908) Sur une infection à corps de Leishman (ou organismes voisins) du gondi. C R Hebd Seances Acad Sci 147:763–766

Olariu TR, Remington JS, McLeod R, Alam A, Montoya JG (2011) Severe congenital toxoplasmosis in the United States: clinical and serologic findings in untreated infants. Pediatr Infect Dis J 30(12):1056–1061

Pappas G, Roussos N, Falagas ME (2009) Toxoplasmosis snapshots: global status of Toxoplasma gondii seroprevalence and implications for pregnancy and congenital toxoplasmosis. Int J Parasitol 39(12):1385–1394

Porter SB, Sande MA (1992) Toxoplasmosis of the central nervous system in the acquired immunodeficiency syndrome. N Engl J Med 327(23):1643–1648

Raeghi S, Akaberi A, Sedeghi S (2011) Seroprevalence of Toxoplasma gondii in sheep, cattle and horses in Urmia North-West of Iran. Iran J Parasitol 6(4):90–94

Robert-Gangneux F, Darde ML (2012) Epidemiology of and diagnostic strategies for toxoplasmosis. Clin Microbiol Rev 25(2):264–296

Saeedi M, Veghari GR, Marjani A (2007) Seroepidemiologic evaluation of anti-toxoplasma antibodies among women in north of Iran. Pak J Biol Sci 10(14):2359–2362

Sanad MM, Al-Ghabban AJ (2007) Serological survey on toxoplasmosis among slaughtered sheep and goats in Tabouk, Saudi Arabia. J Egypt Soc Parasitol 37(1):329–340

Sellami H, Amri H, Cheikhrouhou F, Sellami A, Makni F, Trabelsi H, Trabelsi K, Guermazi M, Ayadi A (2010) Toxoplasmosis in Sfax, Tunisia. Bull Soc Pathol Exot 103(1):37–40

Sharif M, Daryani A, Barzegar G, Nasrolahei M (2010) A seroepidemiological survey for toxoplasmosis among schoolchildren of Sari, Northern Iran. Trop Biomed 27(2):220–225

Sharif M, Gholami S, Ziaei H, Daryani A, Laktarashi B, Ziapour SP, Rafiei A, Vahedi M (2007) Seroprevalence of Toxoplasma gondii in cattle, sheep and goats slaughtered for food in Mazandaran province, Iran, during 2005. Vet J 174(2):422–424

Siala E, Aoun K, Chahed MK, Bouratbine A (2006) Difficulty in dating primary infections by Toxoplasma gondii in pregnant women in Tunisia. Tunis Med 84(2):85–87

Siala E, Ben Abdallah R, Delabesse E, Aoun K, Paris L, Bouratbine A (2007) Contribution of the real time PCR in antenatal diagnosis of congenital toxoplasmosis. Tunis Med 85(5):385–388

Sodqi M, Marih L, Lahsen AO, Bensghir R, Chakib A, Himmich H, El Filali KM (2012) Causes of death among 91 HIV-infected adults in the era of potent antiretroviral therapy. Presse Med 41 (7–8):e386–e390

Tabatabaei S, Soleimani M, Foroutan A, Ahmadabadi M, Zarei R, Piri N, Gordiz A (2011) Ocular toxoplasmosis in Iran: 40 cases analysis. Int J Ophthalmol 4(2):199–201

Tabbara KS, Saleh F (2005) Serodiagnosis of toxoplasmosis in Bahrain. Saudi Med J 26 (9):1383 1387

Villena I, Ancelle T, Delmas C, Garcia P, Brezin AP, Thulliez P, Wallon M, King L, Goulet V (2010) Congenital toxoplasmosis in France in 2007: first results from a national surveillance system. Euro Surveill 15(25)

Yacoub AA, Bakr S, Hameed AM, Al-Thamery AA, Fartoci MJ (2006) Seroepidemiology of selected zoonotic infections in Basra region of Iraq. East Mediterr Health J 12(1–2):112–118

Youssefi MR, Sefidgar AA, Mostafazadeh A, Omran SM (2007a) Serologic evaluation of toxoplasmosis in matrimonial women in Babol, Iran. Pak J Biol Sci 10(9):1550–1552

Youssefi MR, Sefidgar SA, Ghaffari S (2007b) Seroepidemiology of sheep toxoplasmosis in Babol northern Iran 2004. Pak J Biol Sci 10(7):1147–1148

Zia-Ali N, Fazaeli A, Khoramizadeh M, Ajzenberg D, Darde M, Keshavarz-Valian H (2007) Isolation and molecular characterization of Toxoplasma gondii strains from different hosts in Iran. Parasitol Res 101(1):111–115

Rabies

A. Fayaz

Contents

Abstract Rabies is a viral encephalitis caused by the virus belonging to the family Rabdoviridae genus lyssavirus. It is still a public health problem/threat in many countries of the world and the Middle East is no exception. Regarding, rabies is absent in countries like Japan, the UK, Denmark, Greece, Sweden, Qatar and Kuwait (in the Middle East). Rabies exposure in humans or livestock occurs by animal bite or scratch. Concerning this, exposure through non-biting routes

A. Fayaz (✉)
WHO-Collaborating Center for Reference & Research on Rabies, Pasteur Institute of Iran, Tehran, Iran
e-mail: fayaz@pasteur.ac.ir

M.A. McDowell and S. Rafati (eds.), *Neglected Tropical Diseases Middle East and North Africa*, Neglected Tropical Diseases, DOI 10.1007/978-3-7091-1613-5_11, (e.g., Inhalation, transplantation) has been reported. The estimated global incidence of human rabies is around 55,000 deaths per year. Rabies is found in both domestic and wild animals. In most of the developing countries (Asian and African) where domestic animal control programs are not extensively developed, dogs and cats are considered as dominant biting animals, resulting in transmission of the disease to humans; in contrast in Latin America the canine rabies control program has resulted in declined transmission of the disease. In Latin America and also the USA, the frugivorous, insectivorous, and vampire bats could transmit rabies to human. The incubation period of rabies in humans is typically between 1 and 2 months, which might vary from one week, up to 19 years. Two-thirds of the patients present the furious form of the disease and the rest show paralytic (dumb) manifestation. Signs of animal rabies are known as aggressiveness, combative behavior, irritability, etc. The standard laboratory test of rabies for postmortem diagnosis is fluorescent antibody test. In the intra vitam diagnosis of the disease in human using saliva, urine, hair follicle, and cerebrospinal fluid, a positive result would confirm the disease, but the negative might not necessarily rule out the infection. Pre- and post-exposure prophylaxis of humans against rabies is provided by administration of cell culture rabies vaccines (HDCV, PCEC, PHKCV, PVRV) and rabies immunoglobulins (equine or human origin), according to the World Health Organization guidelines.

Keywords Rabies • Post-exposure prophylaxis • Vaccine

Background

Rabies an almost invariably fatal viral disease of animal and man has been known from time immemorial. It is presented in all continents and also endemic in most African and Asian countries. Rabies is transmitted from animal to animal or animal to human by exposure to saliva or other source of infectious virus. After a bite, virus in saliva attaches to peripheral ending nerves and travels to the brain.

In nature, rabies is a disease of mammals and, although a number of carnivores and bat species serve as the natural reservoirs, rabies in dogs is the main source of human infections. Human infection with rabies is nearly always secondary to an animal bite or scratch although exposure through the inhalation of virus, inoculation with improperly inactivated vaccine or transplantation of infected cornea, tissues, and organs have occurred. Louis Pasteur developed a crude-desiccated nervous tissue vaccine for the treatment of rabies. The successful post-exposure treatment of a nine-year old boy, Joseph Meister in 1885, with this nervous tissue vaccine initiated the era of human rabies prevention. Since that time, inactivated virus vaccines derived from cultured cells have been approved for vaccination of humans. For more than the past four decades, concentrated and purified cell culture and embryonated egg rabies vaccines have proved to be safe and effective. These vaccines are intended for pre-exposure prophylaxis as well as post-exposure prophylaxis and have been administered to millions of people worldwide. Rabies is 100 % fatal but preventable with modern vaccines and immunoglobulins.

History

Rabies has been one of the objects of human fascination and fear since the disease was recognized in antiquity. It has been suggested to exist before 2300 BC, because of a description of the disease in the Mesopotamian Laws of Eshnunna (Wiktor 1985). Ancient texts from Egypt, China, Persia, Palestine, and India appear to contain allusions to the disease (Theodorides 1986).

For thousands of years and until the findings of Louis Pasteur regarding the disease, individuals exposed to bite wounds of rabid animals, sought, protection or cure from sorcerers or persons believed to have medical knowledge or magic powers. To avert the onset of the disease in victims exposed to rabid animals, they prescribed and employed procedures that support the belief that they were definitely aware that the cause of the dreadful illness, whatever its nature, entered the victim's body by saliva of the biting animal through the inflicted wounds. Practices varied in different ethnic groups and territories.

Cauterization of wounds by hot iron or by caustic compounds, or the most cruel of all, the amputation of a wounded extremity, dressing of wounds with spider web, the burned hair of the biting animal, or the exceptionally dangerous act of the painting of the wounds with an emulsion of the brain of the biting rabid animal are some procedures practiced in the past.

The first step in correct understanding of rabies was taken by Zinke, who demonstrated experimental transmission of rabies virus in 1804 by inoculation of human saliva into animals. In 1879 Galtier transmitted rabies from dog to rabbits and from rabbit to rabbit and used intravenous injection of rabid material to immunize sheep and goats. The Pasteur group (Emile Roux and the others) established in 1881 that the central nervous system is the principal site for rabies virus replication. They transmitted the disease by inoculation of virus into rabbit and after more than 100 passages Pasteur adopted the term "fixed virus" for it. Emile Roux noticed that the virulence of infected rabbit spinal cord decreased rapidly when they suspended in dry air and was extinguished in 15 days. According to these findings and after vaccination of several dogs against rabies, Pasteur developed a practical method of vaccination. On July 6, 1885 a human patient, Joseph Meister who had been severely bitten by a suspected rabid dog received the preparation of Pasteur consisting of an emulsification of spinal cord derived from rabid rabbits and preserved in a flask of potassium hydroxide for 15 days (Pasteur et al. 1885). Joseph Meister survived the infection and the Pasteur method of treatment aroused great interest in medical circles. Criticism, however, was forthcoming. Occasional failures raised questions about the safety of the vaccine, especially because virulence material was inoculated into the patients by the end of the vaccine series (Plotikin et al. 2008). During the last four decades, the solution to the problem of safety of rabies vaccines lay in development of inactivated rabies vaccines prepared from rabies virus grown in tissue culture free of neuronal tissue.

Etiology

Rabies is caused by the rabies virus belongings to the family Rhabdoviridae genus Lyssavirus with a single, non-segmented, negative-stranded RNA. The RNA genome encodes five viral proteins: nucleoprotein N, phosphoprotein P, matrix protein M, glycoprotein G, and polymerase L. The particle of virus has a bullet shape from 100 to 300 nm in length and 75 nm in diameter (WHO Expert Consultation on Rabies 2005).

Until the 1950s, it was believed that rabies virus was unique. Since 1950, some viruses isolated from animals and humans were shown to be serologically related to rabies virus. These viruses all belong to other known *lyssavirus* genotypes and have been shown or are expected to cause acute progressive encephalitis.

According to the International Committee on Taxonomy of Viruses (ICTV), 12 *Lyssavirus* species are recognized based on genetic and serologic crossreactivity. The *Lyssavirus* genus has been subdivided into two phylogroups (WHO Expert Consultation on Rabies 982, 2013): Phylogroup I includes the rabies virus, European bat lyssavirus type 1 and type 2, Duvenhage virus, Australian bat lyssavirus, Aravan virus, Khujand virus, and Irkut virus.

Phylogroup II includes Lagos bat virus, Mokola virus, and Shimani bat virus. The remaining species of the genus, West Caucasian bat virus cannot be included into either of these phylogroups and suggested to be considered as a representation of an independent phylogroup.

In Phylogroup I, the rabies virus is spread in domestic or wild animals worldwide and has been classified further into street rabies virus and fixed rabies virus. The street rabies virus is derived from one that exists in the nature, and fixed rabies virus denotes to the strains of virus that has been adopted by serial intercerebral passages in rabbit in the laboratory. Fixed rabies virus strains are also used in the vaccine production.

Rabies virus survives at 4 °C for weeks and at −70 °C for years. It is inactivated by CO_2, so when on dry ice, it must be stored in glass-sealed vials. Rabies virus is killed rapidly by exposure to ultraviolet radiation or sunlight, by heat (one hour at 50 °C), by lipid solvents (ether, 0.1 % sodium deoxycholate, by trypsin, by detergents, and by extreme of PH (Jawetz et al. 2007).

Epidemiology

Rabies is a disease of both domestic and wild mammals, particularly dogs, other carnivores, and bat species. In areas in which domestic animal control programs are not extensively developed, dogs and cats account for most of rabid animals reported and cause the majority of human rabies exposures and deaths.

In Latin America canine rabies control programs over the past two decades have resulted in unprecedented success in this region. Official reports of human rabies transmitted by dogs have declined from around 250 in 1990 to less than 10 in 2010, with concomitant in dog rabies; however, official reports likely underestimate the scale of the problem in the few remaining canine rabies foci, particularly in Haiti, Bolivia, Guatemala, and parts of Brazil. Vampire bat rabies-related losses are largely underreporting in this region.

In Asia human rabies from endemic canine rabies was estimated to be around 31,500 in 2003. However since this estimates, the epidemiological situation in many parts of the region has changed, with improvements in rabies control and prevention in many areas particularly with regard to delivery of post-exposure prophylaxis. Nerve tissue vaccines (NTV) have been almost completely phased out of the region, with the exception of Myanmar and Pakistan, still widely using NTVs, Bangladesh discontinued uses these vaccines (WHO Expert Consultation on Rabies 982, 2013).

The vast majority of the estimated 55,000 deaths caused by rabies each year occur in Africa and Asia. In India alone, 20,000 deaths are estimated to occur annually; in Africa the corresponding figure is 24,000. Although all age groups are susceptible, rabies is most common in children under 15 years of age.

The internal market data of vaccine manufacturers suggest that at the global level $\geq$15 million people receive rabies prophylaxis annually, the majority of them live in China and India (World Health Organization 2010). Frugivorous, insectivorous, and vampire bats can transmit the disease in parts of Latin America and the USA. However, dog rabies is the source of 99%of human rabies cases. Some countries like UK, Denmark, Sweden, Greece, Ireland, Iceland, Portugal, New Zealand, Australia, Switzerland, Finland, Norway, France, Belgium, etc., are rabies free (World Health Organization 2010; OIE 2004; Pastoret et al. 2004).

Clinical Description

Rabies is principally a disease of animals and is generally spread to human by bites of rabid animals. Other forms of transmission also have been observed, including by corneal transplant, inoculation of improperly inactivated vaccine, aerosols created in laboratory, or bat-infested caves (Gibbsons 2002; Javadi et al. 1996).

The incubation period in human is typically 1–2 months but may be as short as 1 week or as long as many years (up to 19 years).

Rabies in humans presents either in a furious (two-thirds of the cases) or in a paralytic form. Clinical illness in humans may be divided into following five stages: incubation period, prodromal phase, acute neurologic phase, coma, and death. During the incubation period, there are no symptoms of diagnosis. The prodromal phase lasting 2–10 days may show any of the following nonspecific symptoms: malaise, anorexia, headache, photophobia, nausea and vomiting, sore throat, and fever. Usually there is an abnormal sensation around the wound site. During the acute neurology phase, which last 2–7 days, patients show signs of nervous system, dysfunction, such as nervousness, apprehension, hallucination, and bizarre behavior. General sympathetic overactivity is observed, including lacrimation, papillary dilatation, and increased salivation and perspiration. A large fraction of patients will exhibit hydrophobia (fear of water), the act of swallowing precipitates a painful spasm of throat muscle. This phase is followed by compulsive seizures or coma and death. The major cause of death is respiratory paralysis. In the paralytic form, the patient is conscious but appears to have a neuropathy similar to Guillain Barre Syndrome.

Signs of rabies in animals are known. After a nonspecific prodromal period, a variable proportion of animals develop aggressive or combative behavior, irritability, viciousness, or hyper-reaction to external stimuli and increased salivation. In these cases, the clinical course is described as furious rabies. A paralytic phase that develops is characterized by weakness of one or more limbs and the dysfunction and paralysis of the head and neck muscles. Difficulty in making routine, vocalization leads to alteration of phonation. Cardiac and respiratory failure ultimately leads to death. Alternatively, ataxia and primary paralysis may predominate with no overt aggressive signs, described as "dumb" rabies. Clinical observation of animals leads to a suspicion of rabies and may vary from one animal to another. The way to perform a reliable diagnosis of rabies is to identify the virus or some of its components using laboratory tests (Plotikin et al. 2008).

Diagnosis

A person presenting with an acute neurologic syndrome (encephalitis) dominated by forms of hyperactivity (furious form) or paralytic syndrome (dumb rabies) progress toward coma and death, within 7–10 days after presenting the first symptoms. The history of an associated bite from a known or suspected rabid animal, coupled with the clinical manifestation should provide a reasonably diagnosis of rabies. However such straightforward attributes are not always present and clinical diagnosis may be difficult. At the clinical manifestation stage, saliva, urine, extracted hair follicle, and cerebrospinal fluid (CSF) may be tested by virus isolation or by polymerase chain reaction and CSF may be tested for antibodies to rabies virus. Skin biopsy specimens may be examined for rabies antigens in the cutaneous nerves at the base of follicle (World Health Organization 2010;

Hemachudha et al. 2004). In laboratory tests for intra-vitam diagnosis for human rabies, a positive result is indicative of rabies, a negative result does not necessarily rule out the infection.

Postmortem test is the standard diagnostic technique to search for rabies virus antigen in the brain by fluorescent antibody test (Meslin et al. 1996). A rapid tissue culture isolation test may also be used (Bourhy et al. 1989). More recently a direct immunohistochemical test to detect rabies virus antigen in frozen or glycerol-preserved brain samples has been known to the 100 % sensitive and specific compared to the fluorescent antibody test (Lembo et al. 2006).

Research on Protection of Man Against Rabies in Iran and Other Countries

In 1920, 35 years after the initial Pasteur's attempt to protect the first exposed individual, the Institut Pasteur of Iran was established and from 1923 to 1935, Pasteur's vaccine from rabbit spinal cords was prepared. Criticism was forth coming and occasional failures raised questions about the safety of the Pasteur's vaccine and some cases were sporadically reported from all over the world. There is no record concerning the 12 years using Pasteur's vaccine in Iran. However, in 1936, the spinal cord vaccine was abandoned and Semple type vaccine from the brain tissue was introduced.

Baltazard and Ghodssi of the Institut Pasteur of Iran documented 15 years of observation of postexposure prophylactic treatment and declared that the brain tissue vaccine in common use at that time was almost totally useless in protecting individuals severely exposed to rabies and in whom the disease developed with a short time after exposure (Baltazard and Ghodssi 1953).

To cope with this disastrous situation, the use of a combination of serum and vaccine was recommended by World Health Organization (WHO)-Expert committee on Rabies (1950).

In 1950, Baltazard and Bahmanyar were confronted by a catastrophe that offered a unique opportunity to test the combined serum-vaccine therapy when 29 persons were bitten by a single rabid wolf (Fig. 1).

Of thirteen severely wounded, exposed persons who had multiple and deep wounds on their head and face that were treated with antiserum plus vaccine, only one died. Three of five severely wounded who were treated by vaccine alone died, and of eleven persons mildly exposed and treated with the combined therapy or with vaccine alone, all survived (Baltazard and Bahmanyar 1955).

The results of this Iranian field trial by using the combined treatment (serum + vaccine) was confirmed in 1956 WHO-Expert Committee on Rabies as the best protective procedure to protect the severely exposed persons.

Rabies vaccines of nervous tissue origin triggered severe reactions, resulting in postvaccination paralysis and even death in humans. For resolving this problem, in

Fig. 1 The first exposed persons injured by rabid wolf in Sahneh, Kermanshah, Iran, treated against rabies by serum and vaccine (combined) in Institut Pasteur of Iran, 1954

1950 Robert Kissling in CDC Atlanta succeeded to replicate the rabies virus in non-nervous cells and he prepared an experimental batch of rabies vaccine in hamster kidney cells (Kissling 1958). Koprowski and his coworkers then demonstrated that WI-38 cells, an established line of human diploid cells in wide use in the production of other human vaccines, were the substrate of choice (Koprowski 1967).

The first experimental batch of the rabies vaccine which was produced in human diploid cells, tested in animals (Sickes et al. 1971) and subsequently in human volunteers (Wiktor et al. 1973) was found to be safe and highly immunogenic. Institut Merieux in Lyon, France, agreed then to produce large quantities of Human Diploid Cell Vaccine (HDCV) for clinical trials in Iran. The protocol for a detailed field trial to assess the immunogenecity of the HDCV was established by an international scientific seminar held in Geneva in 1973, and Institut Pasteur of Iran in Tehran was entrusted with the clinical trials of the new HDCV. Immunogenecity trials established that human volunteers vaccinated with one to five doses of HDCV exhibited outstanding specific antibody responses (Bahmanyar 1974).

Convinced by the safety of the HDCV and promptness of the antigenic response in humans and with the consent of Iranian responsible authorities, HDCV was used for the first time in Iran for postexposure prophylaxis (PEP) in human. From 1975 to 1976, 45 persons who were severely bitten by rabid wolves and dogs received rabies immune serum and a complete course of HDCV, consisting of five doses inoculated subcutaneously on days 0, 3, 7, 14, 30, and a booster dose on day 90.

Except for one person, a 90-year-old man who died of a heart attack, these persons have remained healthy when contacted four years after exposure (Bahmanyar et al. 1976; Fayaz et al. 1981). At the end of 2007, the researchers of Institut Pasteur of Iran were able to locate 26 of 45 patients who were exposed and initially received PEP, including the new developed HDCV, 32 years ago and their data confirms the persistence of rabies neutralizing antibody for up to 32 years after vaccination with HDCV (Fayaz et al. 2011).

Rabies Vaccines and Immunoglobulins

Rabies Vaccines

Considering the substantial progress that has been made in the production and use of rabies vaccines in the past three decades, WHO strategy recommends that nerve tissue vaccines should be discontinued. Only cell culture and purified embryonated egg vaccines are now allowed to be used in humans. The first tissue culture vaccine to be approved was Duck Embryo Vaccine (DEV). It gave poor antigenic response in the vaccinated individuals and was subsequently replaced by the new modern cell culture rabies vaccine, the (HDCV). The HDCV is safe with high immunogenicity. Furthermore, two other cell culture vaccines namely Purified Vero Cell Culture Rabies Vaccine (PVRV) and Purified Chicken Embryo Cell Vaccine (PCEC) were developed. PVRV and PCEC are the most widely used cell culture vaccines in millions of patients in the world. In addition, Primary Hamster Kidney Cell Culture Vaccine (PHKCV) is produced in Moscow and in China and the Swiss serum and Vaccine Institute of Bern introduced Purified Duck Embryo rabies Vaccine (PDEV) (Plotikin et al. 2008). All rabies vaccines for human use contain only inactivated rabies virus.

Live Attenuated Vaccines

Live attenuated viruses adapted to growth in chick embryos are used for animals and not for humans.

Rabies viruses grown in various animal cell cultures have been also used as vaccines for domestic animals. A recombinant viral vaccine consisting of Vaccinia virus carrying the rabies surface glycoprotein gene has successfully immunized animals following oral administration. These vaccines may prove valuable in the immunization of both wildlife reservoirs species and domestic animals.

Rabies Immunoglobulins

There are three classes of rabies biological compounds currently available for passive immunization:

- Human Rabies Immunoglobulin (HRIG) which is prepared from the plasma of hyperimmunized humans.
- Equine Rabies Immunoglobulin (ERIG) which is concentrated serum from horses hyperimmunized with rabies virus.
- $F(ab')_2$ products: $F(ab')_2$ fragments are obtained via cleavage of the immunoglobulin by a proteolytic enzyme,(ex; pepsin), followed by separation of the F $(ab')_2$fragments from the Fc portion. Many of the available ERIG are now produced in this way.

Prevention of Rabies in Humans

Although there is no specific treatment for rabies, which is fatal, it is hundred percent preventable by pre- or postexposure prophylaxis.

Pre-exposure Prophylaxis

Pre-exposure prophylaxis is recommended for anyone who will be at continual, frequent, or increased risk of exposure to the rabies virus, either as a result of their residence or occupation (for example, laboratory workers dealing with rabies virus and other lyssaviruses, veterinarians, and animal handlers). Travelers with extensive outdoor exposure in rural high-risk areas where immediate access to appropriate medical care may be limited should also be vaccinated regardless of duration of stay. Children living in or visiting rabies-affected areas are at particular risk. WHO encourages the implementation of carefully designed studies on the feasibility, cost-effectiveness, and long-term impact of incorporating cell culture vaccines (CCVs) into the immunization programs of infants and children, where canine rabies is a public health problem.

Intramuscular Administration for Pre-exposure Prophylaxis

Pre-exposure prophylaxis requires intramuscular doses of 1 ml or 0.5 ml (volume depending on the type of vaccine) to be given on days 0, 7, and 21 or 28. For adults and children aged $\geq$2 years, the vaccine should always be administrated in the deltoid area of the arm; for children aged $<$2 years, the anterolateral area of the thigh is recommended. Rabies vaccine should not be administered in the gluteal area, as the induction of an adequate immune response may be less reliable.

Intradermal Administration for Pre-exposure Prophylaxis

Intradermal administration of 0.1 ml volume on days 0, 7, 21, or 28 is an acceptable alternative to the standard intramuscular route. To lead to significant savings, intradermal immunization sessions should involve enough individuals to utilize all opened vials within 6–8 h.

Requirements for Booster Injections

Booster doses of rabies vaccines are not required for individuals living in or travelling to high-risk areas who have received a complete primary series of pre-exposure or postexposure prophylaxis with a CCV.

Periodic booster injections are recommended as an extra precaution only for people whose occupation puts them at continual or frequent risk of exposure. If available, antibody monitoring of personnel at risk is preferred to the administration of routine boosters. For people who are potentially at risk of laboratory exposure to high concentration of live rabies virus, antibody testing should be done every 6 months. Those professionals who are not at continual risk of exposure through their activities, such as certain categories of veterinarians and animal health officers, should have serological monitoring every 2 years. Because vaccine-induced immunity persists in most cases for years, a booster would be recommended only if rabies virus neutralizing antibody titers fall <0.5 IU/ml.

Postexposure Prophylaxis

Local Treatment of Wounds

Elimination of rabies virus at the site of the infection by chemical or physical means is an effective mechanism of protection. Therefore, the Consultation Committee on rabies emphasized the importance of prompt local treatment of all bite wounds and scratches that might be contaminated with rabies virus. Recommended first-aid procedures include immediate and thorough flushing and washing of the wound for a minimum of 15 min with soap and water, detergent, povidone iodine, or other substances of proven lethal effect on rabies virus (Fig. 2). If soap or an antiviral agent is not available, the wound should be thoroughly and extensively washed with water. People who live in rabies-infected areas should be educated in simple local wound treatment and warned not to use procedures that may further contaminate the wounds. Most severe bite wounds are best treated by daily dressing followed by

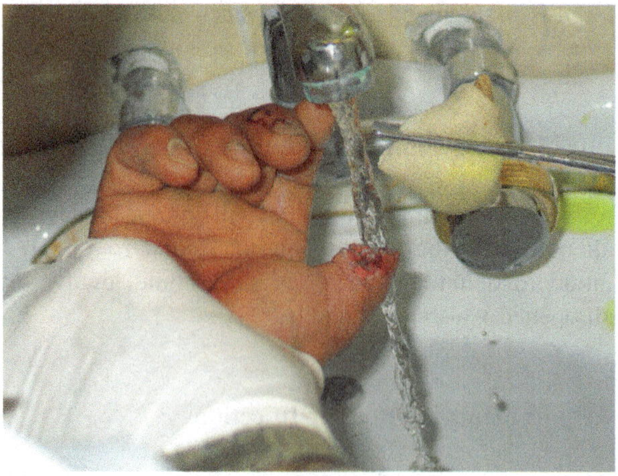

Fig. 2 Washing the fingers (bitten by a rabid dog) with flushing water and iodine containing solution, treated in Institut Pasteur of Iran

secondary suturing where necessary. If suturing after wound cleansing cannot be avoided, the wound should first be infiltrated with passive rabies immunization products and suturing delayed for several hours. This will allow diffusion of the antibody to occur through the tissues before suturing will be performed. Other treatments, such as the administration of antibiotics and tetanus prophylaxis, should be applied as appropriate for other bite wounds.

Categorization of Exposure

The indication for postexposure prophylaxis depends on the type of contact with the suspected rabid animal:

- Category I: touching or feeding animals, licks on intact skin (i.e., no exposure).
- Category II: nibbling of uncovered skin, minor scratches, or abrasions without bleeding.
- Category III: single or multiple transdermal bites or scratches, contaminating of mucous membrane with saliva from licks, licks on broken skin, exposures to bats.

For category I exposures, no prophylaxis is required; for category II, immediate vaccination is recommended; and for category III, immediate vaccination and administration of rabies immunoglobulin are recommended (Figs. 3 and 4).

For categories II and III, local treatments of wounds are indicated.

When it is not possible to complete postexposure prophylaxis with the same CCV, another CCV should be used instead. However, since no study has been performed evaluating vaccine administration (for example, from intramuscular to intradermal) during postexposure prophylaxis, such changes should be the exception.

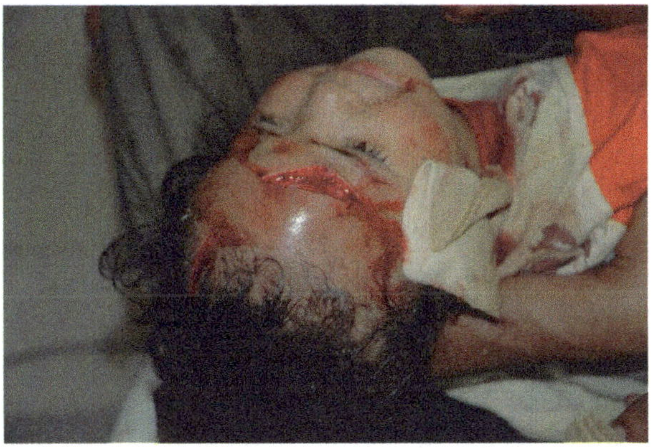

Fig. 3 Two cases of category III exposures: a two-year-old baby and a young man bitten by rabid dog and wolf, respectively, were treated in Institut Pasteur of Iran

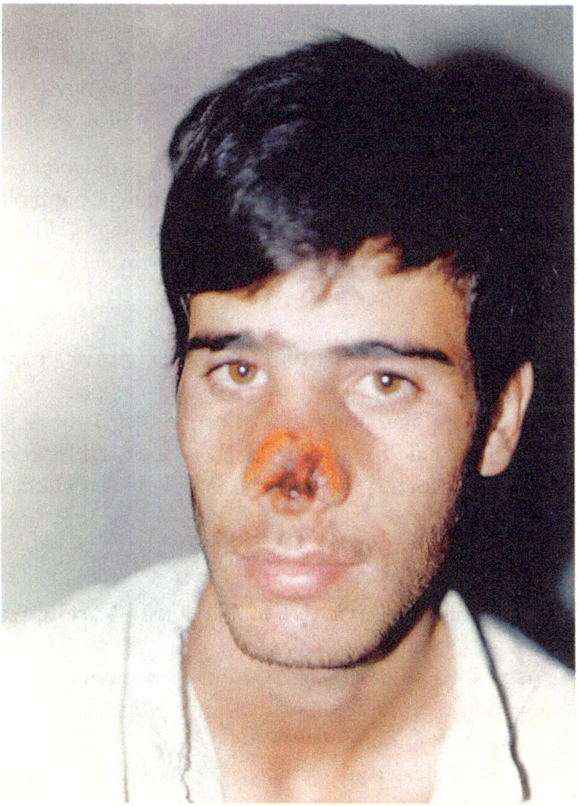

Fig. 4 Two cases of category III exposures: a two-year-old baby and a young man bitten by rabid dog and wolf, respectively, were treated in Institut Pasteur of Iran

Postexposure prophylaxis may be discontinued if the suspect animal is proven by appropriate laboratory examination to be free of rabies or, in the case of dogs or cats, the animal remains healthy throughout a 10-day observation period starting from the date of the bite.

Factors that should be taken into consideration when deciding whether to initiate postexposure prophylaxis include the epidemiological likelihood of the implicated animal being rabid, the category of exposure (I-III) and the clinical feature of the animal, as well as its availability for observation and laboratory testing. In most situations in developing countries, the vaccination status of the implicated animal alone should not be considered when deciding whether to give or withhold prophylaxis.

Intramuscular Administration for Postexposure Prophylaxis

The postexposure vaccination schedule is based on injecting 1 ml or 0.5 ml (the volume depends on the type of vaccine) into the deltoid muscle (or anerolateral thigh in children aged <2 years) of patients with category II and III exposures. The recommended regimen consists of either a 5-dose or 4-dose schedule:

- The 5-dose regimen prescribes 1 dose on each of days 0, 3, 7, 14, and 28.
- The 4-dose regimen prescribes 2 doses on day 0 (1 in each of the 2 deltoid or thigh sites) followed by 1 dose on each of days 7 and 21.

An alternative for healthy, fully immunocompetent, exposed people who receive wound care plus high quality rabies immunoglobulin plus WHO-prequalified rabies vaccines is a postexposure regimen consisting of 4doses administered intramuscularly on days 0, 3, 7, and 14.

Intradermal Administration for Postexposure Prophylaxis

The 2-site regimen prescribes injection of 0.1 ml at 2sites (deltoid and thigh) on days 0, 3, 7, and 28. This regimen may be used for people with category II and III exposures in countries where the intradermal route has been endorsed by national health authorities.

Postexposure Prophylaxis for Previously Vaccinated Individuals

For rabies-exposed patients who can document previous complete pre-exposure vaccination or complete postexposure prophylaxis with a CCV, 1 dose delivered intramuscularly or intradermally on days 0 and 3 is sufficient. Rabies immunoglobulin is not indicated in such cases. This 1-site 2-day intradermal or intramuscular regimen also applies to people vaccinated against rabies who have demonstrated rabies-virus neutralizing antibody titers of $\geq$0.5 IU/ml. As an alternative to this regimen, the patient may be offered a single-visit 4-site intradermal regimen consisting of 4 injections of 0.1 ml equally distributed over left and right deltoids or pre-scapular areas. Vaccination cards recording previous immunizations are invaluable for making correct decisions.

Immunization of Immunocompromised Individuals

In immunocompromised individuals including patients with HIV/AIDS, a complete series of five doses of intramuscular CCV in combination with comprehensive wound management and local infiltration with human rabies immunoglobulin is required for patients with category II and III exposures. When feasible, the rabies-virus neutralizing antibody response should be determined 2–4 weeks following vaccination to assess the possible need for an additional dose of the vaccine.

Rabies Immunoglobulin for Passive Immunization

Rabies immunoglobulin for passive immunization is administered only once, preferably at, or as soon as possible after, the initiation of postexposure vaccination. Beyond the seventh day after the first dose, rabies immunoglobulin is not indicated because an active antibody response to the CCV is presumed to have occurred. The dose of human rabies immunoglobulin is 20 IU/kg body weight; for equine immunoglobulin and $F(ab')_2$ products, it is 40 IU/kg body weight. All of the rabies immunoglobulin, or as much as anatomically possible (but avoiding possible compartment syndrome), should be administered into or around the wound site or sites. The remaining immunoglobulin, if any should be injected intramuscularly at a site distant from the site of vaccine administration. Rabies immunoglobulin may be diluted to a volume sufficient for all wounds to be effectively and safety infiltrated.

Rabies in North Africa

Rabies is endemic in North Africa countries: Morocco, Algeria, Tunisia, Egypt, and Libya. The vast majority of the estimated 55,000 deaths caused by rabies each year occur in rural areas of Africa and Asia. In Africa, the corresponding figure is 24,000 and approximately 4/100,000 of the population is at risk. The incidence of rabies in North Africa is underestimated. According to the most recent reports in the WHO Rabnet database (www.who.int/globalaction/default.asp), RABMED Control data and other sources (Nel and Rupprecht 2007; Biek et al. 2007), many human rabies cases are regularly reported in North Africa.

Despite the substantial committed efforts, rabies is not under control in the North African countries and continues to cause human fatalities and hundreds of animal cases. Dogs remain the main reservoir and transmitter of rabies in North Africa and 85 % of animal rabies cases are from rural areas.

Ruminants and equines are the main victims of rabies among livestock species (Mehdi El Harrak 2011). The annual incidence of human rabies cases varies from 0.02 cases/100,000 population in Tunisia to 0.1 cases/100,000 population in Egypt. The rabies postexposure prophylaxis (PEP) accessibility is subject to large disparities; the lowest rates (0.4 % persons receiving rabies PEP/1000 population) are reported in Sudan, and the highest (3.3 persons receiving rabies PEP/1000 population) are reported in Tunisia (Philippe Gautret et al. 2011).

In Morocco, 90 % of human rabies cases are caused by dog bites, these cases occurs mainly in rural areas in Kenitra (Casablanca and El Jadida). The average of animal rabies cases is 386 per year (Delmas et al. 2008).

In Algeria rabies present a public health problem despite the establishment of a national committee rabies control in 1984. Annually, 950 animal rabies cases are reported. The regions with most affected cases are the center of the country and coastal areas. Dog is the primary disease reservoir and cattle are the main victims of rabies after dogs (Mehdi El Harrak 2011).

Estimating the burden of rabies in Egypt has been associated with a high degree of uncertainty because of the lack of high quality data. Elimination of free-roaming cats and dogs has been conducted in Egypt with little effect. Nervous tissue rabies vaccine use remains widespread in Egypt and Algeria (Aylan et al. 2011; Real and Biek 2007).

In Sudan, animal rabies is reported mainly in dogs. From Libya, no information about human and animal rabies cases has been available; the country has declared itself free of canine rabies, although rabies is present in all neighboring countries (Mehdi El Harrak 2011; Philippe Gautret et al. 2011).

Rabies in the Middle East

Rabies has been known in the Middle East since biblical times (Horton et al. 2013). It is a major public threat in the majority of countries in the region (OIE and Regional representation for the Middle East 2007; Seimenis 2008). According to reported data, a considerable death rate due to rabies is observed in some of these countries, but the true burden of the disease is not clear, due to relative lack of systemic surveillance and reporting (Horton et al. 2013). Cyprus, Kuwait, and Qatar have been nominated as rabies-free countries (OIE and Regional representation for the Middle East 2007). The last rabies case in Kuwait has been reported in 1994 (Country report of the state of Kuwait 2008).

Dogs rank first as the main source of human infection which is followed by cats, cattle, sheep, goat, camels, donkeys, and then wild animals (Horton et al. 2013; Seimenis 2008; Bizri et al. 2000). In the past 15 to 20 years, rabies in wild life has become a problem on the Arabian Peninsula especially in Oman, Saudi Arabia, the United Arab Emirates, Yemen (OIE and Regional representation for the Middle East 2007), Israel, Turkey, and Iran. In some of these countries, red fox and jackals are involved. For control, mass vaccination of dogs has not been established in the Middle East and North Africa and the effective coverage rate in some countries is not exactly determined (Seimenis 2008).

There are approximately 300 human rabies cases annually in the Middle East accompanied by several thousands of (PEP). However, laboratory confirmation has not been performed in all cases, as in Lebanon, during 1991–1999, 11 cases of human rabies was diagnosed clinically without any pathological identification (Bizri et al. 2000). Nevertheless, in most of the countries, there is at least, one central rabies diagnosis laboratory with qualified, well-trained staff.

In Turkey a 3-year national project, supported by the financial cooperation program was initiated in 2005 for the control of rabies. This project involved oral vaccination of dogs and wildlife and parenteral vaccination of farm animals in the Aegean Region (OIE and Regional representation for the Middle East 2007). Dog mediated rabies reduced to restricted foci in urban regions by this effective project. However, re-emergence of rabies in the Aegean region (Horton et al. 2013) due to increased fox rabies (Aylan et al. 2011) reflects the complexity of rabies control in this country (Horton et al. 2013).

Like most of the other countries in the region (Horton et al. 2013) Lebanon, bears the high economic burden of canine mass vaccination (Bizri et al. 2000), unfortunately the wide geographic border with neighboring countries allows continuing in flow of wild animals to the country (Bizri et al. 2000). In spite of a similar situation, in Israel, where the red fox has been the most important reservoir of the disease since 1979, an oral vaccination program against wild animals has been implemented since 1998 and was extended in 2004 to the areas controlled by Israel and Palestinian authority (David et al. 2007).

In contrast, in the rabies-free country of Kuwait, all dogs and cats are vaccinated which is conducted by both the public and private sectors. This country also has regulations for dogs and cats imported in to its territory. All suspected animals which are reported to bite humans and animals are quarantined for a period of 14–21 days (Country report of the state of Kuwait 2008).

Yemen, ranks highest in regards to the spread of rabies due to the lack of strategy to eliminate the disease. Although the Ethical Treatment of Animals Organization on Rabies launched a free vaccination campaign for dogs and cats in Yemen's Sana'a governorate; the number of postexposure prophylaxis has increased largely in recent years and more than 23,000 vaccinations has been reported between the years 2008 to 2012 (Amal Al-Yarisi 2012).

There has been neither a surveillance mechanism for animal bites nor evidence of indigenous transmission of rabies in animals prior to 1990 in Oman. During the years 1991 1997, a total of 6,140 animal bites in humans with a mean of 877 cases per year has been reported. Annual expenditures bits case management was estimated to be more than 250,000 USD during that time period. However evidence of rabies has been found in both wild and domestic animals (camels, sheep, goats, cattle, etc.) (Jaffer and Datta 1998). Seemingly, the surveillance system for rabies in Iraq is at a minimal level, as there is no laboratory for confirmation of rabies diagnosis. Rabies control measures are just restricted to culling of dogs as there are no well-programmed dog vaccination or sterilization campaigns. Rabies in wild life is reported sporadically in western regions of the country but its role in disease transmission to humans or other domestic animals is not clear.

The incidence of human rabies in Iraq, in 2009 was estimated 0.89/one million people.

The reasons for failing to start or complete PEP and control programs might be due to the socioeconomic setting or lack of enough knowledge about the disease, a common phenomenon in most of the countries in the region. Generally, in this country, cattle are known as the dead-end hosts for rabies and it is assumed to be spillover of the virus from undetected cases in dogs or wild life in the area (Horton et al. 2013).

In Iran, rabies is one of the most important zoonosis, which has spread over the entire country, even to central desert parts. However, the north-east, east, and south are reported as the most affected areas. An increasing health budget is spent annually on PEP including HRIG and CCV reflecting the ascending demand for rabies PEP in more than 300 bite management centers in all over the country. The outcome of such trends would explain the doubling number of PEP between the years 1997–2009 and decreased mortality rate from 0.9 per million people in the 1980 to 0.02–0.03 per million people in recent years (Aylan et al. 2011).

In contrast to Iran, the rabies incidence is increasing in Pakistan, where 5,000 deaths are recorded annually and are much higher in India, where more than 20,000 people die (http://rabies.org.in) each year accounting for nearly 36 % of the total deaths due to Rabies world-wide (http://www.who.int/bulletin/volumes/87/12/09-021209/en). The WHO, in collaboration with provincial health authorities, is working to develop dog bite treatment centers to strengthen PEP in Pakistan. In addition, to involve the other sectors such as livestock authorities and veterinary research center, in rabies control, several plans have been developed. The main fields of WHO's' strategies for rabies control in this country are mass awareness on rabies transmission, establishment of rabies treatment centers at each district headquarters hospital, enforcement of laws relating to vaccination of pet animals, providing the modern method of decreasing the stray dogs such as bait vaccination and dog elimination, and development of surveillance system for human and animal rabies over the next two years (Daily Times 2011).

In Afghanistan, rabies can be found in dogs, bats, and other mammals (http://wwwnc.cdc.gov/travel/destinations/traveller). The Disease Early Warning System collects data on dog bite cases through a wireless system, from district hospitals and investigates suspected rabies events. From 2008 to 2012, 144 cases of rabies and five deaths were registered. A National Strategy for Control of Rabies is currently being developed by the Ministry of Public Health and guidelines on rabies vaccination have been updated. The Ministry of Public Health, Ministry of Agriculture and Livestock, the Food and Agriculture Organization of the United Nations, and WHO have established a zoonotic diseases task force for prevention and control of zoonotic diseases (http://www.emro.who.int/afghsnistan-news/rabies-day-2012.html). Additionally, US army helps rabies control in Afghanistan by providing 2,000 doses of dog vaccine making dog catching poles in armies' camp (http://www.rabiescontrol.net/news-archiev).

Programmes for Rabies Control and Elimination

As demonstrated in industrialized countries and in most of Latin America, eliminating rabies from dog populations significantly reduces human exposure to the disease. Mass vaccination of dogs is the single most cost-effective intervention to control and eliminate canine rabies. However, successful rabies control also depends on measures such as managing the dog population, mainly by promoting responsible dog ownership; compulsory notification of rabies in humans and animals; ensuring the availability of reliable diagnostic procedures; conducting post-mortem examination to confirm the cause of death in people suspected to have been infected with rabies; and improving coordination between all public sectors involved in rabies control (WHO Expert Consultation on Rabies 2005, WHO Expert Consultation on Rabies 982, 2013). Programs for rabies control must be multidisciplinary involving the animal and public health services.

Canine rabies is still widespread, occurring in more than 80 countries and territories predominantly in the developing countries (WHO Expert Consultation on Rabies 982, 2013) in Africa and Asia where approximately 55,000 human rabies deaths occur yearly, following contact with rabid dogs (http://www.who.int/bulletin/volumes/87/12/09-021209/en, http://www.who.int/rabies/animal/dogs/en/index.html).

In many of those countries, few activities are underway to prevent rabies occurrence in humans and to control rabies in dogs, even when the number of human deaths is high. The examples of these cases are in Bangladesh, India, Pakistan, Cambodia, Laos, and Nepal in Asia, as well as most African, east Mediterranean, and Arabic peninsular countries (http://www.who.int/rabies/animal/dogs/en/index.html). On the other hand, during the last three decades, a significant reduction in human rabies has been achieved in many parts of the world due to improved preventive exposure treatment delivery systems in conjunction with significant activities for dog rabies control, awareness of appropriate treatment and improved diagnostic facilities (WHO Expert Consultation on Rabies 982, 2013, http://www.who.int/rabies/animal/dogs/en/index.html).

In India, a 2-year project supported by the Global Alliance for rabies Control (GARC) and its partners called "Adopt a Village" has brought medical and veterinary professionals together in a novel intersectoral project for the prevention and control of rabies in a rural community. After the initial assessment of knowledge about rabies, a variety of different educational tools were developed for use in the study villages, including training charts used by rabies volunteers to educate villagers, posters and videos, an educational calendar, book labels for school children, and even a Snakes & Ladder game for educating school children about how to prevent rabies. The use of local TV, school networks, wall paintings, and folk dance performers were all utilized to help spread the message, and World Rabies Day events were held in the study villages. The previously inadequate surveillance systems for rabies in both humans and animals are being improved and a system of reporting animal bite cases from local hospitals is also being established. Pre-exposure vaccination of school children and other risk groups using intradermal vaccination is also in progress (The newsletter of the Global Alliance for Rabies Control 2011).

In Bangladesh, the Bangladesh Anti-Rabies Association (BARA) identified Rabies as one of the major public health problems in this country and it is now taking a very active role in fighting rabies, using advocacy, training, street dog population management, and the vaccination and sterilization of pet dogs. Training on the use of the chemical sterilant, Esterilsol, has also been carried out. Finally a school-based health education project is also just beginning in Dhaka (The newsletter of the Global Alliance for Rabies Control 2011).

In other countries, significant activities for dog rabies control have led to a sustainable reduction in dog rabies as reported for Iran, Thailand, South Africa, and in some Latin American countries. In the countries Morocco, Sri Lanka, and Tunisia these activities led to containment of the rabies situation (http://www.who.int/rabies/animal/dogs/en/index.html).

In general effective animal vaccines providing appropriate immunity and mass parenteral vaccination programs are the backbones of the canine rabies control. In recent years there have been a number of successful rabies control and elimination programs achieved through the mass vaccination of dogs, which has resulted in a reduction of human rabies cases. To achieve long-term rabies control and elimination, at least 70 % of the population should be vaccinated. The required vaccination coverage can be achieved by application of strategies comprising well-designed educational efforts, cooperation at intersectoral and interdisciplinary levels, community participation, local commitment in planning and execution, availability of a recognized quality vaccine, multimedia support, and effective general supervision of activities by appropriate authorities. No, or very low, vaccination coverage in the target population is directly correlated with rabies persistence and hence jeopardizes prospects of elimination over entire regions even when coverage elsewhere is high. Vaccinations may be more effective if carried out comprehensively over a smaller contiguous area than across many separate small areas. A predictive model of rabies transmission would be able to help identify the best strategy for such a situation.

Vaccination programs should consider the local ecology of the dog population including the degree of ownership (dogs may be owned and confined, owned and roaming, community owned, or unowned) to ensure the methods of vaccination delivery maximize access to dogs for vaccination and to support culturally appropriate education efforts. Pilot vaccination campaigns could also provide ideal opportunities to improve dog population estimates, refine strategies in relation to cost and time-scale for future vaccination programs. All dogs and cats should be immunized, regardless of age, weight, and the health state. Many young animals may not be presented for vaccination since their owners and veterinarians mistakenly believe they were too young for vaccination. The aim is vaccination, as many animals as feasible (WHO Expert Consultation on Rabies 982, 2013). The owner of every dog over the age of four months shall ensure that the dog is vaccinated for rabies. Dogs less than four months of age must be confined at home or kept under close leash supervision by the owner when off property. Twenty-eight days after primary vaccination peak rabies antibody levels are reached and a dog is considered vaccinated for one year. While cats could be a vector of rabies to humans there is no evidence that their populations act as reservoirs for the rabies virus. Cats are considered vaccinated from 28 days to one year following primary vaccination, and 1, 3, or 4 years following booster vaccinations, depending on the vaccine used (California Compendium of Rabies Control and Prevention 2012).

Three basic approaches are suggested to access dogs for vaccination in canine rabies-endemic areas: 1. House to house visits 2. Fixed vaccination posts in well-recognized sites and 3. Mobile teams that set up temporary vaccination posts. According to reported experiences, such posts are usually sufficiently attended only distances of less than 500 m or about a 10-min walk from their home. The choice of approach will depend on the specific community and the decision should be made at the local level and even a combination of approaches might be required.

It is reported that, in a few cases, rabies vaccination coupled with sterilization of dogs has resulted in local elimination of human rabies cases (WHO Expert Consultation on Rabies 982, 2013). It has been concluded that the elimination of a reservoir species is impractical, expensive, ecologically unacceptable and ethically unacceptable (Rupprecht et al. 2001). Additionally, there is no evidence that removal of dogs alone has ever had a significant impact on the spread of rabies (WHO Expert Consultation on Rabies 982, 2013). Population reduction using an array of management techniques (e.g., hunting/bounties, trapping, poison baits, gassing, and trapping) has been used to reduce the population of reservoir species to below a threshold density and thereby restrict the opportunity for the virus to spread among the animal population (Debbie 1991). The disadvantages of removing mature individuals from the population, the cost of trapping and euthanasia, and the negative perceptions by the public outweigh any advantages (Winkler and Jenkins 1991). Further, lowering the reservoir population below a "transmission threshold" (Aubert 1994) may not result in slowing of the progression of rabies or increase the possibility of it dying-out in the population (Breitenmoser et al. 1995). Regarding these facts, the number of animal cases and animal bites as well as people seeking and receiving postexposure treatment should be reported in order to provide additional epidemiological information on disease burden and to evaluate the effectiveness and cost–benefit of rabies control programs. Further involvement of frontline veterinarians, animal health workers, in both private and government, are critical as they are the most likely professionals to see a clinically rabid dog and need to know the symptoms for clinical diagnosis, method of sample collection, and process for reporting. Therefore, adequate rabies surveillance both in humans and animals is the other basis for any rabies control programs. In this regard, reactive vaccination is not recommended unless increased surveillance revealed the disease has been reduced to low levels in a few remaining foci. Reactive strategies take much longer to control rabies and are less likely to lead to successful control than systematic vaccination of an entire area.

Communication between the medical and veterinary sectors is recommended to increase rabies diagnosis in countries where outbreak investigators are minimal (WHO Expert Consultation on Rabies 982, 2013).

Other management options attempt to protect victim species from reservoir species. Human public health programs and veterinary programs for domestic animals are integral to this approach. Further, management actions that reduce or discourage opportunities for wildlife to interact or contact humans, pets, or their property can be implemented. Examples include garbage management, modification or elimination of habitat, and the storage or removal of human and pet foods (Hanlon et al. 1999).

The cost of vaccination in many low and middle income countries is high due to the number of dogs to be vaccinated in mass campaign but will differ depending on the approach selected and characteristics of the location, especially where travel is necessary to access dogs. Nevertheless the costs of dog vaccination are recovered through the benefit of saving human lives and the economic savings of reduced demand for postexposure treatment.

Only inactivated and adjuvanted rabies vaccines should be used for mass parenteral vaccination. As these vaccines are susceptible to extreme high temperature, keeping the cold chain should be assured. In endemic countries, vaccination may occasionally fail due to poor cold chain compliance or immunity may wane, therefore, vaccinations should be continued throughout a dog's life (WHO Expert Consultation on Rabies 982, 2013).

Rabies Control in Wild Life

Rabies virus is maintained in populations of wild animals and occasionally spills over into domestic animals and humans (California Compendium of Rabies Control and Prevention 2012). It is assumed that the control of rabies in wild life is complicated by a) ecologic and biologic factors associated with wildlife reservoirs, b) multidisciplinary approaches for management of such public health problem derived from wildlife, c) limited available control methods, and d) different broad range of public attitude toward wildlife (Hanlon et al. 1999). Additionally, rabies is the most significant viral zoonosis in bats.

In Africa, sporadic documented cases of rabies in wild life indicate that rabies virus is circulating in wild carnivores in the southern parts of this continent. Wild canids such as jackals and bat-eared foxes are also assumed to be reservoirs for rabies virus. The reported significant mortality in kudus (*Tragelaphusstrepsiceros*) in Namibia invokes the speculation that direct oral transmission of infective saliva is occurring from kudus to kudus. According to the documentations for Ethiopian wolf (*Canissimensis*) and African wild dog (*Lycaonpictus*), African canids are also threatened by spillover from rabies virus in dogs (WHO Expert Consultation on Rabies 982, 2013).

In the forest steppe and steppe zones of continental Asia, wild canids are the main reservoir of rabies primarily by Red fox and in the Russian Far East, the Raccoon dog. In addition, occasional cases of rabies in wild carnivores were reported in some countries in the Middle East, Central, South, and South-East

Asia that would not clear to date independency of rabies in wild life from the dog rabies transmission cycle. Reported cases of fox rabies from Israel, Saudi Arabia, West bank and Gaza Strip, Aegean part of Turkey, and the majority cases of rabies in cattle kept under free ranged conditions appear the sustained spill over from rabid foxes (WHO Expert Consultation on Rabies 982, 2013).

Three bat associated lyssaviruses have been isolated: Aravan, Khujand, and Irkut which isolated in Kyrgyzestan, Tajikistan, and Irkutsk, respectively (WHO Expert Consultation on Rabies 982, 2013).

Prevention and control of rabies in bats and terrestrial mammals pose considerable challenges. It is generally not possible or desirable to control rabies by reducing the size of wild carnivore or bat populations. Selective population reduction may be attempted in terrestrial rabies outbreaks of limited geographic scope, but these efforts can be labor and resource intensive and provide effective control only until immigration or re-introduction of the incriminated species.

Principles of rabies prevention should focus on excluding wild animals from areas of human and domestic animal habitation and activity and avoidance of contact with possibly rabid wild animals. Public education on the risks of rabies transmission from wild animals is paramount to effective disease prevention (California Compendium of Rabies Control and Prevention 2012).

The methods for rabies control involve vaccination of wild life reservoirs (Hanlon et al. 1999). In trap-vaccinated-release (TVR) programs, targeted reservoir species are live-trapped and manually injected with liquid vaccine (i.e., parenteral vaccination) (Rosatte et al. 1992). However the efficacy of parenteral rabies vaccination in wildlife has not been established and no parenteral vaccine is licensed for use in wildlife in the USA.

Conversely, an oral rabies vaccine is currently licensed for use in raccoons in the USA (Jenkins et al. 2001) and Europe (WHO Expert Consultation on Rabies 982, 2013). Oral rabies vaccination (ORV) also may be a more economically and technically feasible alternative for use on a large scale. Although ORV is less invasive to individual animals than TVR, ORV may be more intrusive on the landscape. ORV utilizes baits attractive to targeted reservoir species that once taken (bitten) release an encapsulated, attenuated rabies virus vaccine into the mouth or pharyngeal tissues to elicit an immune response (Hanlon et al. 1998). The ORV strategy developed for elimination of fox rabies has been successfully used in large parts of Western and Central Europe, Canada, and the USA, but it should be keep in mind that the ORV strategy which works for one wild life reservoir species may not necessarily work for others (WHO Expert Consultation on Rabies 982, 2013), and no rabies vaccines are licensed for use in animal species other than dogs, cats, cattle, horses, sheep, and ferrets in the USA. Although the effectiveness of rabies vaccination in other species is unknown, due to their susceptibility to rabies, wild carnivores and bats should not be kept as pets (Compendium of Animal Rabies Prevention and Control 2011). Vaccines used in the field must fulfill the requirements for biologics such as efficacy, safety, and stability which are determined in laboratory experiments and in the field. The risks which should be assessed with ORV are recombinant vaccines, recombination with similar animal viruses circulating in nature, attenuated rabies vaccines, and reversion to virulence. According to the susceptibility of target tissue (oral cavity, oropharygeal mucosa, tonsils, or small intestine), the baits are designed for each target wild animal to elicit an immune response and inactivation by degrading stomach environment should be avoided. The basic elements of evaluation of the ORV success are rabies surveillance and monitoring of oral vaccination campaigns which require sustained, constant, and intensive approaches (WHO Expert Consultation on Rabies 982, 2013).

References

Amal Al-Yarisi (2012) More Yemenis at-risk of rabies. (Health & Environment Section) Yemen Times, 24 Sept 2012

Aubert M (1994) Control of rabies in foxes: what are the appropriate measures? Vet Rec 134:55–59

Aylan O, Mohamed EI-Seyed AF, Farahtaj F, Janani AR, Lugach O, Tarkhan-Mouravi O, Usluer G, Vodopija R, Vranjes N, Tordo N, Dodet B (2011) Report of the first meeting of the Middle East and Eastern Europe Rabies Expert Bureau, Istanbul, Turkey (8–9 June 2010), Advances in Preventive Medicine (2011), Article ID: 812515. doi:10.4061/2011/812515

Bahmanyar M (1974) Results of antibody profile in man vaccinated with HDC-S vaccination with various schedules. In: Regamery RH (ed) Symposium series on immunobiological standards, Basil, Karger, vol 21, pp 231–239

Bahmanyar M, Fayaz A, Nour-Salehi SD et al (1976) Successful protections of humans exposed to rabies infection: post-exposure treatment with the new human diploid cell culture vaccine and anti-rabies serum. JAMA 236:2751–2754

Baltazard M, Bahmanyar M (1955) Essaipratique du serum antirabique chez les mordus par loups enrages. Bull World Health Org 13:747–772

Baltazard M, Ghodssi M (1953) Prevention de la rage humaine. Rev Immunol 17:366

Biek R, Henderson JC, Waller IA, Rupprecht CE, Real LA (2007) A high resolution genetic signature of demographic an spatial expansion in epizootic rabies virus. Pro Natl Acad Sci USA 104:7993–7998

Bizri A, Azar A, Saam N, Mokhbat J (2000) Human rabies in Lebanon. Epidemiol Infect 125:175–179

Bourhy et al (1989) Comparative field evaluation of the fluorescent antibody test versus isolation from tissue culture and enzyme immuno diagnosis for rapid laboratory diagnosis of rabies. J Clin Microbiol 27:519–552

Breitenmoser U, Kaphegyi T, Kappeler A, Zanoni R (1995) Significance of young foxes for the persistence of rabies in northwestern Switzerland. In: Proceedings of the third Congress European Society of veterinary virology, pp 391–396

California Compendium of Rabies Control and Prevention (2012) Part I, pp 1–5

Compendium of Animal Rabies Prevention and Control (2011) National Association of State Public Health Veterinarians. http://www.nasphv.org/documentsCompendia.html

Country report of the state of Kuwait (2008) WHO/OIE inter country expert workshop on protecting humans from domestic and wildlife rabies in the Middle East, 23–25 June 2008, Amman, Jordan

Daily Times (2011) Friday, 29 July 2011 (in Pakistan)

David D, Hughes GJ, Yakoubson BA, Davidson I, Un H, Aylan O, Kuzim IV, Rupprecht CE (2007) Identification of novel canine rabies virus clades in the Middle East and North Africa. J Gen Virol 88:967–980

Debbie JG (1991) Rabies control of terrestrial wildlife by population reduction. In: Baer GM (ed) The natural history of rabies, 2nd edn. Academic, New York, pp 477–484
Delmas O, Holmes EC, Talbi C, Larrous F, Dacheux L et al (2008) Genomic diversity and evolution of the lyssaviruses. PLoS One 3:e2057
Expert Committee on Rabies (1950) First report WHO, 1950, WHO technical report series 28:1–26
Fayaz A, Simani S, Nour-Salehi S, Bahmanyar M (1981) Booster effect of human diploid cell anti-rabies vaccine in previously treated persons. JAMA 246:2334–2335
Fayaz A, Simani S, Janani A, Farahtaj F, Biglari P, Howaizi N, Eslami N (2011) Antibody persistence, 32 years after post-exposure prophylaxis with human diploid cell rabies vaccine (HDCV). Vaccine 29:3742–3745
Gautret P, Ribadeau Dumas F, Parola P, Broukui P, Bourhy H (2011) Risk for rabies importation from North Africa. Emerg Infect Dis 17(12):2187–2193, http://www.cdc.gov/eid
Gibbsons RV (2002) Cryptogenic rabies, bats and the question of aerosol transmission. Ann Emerg Med 39(5):528–536
Hanlon CA, Niezgoda M, Hamir AN, Schumacher C, Koprowski H, Rupprecht CE (1998) First North American field release of a vaccinia-rabies glycoprotein recombinant virus. J Wildl Dis 34:228–239
Hanlon CA, Childs JE, Nettles VF et al (1999) Recommendations of the Working Group on Rabies. Article III: Rabies in wildlife. J Am Vet Med Assoc 215:1612–1618
Hemachudha T et al (2004) Antemortem diagnosis of human rabies. Clin Infect Dis 39:1085–1086
Horton DL, Ismail MZ, Sirvan ES, Wali AR, Ab-Dulla HE, Wise E, Voller K, Harkess G, Marston DA, McElhinney LM, Abbas SF, Fooks AR (2013) Rabies in Iraq: trends in human cases 2001–2010 and characterisation of animal rabies strains from Baghdad. PLoS Negl Trop Dis 7 (2):e2075. doi:10.1371/journal.pntd.0002075
http://rabies.org.in/
http://www.emro.who.int/afghsnistan-news/rabies-day-2012.html
http://www.rabiescontrol.net/news-archiev/
http://www.who.int/bulletin/volumes/87/12/09-021209/en/
http://www.who.int/rabies/animal/dogs/en/index.html
http://wwwnc.cdc.gov/travel/destinations/traveller/
Jaffer Y, Datta KK (1998) Dept. of Surveillance & Disease Control, Dept. of Family & Community Health, vol VII, no.3, Jul–Sep 1998. Sultanate of Oman
Javadi MA, Fayaz A et al (1996) Transmission of rabies by corneal graft. Cornea 15(4):431–433
Jawetz E et al (2007) Review of medical microbiology. McGraw-Hill, New York
Jenkins SR, Auslander M, Conti L, Johnson RH, Leslie MJ, Sorhage FE (2001) Compendium of animal rabies prevention and control, 2001. J Am Vet Med Assoc 218:26–31
Kissling RE (1958) Growth of rabies virus in non-nervous tissue culture. Proc Soc Exp Biol Med 98:223–225
Koprowski H (1967) Vaccines against rabies: present and future. In: Proceeding of the first international conference on vaccines against viral and rickettsial diseases of man, Pan American Health, Washington DC, pp 488–493
Lembo T et al (2006) Evaluation of a direct rapid immunohistochemistry test for rabies diagnosis. Emerg Infect Dis 12:310–313
Mehdi El Harrak (2011) OIE global conference on rabies control, Incheon, Republic of Korea, 7–9 September 2011
Meslin FX, Kaplan MM, Koprowski H (1996) Laboratory techniques in rabies, 4th edn. WHO, Geneva
Nel LH, Rupprecht CE (2007) Emergence of Lyssaviruses in the old world: the case of Africa. Curr Top Microbiol Immunol 315:161–193
OIE (World Organization for Animal Health) (2004) Rabies in: OIE manual of standards for diagnostic tests and vaccines, 5th edn. OIE, Paris, pp 276–278

OIE, Regional representation for the Middle East (2007) Situation of rabies in the Middle East, updated November 2007

Pasteur L, Chamberland C-E, Roux E (1885) Method pour prevenir la Rage après morsure. C R Acad Sci 101:765–772

Pastoret PP, Kappeler A, Auler M (2004) European rabies control and its history. In: King AA, Fooks AR, Aubert M, Wandeler AI (eds) Historical perspective of rabies in Europe and the Mediterranean Basin. OIE, Paris, pp 337–347

Plotikin SA, Koprowski H, Rupprecht CE (2008) Rabies vaccines. In: Plotkin SA, Orenstein WA (eds) Vaccine, 5th edn. Elsever, Philadelphia, PA, pp 687–714

WHO Expert Consultation on Rabies (2005) WHO technical report series 931, First report, Geneva

WHO Expert Consultation on Rabies (2013), Second report, Geneva

Real LA, Biek R (2007) Spatial dynamics and genetics of infectious diseases on heterogeneous landscapes. J R Soc Interface 4:935–948

Rosatte RC, Power MJ, MacInnes CD, Campbell JB (1992) Trap-vaccinate-release and oral vaccination for rabies control in urban skunks, raccoons, and foxes. J Wildl Dis 28:562–571

Rupprecht CE, Stohr K, Meredith C (2001) Rabies. In: Williams ES, Barker IK (eds) Infectious diseases of wild mammals. Iowa State University Press, Ames, IA, pp 3–36

Seimenis A (2008) The rabies situation in the Middle East. Dev Biol (Basel) 131:43–53

Sickes RK, Cleary WF, Koprowski H et al (1971) Effective protection monkeys against death from street virus by post-exposure administration of tissue culture rabies vaccine. Bull World Health Org 45:1–11

The newsletter of the Global Alliance for Rabies Control (2011) (21):2–3, February 2011

Theodorides J (1986) Histoire de la Rage. Masson, Paris, pp 1–289

Wiktor T (1985) Historical aspects of rabies treatment. In: Koprowsky H, Platkins SA (eds) World's Debt to Pasteur. Alan R. Liss, New York, pp 141–151

Wiktor TJ, Plotkin S, Grella DW (1973) Human cell culture rabies vaccine: antibody response in man. JAMA 224:1170–1171

Winkler WG, Jenkins SR (1991) Raccoon rabies. In: Baer GM (ed) The natural history of rabies, 2nd edn. Academic, New York, NY, pp 325–340

World Health Organization (2010) Weekly epidemiological record, no. 32, 6 August 2010, pp 309–320

Index

M.A. McDowell and S. Rafati (eds.), *Neglected Tropical Diseases - Middle East and North Africa*, Neglected Tropical Diseases, DOI 10.1007/978-3-7091-1613-5,

Made in the USA
Monee, IL
27 January 2025

11101167R00162